UMBELLIFERAE (APIACEAE) OF INDIA

Prasanta Kumar Mukherjee (Mukhopadhyay)
Department of Botany, University College of Science
University of Calcutta, Calcutta, India

and

Lincoln Constance
University Herbarium & Department of Integrative Biology
University of California
Berkeley, California, U.S.A.

AMERICAN INSTITUTE OF INDIAN STUDIES

and

OXFORD & IBH PUBLISHING CO. PVT. LTD.
New Delhi Bombay Calcutta

ISBN 81-204-0807-1

Published by Mohan Primlani for Oxford & IBH Publishing Co. Pvt. Ltd., 66 Janpath, New Delhi 110 001. Typeset by Laserwords, Madras, Processed and printed at Baba Barkha Nath Printers, 26/7 Najafgarh Road Industrial Area, New Delhi 110 015.

1-Y3-7

CONTENTS

PREFACE

This treatment of the Umbellifers of India is an updated version of material contained in the Doctoral thesis I submitted to the University of Calcutta in 1972. The study then as well as now emphasizes morphological and distributional considerations, to provide a useful framework for more detailed studies of anatomy, cytology, phytochemistry, and other branches, which may lead to a better understanding of individual taxa as well as of the group as a whole.

Most Indian species were worked out by British and other European botanists, whose materials are now deposited in various European herbaria, notably those of the Royal Botanic Gardens, Kew and the British Museum of Natural History, London. Fortunately, however, there are numerous duplicates available in India, particularly in the Central National Herbarium (CAL) and the herbarium of the Forest Research Institute, Dehra Dun (DD). Apart from these two paramount repositories, other herbaria in India which served as source banks were the herbaria of the Botanical Survey of India, Northern Circle (BSD), Eastern Circle (ASSAM), Southern Circle (MH), Blatter Herbarium, Bombay (BLAT), and the herbaria of the University of Calcutta (CUH) and the Industrial Section, Indian Museum, Calcutta (ISIM). These were later supplemented by field collections.

In recent years, I have studied specimens at Kew (K) and the British Museum (BM), the Komarov Botanical Institute, Leningrad (LE), Conservatoire et Jardin botaniques, Geneva (G), Muséum National d'Historie Naturelle, Paris (P), Station d'Ecologie Végétale "La Jaysinia", Samoëns (JAY), and the Herbarium of the University of California, Berkeley (UC). I am most grateful to the authorities of these herbaria, for their kind cooperation. My visit to Berkeley in 1983 was made possible by grants from the Smithsonian Institution, Washington, D. C., and the Graduate Division of the University of California, Berkeley. Through the good offices of Dr. Vernon H. Heywood, the British Council hosted my stay in the United Kingdom, enabling me to study briefly at Kew and the British Museum. An exchange programme of the Indian National Science Academy and the Academy of the USSR in 1986 enabled me to visit Moscow, Leningrad, and Tbilisi. In Tbilisi I consulted with Dr. Ida Mandenova and worked for some weeks in the very active laboratory of the State University of Moscow with Drs. M.G. Pimenov, E.V. Kluykov, and their associates. For my subsequent visit to Geneva, Samoëns, and Paris, I am most grateful to Dr. M.A. Farile for his friendly cooperation.

The original dissertation topic was suggested by Dr. S.K. Mukerjee, formerly Keeper of the Herbarium, Central National Herbarium, Sibpur, under whose guidance the thesis work was successfully completed. To this revered teacher, I express my indebtedness. Professor A.K. Sharma, former head of the Department of Botany, University of

Calcutta and a "guru" in every sense, not only extended facilities for my study but also inculcated in me the motivation for research. His encouragement sustained me at every stage of the preparation of this work. With deep respect, I express my gratitude to him. Dr. U.C. Bhattacharyya, Ex-Joint Director of the Botanical Survey of India, has been of constant help since the initiation of this research. I owe special thanks to him. Drs. R.B. Mazumdar and S. Basak helped materially during preparation of the thesis and are cordially thanked.

I consider it an honor to have Professor Lincoln Constance coauthor this volume with me. Not only did he act as an adjudicator for my Ph.D. thesis, but he has also been a motivating force in bringing it to publication. He has long been patiently engaged in various aspects of the work, from checking the nomenclature, searching the literature, and revising descriptions, to typing the manuscript.

Lastly, I should be failing in my duty if I did not acknowledge the forebearance and concern of my wife Jayanti and our children, Anindita and Rajarshi, during my long immersion in this project.

PRASANTA K. MUKHERJEE

PREFACE

My involvement with Indian Umbelliferae began early in 1968, when I received a letter from Mr. Mukherjee outlining his study plans and asking for assistance with literature and inviting general guidance. A few years later I acccpted the invitation to serve as a member of the Board of Examiners for the University of Calcutta to judge the quality of Mr. Mukherjee's thesis, "Taxonomic Studies on Indian Umbelliferae" (Mukherjee, 1972). My general conclusion was that: "That is an important work and helps to fill a very large gap in our knowledge of the family in Asia. It is to be hoped that the information it contains will be published, in some form, without delay." Even a year before the thesis, Mr. Mukherjee had interested me in the problem of *Heracleum* ? *pedatum* Wight, which led to our joint proposal of the genus *Vanasushava* (Mukherjee and Constance, 1974).

Dr. Mukherjee advised me that he did not think he could expect to publish his thesis as a whole, but hoped to establish nomenclatural priority for parts of it. I was able to put him in touch with the organizers of the 2 *ème* Symposium International sur les Ombellifères, held at Perpignan, France, in May, 1977, to help accomplish the latter (Mukherjee, 1978). Even earlier, however, I had conceived the hope of arranging printing of the thesis contents under Public Law 480. I was given substantial encouragement and continuing essential assistance in this course by Dr. David B. Lellinger of the Smithsonian Institution and by my colleague Vice Chancellor Watson M. Laetsch, who put me in touch with Miss Francine Berkowitz of the Office of Fellowships and Grants. The participants in the Perpignan Symposium signed a petition expressing their strong interest that the work be printed.

By 1979 I had received assurance that the project of having the treatment printed in India could go forward, with the proviso that I edit the manuscript. I have been engaged in "editing" for the past decade and it is now generally recognized that my participation may more accurately be characterized as that of co-author than that of "editor".

My special thanks are due to Dr. Lellinger and Miss Berkowitz of the Smithsonian Institution for initial encouragement and acceptance of the proposal, and again to Miss Berkowitz, and to Vice Chancellor Laetsch, Dean William A. Shack of the Graduate Division, and Professor Russell L. Jones, Chairman of the Department of Botany, all of the University of California, Berkeley, who made it possible for Dr. Mukherjee of spend a couple of weeks in May, 1983, working directly with me.

I am grateful also to the authorities of the Herbarium, Royal Botanic Gardens, Kew, the British Museum (Natural History), the Royal Botanic Garden, Edinburgh, and the

Gray Herbarium and Arnold Arboretum of Harvard University, for the generous loan of certain critical material together with helpful comments.

LINCOLN CONSTANCE

INTRODUCTION

Although some aromatic species of Umbelliferae have been known to Indian herbalists since remote antiquity and appear in the Sanskrit writings of Charaka and Sushurata in the pre-Buddhist period, Rheede tot Draakestein (1678–1703) was perhaps the first Westerner to treat an Indian umbellifer. He included the ubiquitous *Centella asiatica* in his seventeenth century *Hortus Indicus Malabaricus*. In the early nineteenth century, Roxburgh (1814) enumerated seven wild and six cultivated species from the plains, D. Don (1825) described 15 species from the Himalaya of Nepal, and, most importantly, Wallich (1829) included 66 chiefly Himalayan species in his *Catalogue*, many of which were subsequently published formally by de Candolle in the *Prodromus* (1830). Wight and Arnott (1834) reported 20 species mainly from the mountains of Peninsular India. Royle's collection of 90 species from the northwestern Himalaya included several new taxa that were published by Lindley (1835). Four new genera and 11 new species were added by Edgeworth (1846, 1851) and Dalzell (1861) reported other new taxa from Bombay Presidency.

The last and only comprehensive treatment of the Umbelliferae of India was that drawn up a century ago by Charles Baron Clarke for Joseph D. Hooker's *Flora of British India* (1879). Clarke's assessment of the family in India gave recognition to 188 species in 39 genera. Since this classic revision, members of the family have been treated in local floras and a number of new taxa have been proposed. Notable contributors include Clarke (1884–1889), Gamble (1919), Haines (1922), and H. Wolff (1910–1930). Chatterjee, in 1939, estimated a total of 180 species distributed in 43 genera, of which, 49 species were exotic and 131 indigenous. In three papers under the general heading of "Apiaceae himalayenses" (1984–1985), Farille and associates have proposed as new—four genera, five subgenera, 20 species, 1 variety, and seven combinations, predominantly from Nepal. (More are promised!) Numerous changes in generic alignment, description of new taxa, and differing assessments of others have made a reexamination of the family long overdue.

Umbelliferae has been treated for Indonesia by Buwalda (1936), for Japan by Hiroe and Constance (1958), for Afghanistan by Kitamura (1960) and by Rechinger f. and Riedl (1963), and for Cambodia, Laos, and Vietnam by Tardieu-Blot (1967). These works provide valuable information. Nasir's excellent illustrated treatment of the family in *Flora of West Pakistan* (1972) covers, however, only the northwestern Himalaya of India. Although Hiroe (1958) purported to survey all Asian Umbelliferae, his omission of many Indian species leaves a substantial gap; extensive realignment of taxa appeared in his more comprehensive work (1979). Cannon (1979) has provided a critical and most helpful list of Umbelliferae of Nepal. A most favorable recent development is the appearance of the treatment of Umbelliferae in the *Flora Reipublicae Popularis Sinicae* under the supervision of Shan and Shen, of which two of three projected volumes have appeared, in 1979 and 1985 respectively. Finally, the long-awaited account of Umbelliferae in *Flora Iranica* by Hedge, Lamond, Rechinger f. *et al.* (1987) has provided many helpful reference points.

GEOGRAPHIC DISTRIBUTION AND AFFINITIES

The scope of the present treatment is primarily the Indian Union (Fig. 1), but Nepal and Bhutan are included as well to provide a better understanding of Himalayan plants, which recognize no political boundaries. Because Bangladesh, although politically independent, is surrounded by the Union and has a closely allied flora that is well represented in Indian herbaria, it also is included.

The general distribution pattern of Indian Umbelliferae reflects the temperate nature of the family and the fact that the species are largely confined to montane areas. More than one-half the area of India lies north of the Tropic of Cancer and, although the climate is generally tropical, above an elevation of 1,250–2,000 m the vegetation is more or less temperate, passing upward into an alpine flora over a large extent of still loftier mountains.

The present study deals with 68 genera and approximately 240 species, of which 95% of both genera and species are indigenous to India and nearly 25% of the species are endemic or restricted in range. In view of the relatively large number of genera, it is at first surprising that only five, *Vanasushava*, *Karnataka*, *Pinda*, and *Polyzygus*, all monotypic taxa of peninsular India, and *Kedarnatha*, from the Himalaya of Uttar Pradesh, appear to be endemic to India. If one considers the whole area of this study and thus adds Bhutan, Nepal, and Bangladesh, the number of endemic genera rises only to eight, with *Pleurospermopsis*, *Keraymonia*, and *Lalldhwojia*, and the total of endemic species attains about 85, or 28%. Thus, most of the genera occurring in India exhibit greater differentiation and development elsewhere.

Of the approximately 240 species of indigenous Umbelliferae, some 160 are Himalayan, 25 are found in peninsular India (including Bombay Presidency), and 30 occur in east and northeast India. These represent the three principal foci of the family in the subcontinent. The Himalaya afford a unique corridor between eastern and western Asia and, as might be expected, have the greatest diversity of Umbelliferae. The largest number is found in Central and Western Himalaya including Nepal (122), while Eastern Himalaya (from Sikkim eastward) has 50; some 20 are common to both sections of the range. Most of the characteristically Himalayan genera, such as *Chamaesium*, *Bupleurum*, *Acronema*, *Sinocarum*, *Trachydium*, *Schulzia*, *Keraymonia*, *Vicatia*, *Physospermopsis*, *Cortia*, *Cortiella*, *Ligusticum*, *Selinum*, and *Tordyliopsis*, extend to China (Tibet, Sichuan, Yunnan) or Burma, on the one hand, or to Pakistan and/or Afghanistan, on the other, or considerably beyond India in one direction or another. Temperate forms may, under local circumstances, be found at lower elevations in the hills and mountains of northeastern India and southern India (Hooker and Thomson, 1855). Where the same genus occurs both

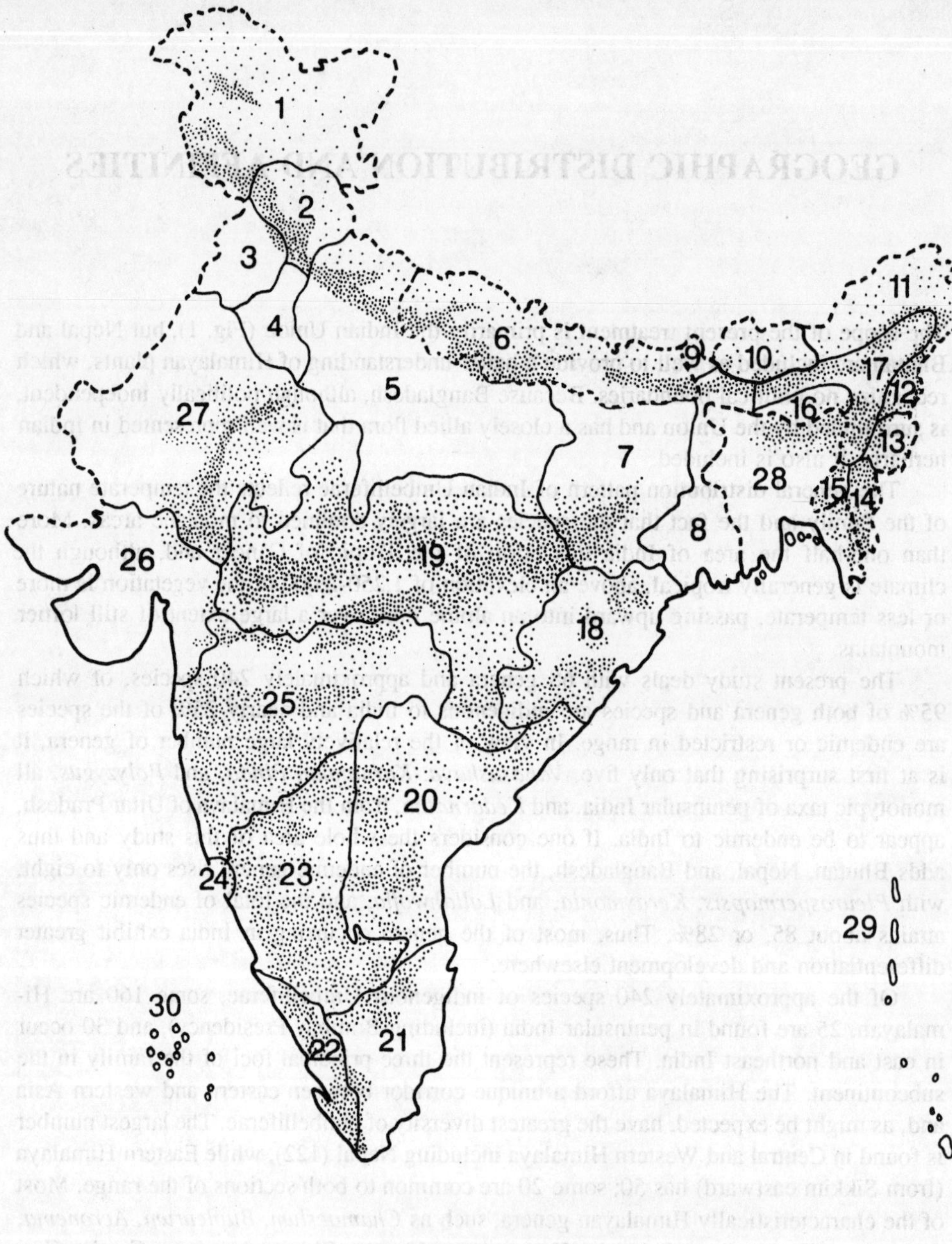

Figure 1: 1. Jammu and Kashmir; 2. Himachal Pradesh; 3. Punjab; 4. Haryana; 5. Uttar Pradesh; 6. Nepal; 7. Bihar; 8. West Bengal; 9. Sikkim; 10. Bhutan; 11. Arunachal Pradesh; 12. Nagaland; 13. Manipur; 14. Mizoram; 15. Tripura; 16. Meghalaya; 17. Assam; 18. Orissa; 19. Madhya Pradesh; 20. Andhra Pradesh; 21. Tamil Nadu; 22. Kerala; 23. Karnataka; 24. Goa Daman and Diu; 25. Maharashtra; 26. Gujarat; 27. Rajasthan; 28. Bangladesh; 29. Andaman and Nicobar Island; 30. Lakshadweep and Minicoy Island.

in the Himalaya and in southern India, the number of Himalayan species is usually the greater. Only five genera, *Karnataka*, *Polyzygus*, *Vanasushava*, *Pinda*, and *Zosima*, that occur naturally elsewhere in India, are unrepresented in the Himalaya. Only a few native species, such as *Pycnocycla glauca*, *Sesseli diffusum*, and *Peucedanum nagpurense*, and a number of mostly annual weeds from the Mediterranean area or central or southwestern Asia, are represented in the central hills and plains.

The most species-rich genera of Indian Umbelliferae are the circumboreal or largely Eurasian *Bupleurum*, *Pimpinella*, and *Heracleum*, each with 20 or more taxa. All three genera have endemic species in each of the three areas of umbellifer endemism — Himalaya, northeastern India, and peninsular India. *Pleurospermum*, with ± 20 species, is almost confined to the Himalaya. *Acronema* and the closely related *Sinocarum*, collectively with some 18 species in India, are essentially restricted to the Himalaya, but expand into an even greater number of taxa in China (Liou and Shan, 1980; Shan, 1985).

All these numbers, it should be emphasized, are to be taken *cum grano salis* since many generic boundaries are debatable, a number of described taxa are of questionable status, and correlation with the floras of surrounding areas is very imperfect. However, the generalized distributional patterns indicated are probably valid.

LECTOTYPIFICATION

An admonition from Hooker (1855) bears repeating: "And here I must caution botanists against an over-reliance upon the names attached to the Indian collections which have been distributed from Kew, first by myself, and latterly by the keeper of the herbarium, between 1855 and 1870. These collections originally comprised about half a million of specimens, which had been accumulating for upwards of thirty years, principally in the India House (where a great number were wholly destroyed by damp and vermin), at Kew, and at the Linnean Society, and consisted chiefly of the collections of Griffith, Falconer, Helfer, Royle, G. Thomson, T. Thomson and myself, Law, Stocks, Dalzell, and Gibson, together with the remainder of Wallich's, and latterly Wight's. From these, after arrangement, upwards of 380,000 specimens were distributed in sets to public and private museums in Europe, India, and America, every specimen bearing a ticket with the name of the locality and collector, and that of the plant, as far as it could be approximately determined. I have no reason to suppose that these collections contain more errors in nomenclature than do similar ones; but, as was explicitly stated when they were sent out, such names are not to be regarded as authoritative" (Preface, p. viii).

Since such authors as Lindley, Clarke, and Edgeworth rarely cited collection numbers in describing new taxa, problems of lectotypification are very formidable. Moreover, de Candolle, who received the initial assignment to work up Umbelliferae, did not cite Wallich's catalogue numbers in publishing a number of his names. One would assume that the types of all Candollean taxa would be in the *Prodromus Herbarium,* lacking any evidence to the contrary. Nasir designated holotypes (at Kew) for *Sium latijugum* Clarke, *Eriocycla thomsonii* (Clarke) H. Wolff, *Selinum papyraceum* Clarke, *Ligusticum marginatum* Clarke, *L. elatum* Clarke, *Heracleum cachemiricum* Clarke, *H. pinnatum* Clarke, *Platytaenia lasiocarpa* ssp. *thomsonii* (Clarke) Rech. f. & Riedl, and *Pleurospermum stylosum* Clarke. He also attributed holotype status to 12 taxa of Wallich/Candolle at Kew, but we think this must be incorrect. Tardieu-Blot designated at least four more Kew collections by Wallich as being holotypes. Leute (1970) indicated Kew holotypes for *Ligusticum elatum* (Edgew.) Clarke, *L. marginatum* Clarke, and *L. thomsonii* Clarke, but refrained from doing so for *Cortia* (= *Selinum*, in large part). Rechinger f. and Riedl designated Kew lectotypes for *Aulacospermum stylosum* (Clarke) Rech. f. & Riedl and *Platytaenia lasiocarpa* ssp. *thomsonii*; they also included photographs of the Kew specimens of *Ligusticum elatum, L. marginatum, L. thomsonii*, and *Selinum papyraceum*.

In view of this somewhat chaotic situation, it seems preferable to refrain from designating further holotypes, lectotypes and/or neotypes until each genus can be studied intensively, with all pertinent materials at hand.

UMBELLIFERAE

Apiaceae Lindley, Nat. Syst. ed. 2, 21. 1836, *nomen conserv.*
Ammiaceae Sprengel ex Pfeiffer, Nom. Bot. 1: 145. 1873.
Eryngiaceae Nevski, Acta. Inst. Bot. Acad. Sci. ser. 1, fasc. 4: 275. 1937, *nomen illeg.*
Hydrocotylaceae Hylander, Uppsala Univ. Srskr. 1945, 7: 20. 1945.
Saniculaceae A. Löve & D. Löve, Opera Bot. 5: 263. 1961, *nomen illeg.*

Perennial, biennial, or annual mostly terrestrial herbs, shrubs, or rarely small trees, caulescent or acaulescent from taproots, fibrous roots, or tubers, glabrous to variously pubescent, sometimes prostrate and creeping or floating, or forming mats or cushions. Leaves alternate, or rarely opposite; leaf blades usually much dissected to divided or decompound, but sometimes simple and even entire, with sheathing petioles, or less commonly stipulate and nonsheathing. Flowers perfect or staminate, sometimes polygamous, small, regular or the marginal petals radiant, in lax compound or simple umbels, or the umbels capitate. Umbels sometimes subtended by bracts to form an involucre; umbellets often subtended by bracteoles (bractlets) to form an involucel; bracts and bracteoles simple to compound, persistent or caducous, or frequently one or both lacking. Calyx tube wholly adnate to ovary, the free lobes or teeth conspicuous and persistent to obsolete or lacking. Petals 5, with or without a narrower inflexed apex. Stamens 5, inserted alternately with the petals on an epigynous disk, the bilocular anthers basi- or dorso-fixed, dehiscent longitudinally. Ovary bicarpellate, inferior, bilocular, with one anatropous pendulous ovule in each locule; styles 2, usually swollen at base into a stylopodium. Fruit a schizocarp consisting of 2 mericarps united by their faces (commissure), variously compressed or terete, each mericarp with 5 primary ribs and less commonly also with 5 secondary ribs, the ribs filiform to strongly winged; mericarps l-seeded, usually separating readily at maturity and suspended from the summit of a slender, persistent, entire to bipartite axis (carpophore). Vittae (oil tubes) usually present in the pericarp between the ribs and on the commissural surface, which is plane to deeply sulcate. Embryo small, with abundant cartilaginous endosperm.

Type genus: *Apium* Linn.

A largely temperate and montane family consisting, according to various estimates, of 200–300 genera and 2,000–3,500 species.

ARTIFICIAL KEY TO GROUPS OF GENERA

A. Flowers in simple umbels or heads, these occasionally aggregated into branched inflorescences. .... Group I, page 9.

A. Flowers in definitely and regularly compound umbels, these sometimes combined into more complicated inflorescences.

B. Each umbellet bearing a single sessile central female flower surrounded by pedicellate male flowers, the pedicels inflated at maturity. .... 5. *Pycnocycla*, Page 22.

B. Each umbellet bearing perfect or male flowers, or polygamous, the flowers usually all pedicellate and pedicels not inflated.

C. Fruit with both primary and secondary ribs equally or unequally developed. .... Group II, page 23.

C. Fruit with only primary ribs evident, secondary ribs lacking.

D. Fruit terete or compressed laterally if at all.

E. Fruit elongate, linear to oblong and at least twice as long as broad, often several times as long, or if rarely oblong-ovoid, the fruit beaked. .... Group III, page 35.

E. Fruit oblong to globose or elliptic, less than twice as long as broad (cf. also 17. *Foeniculum* and 18. *Bupleurum*). .... Group IV, page 66.

D. Fruit compressed or strongly flattened dorsally, some or all ribs winged.

F. Fruit with lateral and at least some dorsal ribs equally or unequally winged (cf. also Group VI). .... Group V, page 168.

F. Fruit with lateral ribs winged, dorsal ribs low, filiform, or obsolete (cf. also Group V). .... Group VI, page 208.

GROUP I

1. Low, delicate, creeping herbs; fruit strongly flattened laterally, never compressed dorsally; calyx teeth obscure or lacking.
   2. Leaves with paired stipules, petioles without sheaths; secondary ribs lacking .... 1. *Hydrocotyle.*
   2. Leaves lacking stipules, petioles sheathing at base; fruit with evident secondary ribs and prominent reticulations. .... 2. *Centella.*
1. Stout to slender, usually erect or ascending herbs; fruit terete to strongly flattened dorsally, never compressed laterally; calyx lobes prominent, persistent on fruit.
   3. Flowers polygamous and some usually pedicellate, in simple umbels or heads; foliage not spinose; fruit with prickles or tubercles. .... 3. *Sanicula.*
   3. Flowers all perfect and sessile in heads; foliage usually spinose; fruit squamate or papillose, never prickly. .... 4. *Eryngium.*

1. HYDROCOTYLE Linn.

Sp. Pl. 234. 1753; C. B. Clarke in Hook. f., Fl. Brit. Ind. 2: 667. 1879.

Prostrate, ascending, or rarely erect perennial herbs, glabrous or variously hairy, usually rooting at nodes. Leaves petiolate, simple, cordate, the margins crenate to deeply lobed, palmately nerved; stipules present, small, membranaceous; petioles not sheathing. Umbels simple, small, headlike, opposite the leaves or axillary. Involucre present or lacking. Flowers white, greenish-white or yellow; petals ovate, entire and without an inflexed apex; calyx teeth small or obsolete; stylopodium conic or depressed. Fruit orbicular or ellipsoid, compressed or flattened laterally, glabrous or variously hairy or papillate, the 5 primary ribs evident; vittae obscure; commissural face plane or concave. Carpophore usually wanting.

Type species: *Hydrocotyle vulgaris* Linn.

Distribution: About 75 species, widely distributed in tropical, subtropical, and temperate regions, especially abundant in the Southern Hemisphere. Of the 6 species occurring in India, *H. himalaica* is endemic to the Eastern Himalaya and *H. conferta* is confined to peninsular India; the others are more widely distributed, especially in the mountains of India and beyond.

KEY TO SPECIES

1. Umbels long, pedunculate.
    2. Pedicels slender, 3–6 mm long.
        3. Fruit reddish-brown to blackish, dorsal ribs indistinct; leaves orbicular, shallowly 9-lobed or unlobed. .... 1. *H. himalaica.*
        3. Fruit yellowish-brown, often with red spots, dorsal ribs slender; leaves pentagonal or 5-lobed .... 6. *H. hookeri.*
    2. Pedicels short or obsolete; rarely more than 1.5 mm long.
        4. Leaves 2–15 cm in diameter, lobes usually triangular; upper peduncles fascicled; fruit reddish-brown to blackish. .... 2. *H. javanica.*
        4. Leaves 0.5–2.5 cm in diameter, lobes rounded or obtuse; peduncles solitary; fruit yellowish-brown.
            5. Fruit distinctly pedicellate; peduncles longer than subtending leaves; leaf sinuses narrow, sometimes closed. .... 3. *H. ramiflora.*
            5. Fruit sessile or subsessile; peduncles shorter than subtending leaves; leaf sinuses usually open, often broad. .... 4. *H. sibthorpioides.*

1. Umbels subsessile to sessile; flowers subsessile.
    5. Leaves 0.5–2 cm in diameter, usually glabrous; petioles 1–2.5 cm long. .... 4. *H. sibthorpioides.*
    5. Leaves 2–6 cm in diameter, setulose; petioles 4–18 cm long. .... 5. *H. conferta.*

1. **Hydrocotyle himalaica** P. K. Mukherjee,
Indian Forester 95: 470. 1969

Hydrocotyle javanica var. *podantha* C. B. Clarke in Hook. f., Fl. Brit. Ind. 2: 668. 1879, non *H. podantha* Molkenb. 1851.

Stems 10–30 cm tall, softly ferrugineus-tomentose. Leaf lamina orbicular or reniform, 2–8 cm in diameter, obtusely repand-crenate, chartaceous, ferrugineus-tomentose, the sinus broad, the principal nerves 9, raised on ventral surface, rough and bristly, stipules ovate, membranaceous, 2–3 mm long; petioles 2.5–18 cm long, tomentose. Umbels globose, ca 40-flowered, 1–3 cm in diameter. Peduncles leaf opposed, 3–7 cm long, ferrugineus. Petals 1–1.3 mm long, triangular, the apex slightly incurved, the base broad; calyx teeth minute, ± 1 mm long, hairy dorsally; stamens 1.7 mm long, exceeding petals. Pedicels 0.5–0.6 mm long, puberulent. Fruit suborbicular, broader than long, reddish-brown to blackish, 1.5 mm in diameter, the primary ribs indistinct. June–Oct.

Type locality: Khursiong, West Bengal, *Clarke 38, 825A* (CAL: holotype).

Distribution: Himalaya, from Nepal to Arunachal Pradesh, between 1,500 and 3,000 m.

Representative specimens: Nepal: Arun Valley, *Stainton 492* (CAL). Sikkim: Great Rangeet Valley, *Kurz* (CAL); Choongthang, Sept. 1903, *Prain* (CAL). West Bengal: Birch Hill, Darjeeling, *Mukherjee* (CUH). Bhutan: Dorokha, *Thothathri 10,500* (CAL). Arunachal Pradesh: Sonegoodam to Dreyi, Lohit, F. D., *R. S. Rao 10,359* (CAL).

2. **Hydrocotyle javanica** Thunb., Diss. Hydrocot. no. 17,
pl. 2. 1798; C. B. Clarke in Hook. f., Fl. Brit. Ind. 2: 667. 1879.

Hydrocotyle hispida Buchanan-Hamilton ex D. Don, Prodr. Fl. Nepal. 182. 1825; *H. nepalensis* var. *longipes* DC., Prodr. 4: 65. 1830. Based on *Buchanan-Hamilton*, Nepalia.

Hydrocotyle nepalensis Hook, Exot. Bot. 1: pl. 30, 1823; *H. nepalensis* var. *brevipes* DC., Prodr. 4: 65. 1830. Neither locality nor collector indicated.

Hydrocotyle heyneana Wallich, Cat. 563. 1829, *nomen nudum; H. strigosa* Buchanan-Hamilton ex Wallich, Cat. 7219. 1829, *nomen nudum; H. zeylanica* DC., Prodr. 4: 67. 1830; *H. zeylanica* var. *heyneana* DC., loc. cit. Based on *Leschenault*, "In Zeylona circa Kandy."

Hydrocotyle polycephala Wight & Arn., Prodr. Fl. Ind. Orient. 366. 1834. Based on *Wight 1187*, "Peninsular India Orientalis" (K).

Stems procumbent to erect, 10–50 cm high, terete, glabrous, hispid, or hirsute. Leaf lamina orbicular-reniform, 2.5–15 cm in diameter, glabrous or variously hairy, the sinus broad or narrow, 7-lobed, the lobes triangular, crenate with sharply acute or rounded teeth very variable; stipules broadly lanceolate, 3–6 mm long; petioles 2–15 cm long, reduced upward, usually with reflexed hairs. Umbels 15–50-flowered. Peduncles solitary or in groups of up to 15, often opposite the leaves, unequal, 1–7 cm long, glabrous or variously hairy, shorter or longer than petioles. Involucre of several ovate, acute, entire or lobed bracts ca 1 mm long. Petals lanceolate, acute; calyx teeth obsolete. Pedicels ± 1.5 mm, rarely 2.5 mm long. Fruit broader than long, flattened laterally, reddish-brown to blackish and often mottled purple, 1–1.25 mm long, 1.25–2 mm broad, the ribs very slender. Throughout the year.

Type locality: "Java Volgens," *Thunberg*.

Distribution: Tropical Asia: Malayasia, Australia, China, Korea, and Vietnam; in India, widely distributed on the major hills, from 30 to 3,000 m.

Representative specimens: Uttar Pradesh: Mussoorie, *Mackinnon* (CAL). Nainital, *Champion* (CAL). Nepal, *Buchanan-Hamilton* (CAL). Sikkim: Rangeet Valley, *Anderson 144* (CAL). West Bengal: Darjeeling, *Gamble 3926B* (CAL). Bhutan: Labha, *Sen Gupta 382* (CAL). Arunachal Pradesh: Paya-Nava, Lohit, F. D., *R. S. Rao 10,682* (CAL). Nagaland: Kohima, *Clarke 40,991* (CAL). Tripura: Panghman, *Deb 27,067* (CAL). Mizoram: South Lushai Hills, *Gage 234* (DD). Manipur: On way to Muku, *Watt 5037* (DD). Meghalaya: Cherapunji, *Panigrahi 3711* (ASSAM, CAL). Tamil Nadu: Nilgiri, *King* (CAL); Pulney, *Bourne 1097* (CAL). Maharashtra: Bombay Presidency, *Gibson* (CAL, DD).

Note. The Indian plants of this alliance, described under as many as 6 or 7 names, were, in my opinion, correctly reduced to a single species by Clarke. Collections examined from all parts of India show variation in pubescence, shape of leaf lobes, width of sinuses, and number and length of petioles and peduncles. Separation of Indian and Japanese plants from Indonesian ones, as *H. nepalensis*, on the basis of such characters does not seem well founded. Although intensive morphological, cytological, and geographical analysis of this species complex may result in the recognition of smaller taxa, I believe it best to emphasize the evident similarities rather than the possible differences at this time. (P.K.M.)

3. **Hydrocotyle ramiflora** Maxim. Bull. Acad. St. Petersb. 31: 46. 1887.

Hydrocotyle wilfordii Maxim., Bull. Acad. St. Petersb. 31: 46. 1887. Based on *Wilford*, Korea.

Hydrocotyle maritima Honda, Bot. & Zool. 2: 1825. 1934; *H. ramiflora* var. *maritima* (Honda) Hiroe, Acta Phytotax. Geobot. 40. 1950. Based on *Jyotani*, Oshima Island, Japan.

Stems decumbent creeping or floating, 10–30 cm long, glabrous or foliage very sparingly hirtellous. Leaf lamina suborbicular to reniform, 1–2(–3) cm in diameter, shallowly crenately 5–9-lobed, the lobes crenate, the crenations low and subequal, the sinus usually narrow, sometimes closed, the principal nerves usually 5, evident; stipules broadly ovate, scarious, brownish; petioles 1–12 cm long, slightly hirtellous at summit. Umbels globose, 5–20-flowered, 2–5 mm in diameter. Peduncles slender, 1–5 cm long. Petals ca 1 mm long, the apex acute, plane, the base broad; calyx teeth 0; stamens not exceeding petals. Pedicels 0.5–2 mm long, glabrous. Fruit suborbicular, 1–1.5 mm in diameter, yellowish-brown, often dark spotted, the ribs distinct, acute.

Type locality: "Yezo [Hokkaido], circa Hakodate," June 1861, *Maximowicz*.

Distribution: Japan, Korea, China, to northeastern India; adventive in southern USSR.

Specimen examined: Arunachal Pradesh: Tawang, 1988, *Mukherjee* (CUH).

4. **Hydrocotyle sibthorpioides** Lam., Encyc. 3: 153. 1789; C. B. Clarke in Hook. f., Fl. Brit. Ind. 2: 669. 1879.

Hydrocotyle nitidula A. Rich., Ann. Gen. Sci. Phys. 4: 60, pl. 63, fig. 33. 1820. Neither locality nor collector indicated.

Hydrocotyle tenella Buchanan-Hamilton ex D. Don, Prodr. Fl. Nepal. 183. 1825. Based on *Buchanan-Hamilton*, Nepalia.
Hydrocotyle rotundifolia Roxb., Hort. Beng. 21. 1814, *nomen nudum*; DC., Prodr. 4: 64. 1830; C. B. Clarke in Hook. f., Fl. Brit. Ind. 2: 668. 1879. Based on *Roxburgh*, Botanic Garden, Calcutta.

Stems prostrate, filiform, glabrous or sparsely hairy, with long internodes. Leaf lamina orbicular, 0.5–2.5 cm in diameter, shallowly 5–7-lobed, deeply cordate, the lobes crenate, pilose or hirsute on both surfaces, rarely glabrous; stipules ovate-elliptic, membranaceous, ca 0.5 mm long; petioles 1–2.5 cm long, reduced above, filiform, hairy, with spreading or reflexed hairs. Umbels axillary, 10- or usually fewer-flowered. Involucre of 4–8 ovate-lanceolate bracts. Peduncles 2 cm long to obsolete. Petals ovate, acute; calyx teeth 0; styles reflexed. Pedicels very short to obsolete. Fruit suborbicular, yellow to brown, sometimes red punctate, glabrous, 0.8–1.5 mm in diameter, the mericarps pentagonal in transection, the lateral ribs more prominent than the dorsal ribs. Feb.–Sept.

Type locality: "A l'Isle de France" [Mauritius], no collector indicated.

Distribution: Tropical Asia, Africa, China, Japan; introduced into America, Australia, and India, where it is adventive from the plains to 3,000 m.

Representative specimens: Uttar Pradesh: N of Almorah, *Anderson* (CAL). West Bengal: Teesta Bank, Rungpoo, *Clarke 37,090* (CAL). Bhutan: Samkhara, *Sen Gupta 1108* (CAL). Arunachal Pradesh: Makum, *Gammie 95* (CAL). Tripura: Agartala, *Debbarman 686* (CAL). Bangladesh: Sylhet, *Wallich 562* (CAL, K).

5. **Hydrocotyle conferta** Wight, Ic. Pl. Ind. Orient. 3: 11, pl. 1002. 1846; C. B. Clarke in Hook. f., Fl. Brit. Ind. 2: 668. 1879.

Stems slender, procumbent to suberect, villous. Leaf lamina orbicular-reniform, 2–6 cm long, 2.5–5 cm broad, mostly 7-lobed, the lobes triangular, crenate, the crenations subequal, setulose; petioles 4–18 cm long. Umbels axillary and opposite the leaves, 5–15-flowered, sessile or shortly pedunculate. Involucre obscure. Peduncles glabrous, up to 2.5 mm long. Pedicels less than 1 mm long. Fruit suborbicular, subcompressed laterally and subpentagonal in transection, smooth or obscurely punctate, 1 mm long, 1.5 mm broad, the pericarp hard. Vittae lacking. June.

Type locality: "Peninsula India Orientalis," *Wight 1158* (K).

Distribution: Endemic in Nilgiri and Pulney hills of peninsular India, at ca 2,000 m.

Representative specimens: Tamil Nadu: Bear Shola, Pulney, *Bourne 1505* (CAL, MH); Nilgiri, *Schmidt* (CAL). Kerala: Kondalur Forest, *Sebastine 18438* (MH).

6. **Hydrocotyle hookeri** (C. B. Clarke) Craib, Kew Bull. 1911: 9. 1911.

Hydrocotyle javanica var. *hookeri* C. B. Clarke in Hook. f., Fl. Brit. Ind. 2: 668. 1879.

Stems slender, procumbent with ascending branches, glabrous to puberulent, up to 20 cm long. Leaf lamina orbicular-reniform, 3–8 cm long, 4–8 cm broad. 5-lobed or pentagonal, crenate or crenate-serrulate, glabrous or hirsute; stipules oval, ca 3 mm long; petioles 4–12 cm long, hirsute. Umbels axillary or leaf opposed, ca 50-flowered. Involucral bracts several, ovate, acute, 1 mm long. Peduncles 3–12 cm long, glabrous or pubescent.

Pedicels 3–6 mm long. Fruit yellowish-brown, with or without red spots, pentagonal in transection, ca 1 mm long, 1.5 mm broad; the lateral ribs more prominent than the slender dorsal ribs. Aug.–Nov.

Type locality: "Khasia, alt. 5,000–6,000 ft. [Mowflong], *Hooker & Thomson* (Herb. Ind. Orient., *Hydrocotyle* 5) (CAL, K).

Distribution: In India, occurring in Arunachal Pradesh and on the Khasia and Naga hills, from 1,500–2,000 m; China, Vietnam.

Representative specimens: Arunachal Pradesh, Koppu-Gelling, Siang F. D., *Panigrahi 17,450* (ASSAM, CAL). Nagaland: Kohima, *Bor 16,161* (ASSAM).

2. CENTELLA Linn.

Pl. Rar. Afr. 20. 1760, *nomen conserv.*

Solandra Linn., Syst. Nat. ed. 10, 1269. 1757, *nomen rejec.*
Hydrocotyle sect. *Centella* DC., Prodr. 4: 68. 1830; C. B. Clarke in Hook. f. Fl. Brit. Ind. 2: 669. 1879.

Prostrate perennial herbs, glabrous or pubescent, producing runners and rooting at nodes. Leaves petiolate, simple, the margins entire to repand-dentate, palmately nerved; stipules lacking; petioles sheathing at base. Umbels simple, subcapitate, the peduncles axillary. Involucre of 2 or 3 conspicuous bracts. Flowers white or rose tinged; petals orbicular, with a narrower inflexed apex; calyx teeth obsolete; styles short; stylopodium obsolete. Pedicels slender to obsolete. Fruit ovoid to orbicular, flattened laterally and constricted at commissure, both primary and secondary ribs filiform and connected by transverse ridges, forming a reticulum; an oil-bearing layer beneath epidermis; commissural face plane. Carpophore present, entire.

Type species: *Centella villosa* Linn.

Distribution: About 20 species, mostly South African. The *Centella asiatica* complex, however, is circumtropical and extends into the temperate zones in both hemispheres. *Centella asiatica* occurs throughout India, from the plains to 2,000 m.

1. **Centella asiatica** (Linn.) Urban in Martius, Fl. Bras. 11: 287, pl. 78, fig. 1. 1879.

Hydrocotyle asiatica Linn., Sp. Pl. 234. 1753; C. B. Clarke in Hook. f., Fl. Brit. Ind. 2: 668. 1879.

Stems filiform, with elongated internodes, rooting at nodes. Leaves orbicular-reniform, broader than long, 0.5–5 cm in diameter, entire or crenate-repand, glabrous or nearly so, the nerves numerous; petioles 1–25 cm long, glabrous or nearly so. Umbels 3-4-flowered. Involucre of 2 ovate, acute bracts ca 1.5 mm long. Peduncles pubescent or glabrate, ca 2.5 cm long. Flowers pink, 1 or 2 central flowers pedicellate, the others sessile; petals ovate. Fruit broader than long, ca 2 mm long, 2–3 mm broad. Throughout the year.

Type locality: "Indiae"; Linnean Herbarium, Savage Catalog 3325.

Distribution: Tropical and subtropical Asia and the Pacific, extending into the temperate zones. Found throughout India as a weed, from the plains to 2,000 m.

Representative specimens: Himachal Pradesh: Neerath, *Nair 29,850* (BSD). Uttar Pradesh: Dehra Dun, *Duthie 2624* (DD). Nepal: Pokhra, *Pande 232* (DD). West Bengal: Behala, *Sarkar* (CUH). Bhutan: Tama, *Sen Gupta 1162* (CAL). Arunachal Pradesh: Baha, Kameng F. D., *Panigrahi 15,252* (ASSAM). Assam: North Lakhimpur, *Panigrahi 27,842* (ASSAM). Orissa: Puldumer, Kalahandi Estate, *Mooney 2438* (DD). Andhra Pradesh: Anantagiri, Vishakapatnam District, *Subbarao 19,569* (MH). Tamil Nadu: Kodaikanal, *Bourne 27* (CAL); Coonor, *Gamble 12,016* (CAL). Kerala: Anathod, Kottayam District, *Deb 30,416* (MH). Maharashtra: w/o locality, *Dalzell* (DD).

Note. Mittall (1955) has suggested that the inflorescence of *Centella asiatica* is actually a cyme rather than a simple umbel, as usually described. A thorough study of the inflorescences of the entire genus by Froebe (1979, pp. 39–49, figs. 16–19) indicates it to be a "trunkate pleiosciadiod" inflorescence, that is, a greatly simplified and reduced form of a vastly more complicated branching system.

3. SANICULA Linn.

Sp. Pl. 235. 1753; C. B. Clarke in Hook. f., Fl. Brit. Ind. 2: 670. 1879.

Erect or ascending perennial herbs, essentially glabrous, usually from a short rootstock or tuberous roots. Leaves petiolate to subsessile, membranaceous to subcoriaceous, palmately or pinnately lobed or divided to compound, rarely entire, the divisions toothed, dissected, or entire; petioles sheathing. Umbels simple, capitate, usually cymosely arranged. Involucre present. Flowers polygamous in the umbel, greenish-white to yellow, red, or purple; petals spatulate, with a narrower inflexed apex; calyx lobes prominent, setiform to ovate, distinct to connate; stylopodium obsolete. Staminate flowers pedicellate, deciduous; perfect flowers few, shortly pedicellate to subsessile. Fruit ovoid-oblong to globose, somewhat compressed laterally, densely covered with prickles or tubercles, the mericarps subterete in transection, the ribs obsolete. Vittae regular or irregular, few to many, large or small. Carpophore wanting. Commissural face plane to sulcate.

Type species: *Sanicula europaea* Linn.

Distribution: About 40 species, semicosmopolitan but absent from Australia and New Zealand. In India, a single widespread species.

1. **Sanicula elata** Buchanan-Hamilton ex D. Don, Prodr. Fl. Nepal. 183. 1825.

Sanicula europaea var. *elata* (Buchanan-Hamilton ex D. Don) Boiss., Bull. Soc. Bot. France 53: 421. 1906.

Sanicula hermaphrodita Buchanan-Hamilton ex D. Don, Prodr. Fl. Nepal. 183. 1825. Based on *Buchanan-Hamilton*, Nepal.

Sanicula javanica Blume, Bijdr. Fl. Ned. Ind. 15: 1882; *S. europaea* var. *javanica* (Blume) H. Wolff in Engl., Pflanzenreich 61 ($4^{\underline{228}}$): 64. 1913. Based on *Blume* (?), "Montis Gede," Java.

Sanicula europaea sensu C. B. Clarke in Hook. f., Fl. Brit. Ind. 2: 670. 1879, non Linn. 1753.

Sanicula europaea var. *pseudoyunnanensis* H. Wolff in Feddes, Repert. 27: 113. 1929. Based on *Duthie 10,406*, Panchmarhi, Water's meet, 23 Feb. 1891.

Plants slender, ascending to erect, 15–100 cm tall, the solitary stem dichasially branched above, from a horizontal rootstock bearing fibrous roots. Basal leaves longly petiolate, broadly ovate to orbicular, cordate, 3.6–5 cm in diameter, palmately 3–5-lobed, the principal segments lanceolate to rhombic-ovate, entire to pinnately lobed, the margins usually crenate-serrate with spinulose teeth, the petioles up to 18 cm long; cauline leaves gradually reduced upward, shortly petiolate to sessile, ternate or 3-parted. Involucre of 2–6 lanceolate bracts, 1–4 mm long. Inflorescence of cymosely arranged, simple umbels, each containing 3–6 deciduous, pedicellate staminate flowers and usually 3 subsessile perfect flowers; petals whitish to yellowish-green; calyx lobes linear, ca 1 mm long; styles twice as long as calyx. Fruit globose-ovoid to ovoid-oblong, 2–3 mm long, 1.5–2.5 mm broad, densely covered with uncinate bristles; vittae several and minute in intervals, 2 large ones on the commissure. Commissural face slightly concave. April–Oct.

Type locality: Narainhatty, Nepal, *Buchanan-Hamilton*.

Distribution: Southeastern Asia from Pakistan to western China and southern Japan, to southeastern Africa; in India, in the Himalaya and the hills of Assam and peninsular India.

Representative specimens: Jammu & Kashmir: Lidder Valley, *Duthie 13,459* (CAL, DD). Uttar Pradesh: Bansor Valley, *Duthie 14,501* (CAL, DD); Mussoorie, *Santapau 28,520* (BSD, CAL); Nainital, *Duthie 4044* (DD). Himachal Pradesh: Namali, *Bhattacharyya 40,340* (BSD). Nepal: Narainhatty, *Buchanan-Hamilton* (CAL). Sikkim: w/o locality, *Hooker* (CAL). West Bengal: Darjeeling, *Gamble 3921C* (DD); Bhanjan, Darjeeling District, *Mukherjee s. n.* (CUH). Arunachal Pradesh: Sonegoodam-Dreyi, Lohit, F. D., *R. S. Rao 10,396* (CAL). Nagaland: Naga Hills, *Bor 2687* (DD). Meghalaya: Khasia, *King's Collector* (CAL). Assam: w/o locality, *Griffith 2595/1* (CAL). Madhya Pradesh: Panchmarhi, *Duthie 10,406* (DD). Tamil Nadu: Ooty, collector not known (MH); Kodaikanal, *Bourne* (CAL).

4. ERYNGIUM Linn.

Sp. Pl. 232. 1753; C. B. Clarke in Hook. f., Fl. Brit. Ind. 2: 669. 1879.

Prostrate to erect, usually caulescent perennial or biennial herbs, mostly spinose and essentially glabrous, arising from taproots or rootstocks bearing fibrous roots. Leaves petiolate to sessile, coriaceous to membranaceous, entire or pinnately or palmately lobed or dissected, often spinose or ciliate; petioles sheathing. Umbels simple, capitate, solitary or more commonly cymosely arranged in branched inflorescences. Involucre usually conspicuous, of one or more series of bracts. Flowers white, bluish, or purple, all perfect, sessile, each subtended by a bracteole; petals ovate to oblong with a narrower inflexed apex; calyx lobes prominent, lanceolate or ovate, acute or mucronate, rarely obtuse; stylopodium lacking. Fruit globose to obovoid, the mericarps terete to flattened dorsally in transection, densely squamate or papillose, the ribs obsolete. Vittae usually inconspicuous. Carpophore wanting. Commissural face plane or slightly concave.

Type species: *Eryngium maritimum* Linn.

Distribution: About 200 species, cosmopolitan, but especially well developed in tropical and subtropical America and in the Mediterranean region. Of the Indian species, three are Eurasian, the fourth was introduced as a culinary herb from tropical America.

KEY TO SPECIES

1. Heads cylindrical, comate; inflorescence green at maturity; plants strongly *Coriandrum*-scented. .... 1. *E. foetidum*.
1. Heads ovoid, globose, or hemispherical, lacking a coma; inflorescence bright blue at maturity.
   2. Basal leaves oblong to oval, cordate, undivided, the margins crenate-serrate.
      3. Bracteoles both entire and tricuspidate; styles greatly exceeding calyx lobes. .... 2. *E. planum*.
      3. Bracteoles all entire; styles equaling or only slightly exceeding calyx lobes. .... 3. *E. coeruleum*.
   2. Basal leaves orbicular, ternately divided, the margins spinose-dentate. .... 4. *E. billardieri*.

1. Eryngium foetidum Linn., Sp. Pl. 232. 1753.

Plants slender, erect, biennial, the stem up to 30 cm tall, from a short rootstock bearing fibrous roots. Basal leaves thin, oblanceolate, obtuse, cuneate at base, 3–30 cm long, 1–5 cm broad, callous margined, crenate to finely spinose-serrate, the venation pinnately reticulate, the petioles broadly sheathing, up to 3 cm long; upper cauline leaves sessile, opposite, deeply spinose-serrate to parted. Inflorescence 3–5-furcate, the lateral branches continuing as monochasia. Heads cylindrical, 7–11 mm long, 3–5 mm broad, subtended by an involucre of ca 6 foliaceous lanceolate, usually spinose bracts 1–2.5 cm long, and with a conspicuous coma. Flowers greenish; petals oblanceolate; calyx lobes ovate-lanceolate, acute, 0.5 mm long; styles exceeding calyx. Fruit globose, ca 1.5 mm in diameter, the mericarps subcompressed dorsally, densely papillose. March–May.

Type locality: Mexico, collector not known.

Distribution: Indigenous to Central America, but introduced and sometimes cultivated in tropical Africa and tropical and subtropical Asia; in India, occurs widely at low altitudes.

Representative specimens: Uttar Pradesh: Dehra Dun, *Raizada* (DD). West Bengal: Jalpaiguri, *Mukherjee 1029* (CAL). Arunachal Pradesh: Cheng-Pang to Khela, *Panigrahi 14,460* (ASSAM, CAL). Mizoram: Aizal, *Godfrey 516* (CAL). Meghalaya: Burui Forest, K. & J. Hills, *P. C. Kanjilal 9837* (ASSAM). Assam: N. Cachar Hills, *U. N. Kanjilal 5621* (ASSAM, CAL). Orissa: Sambalpur, *Mooney 3795* (DD). Kerala: Chandanathode, Cannanore District, *Ellis 25,268* (MH).

2. **Eryngium planum** Linn., Sp. Pl. 232. 1753.

Eryngium coeruleum sensu C. B. Clarke in Hook. f., Fl. Brit. Ind. 2: 669. 1879, *pro parte*, non M. Bieb. 1798.

Plants stout, perennial, up to 1 m tall, from a woody taproot. Basal leaves coriaceous, oblong-oval, cordate, 10–15 cm long, 3–8 cm broad, crenate to spinulose-serrate, palmately 7–9-veined, the petioles stout, 10–15 cm long; upper cauline leaves sessile or subsessile, entire or 5–7-lobed. Inflorescence repeatedly cymosely branched. Heads usually ovoid, 1.5–2 cm long, 1–1.5 cm broad, subtended by an involucre of 6 rigid, linear, spinose-dentate bracts 1.5–3.5 cm long. Flowers bluish; petals ovate-oblong; calyx lobes lanceolate, acute, 2–2.5 mm long; styles greatly exceeding calyx. Fruit ovoid, 3–4 mm long, the mericarps compressed dorsally, densely covered with subequal linear white scales up to 1 mm long. June–Sept.

Type locality: "In Russia, Polonia, Austria, Helvetia," collector not known.

Distribution: Central Europe and temperate Asia to Siberia; in India occurs only in the Kashmir Valley at 1,500 to 2,000 m.

Representative specimens: Jammu & Kashmir: Kashmir, *Thomson* (CAL), *Falconer 490* (CAL), *Tanner 265* (DD).

3. **Eryngium coeruleum** M. Bieb., Tabl. Prov. Casp. 112. 1798; C. B. Clarke in Hook. f., Fl. Brit. Ind. 2: 669. 1879, *pro parte*.

Eryngium biebersteinianum Nevski ex Bobrov. Fl. U.R.S.S. 16: 86. 1950.

Plants stout, perennial, up to 1 m tall, from a taproot. Basal leaves subcoriaceous, oblong to orbicular, 4–5 cm long, 3–4 cm broad, undivided, the margins crenate or crenate-serrate; lower cauline leaves trisect with oblong divisions, the upper cauline leaves palmately 5-lobed or 5-parted, their divisions narrow, spinose-serrate and sometimes lobed. Inflorescence divaricately branched. Heads subovoid to hemispherical, ca 1 cm in diameter, subtended by an involucre of 4–6 rigid, spinose, linear bracts 2–3 cm long, 2–4 mm broad. Flowers purplish; petals oblong, fimbriate at apex; calyx lobes lanceolate, mucronate, ca 2.5 mm long; styles slightly exceeding calyx. Fruit turbinate, ca 0.5 mm long, the mericarps angled, densely covered with subequal, lanceolate, long-acuminate scales. June–Sept.

Type locality: Described from the "Shirvan Steppe."

Distribution: Mediterranean area to central Asia; in India, found only in the Kashmir Valley.

Representative specimens: Jammu & Kashmir: Kashmir, *Falconer 489* (CAL); Srinagar, 1700 m, 14 July 1891, *Gammie* (DD), 7 Oct. 1889, *Gallar 9032* (DD).

Note. *E. biebersteinianum* Nevski was apparently created because *E. coeruleum* M. Bieb. was thought to be preempted by the existence of *E. coeruleum* Gilib. Fl. Lithuan. 1785. However, Gilibert's names are to be disregarded for purposes of priority. (L. C.)

4. **Eryngium billardieri** Delaroche f. ssp. **nigromontanum** (Boiss. & Buhse) H. Wolff in Engl., Pflanzenreich 61 (4^{228}): 150. 1913.

Eryngium nigromontanum Boiss. & Buhse, Aufz. Pfl. Reise Transk. Pers. 95. 1860.
Eryngium billardieri sensu C. B. Clarke in Hook. f. Fl. Brit. Ind. 2: 670. 1879, non Delaroche f. 1808.
Eryngium billardieri γ *meiocephalum* Boiss. Fl. Orient. 2: 875. 1872. Based on *E. nigromontanum* Boiss. & Buhse.

Plants stout, perennial, erect, 15–80 cm tall, from a taproot. Basal leaves coriaceous, 5–20 cm long, palmately 3-lobed, the segments trifid or pinnatifid, spinose-dentate; lower cauline leaves similar but shortly petiolate; upper cauline leaves sessile, bipinnatifid, with short and narrow spinose-dentate divisions. Inflorescence widely branched above. Heads ovoid-globose, 5–15 mm in diameter, subtended by an involucre of 5–8 linear, rigid, entire or spinose bracts 1–3 cm long. Flowers bluish; petals oblong, fimbriate at apex; calyx lobes lanceolate, acuminate. Fruit turbinate, the mericarps compressed dorsally, bearing a few curved lanceolate lateral scales and numerous rounded dorsal ones. June–July.

Type locality: Lebanon.

Distribution: Eastern Mediterranean to Central Asia; in India, only in the Kashmir Valley.

Representative specimens: Jammu & Kashmir: Kashmir, *Meebold 3387* (CAL); Ladak, *Thomson* (CAL); Srinagar, *T. A. Rao 784* (BSD).

5. **PYCNOCYCLA** Lindley

In Royle, Ill. Bot. Himal.
Mount. 232. 1835; C. B. Clarke in Hook. f.,
Fl. Brit. Ind. 2: 694. 1879.

Rigid, erect, caulescent, perennial, subshrubby herbs, pubescent or glabrous, from taproots or woody rootstocks, the nearly naked branches often rushlike. Leaves longly petiolate, pinnately dissected, the divisions rigid, spinescent, filiform. Inflorescence of terminal or axillary compact compound umbels. Involucre of several linear to lanceolate spinescent bracts; bracteoles of the involucel similar. Rays few and very short. Umbellets consisting of a single sessile central perfect flower surrounded by several male flowers borne on pedicels which become enlarged and inflated at maturity. Flowers white to purplish; petals obovate with a narrower inflexed apex, hairy dorsally; calyx lobes prominent, unequal; ovary pubescent; stylopodium conical. Fruit oblong to ovoid, narrowed at apex, slightly compressed laterally, the mericarps subterete (frequently only one developed), pubescent, the ribs filiform, the lateral situated on margin of commissure. Vittae minute, one to several in intervals and on commissure. Commissural face concave or sulcate. Carpophore obsolete.

Type species: *Pycnocycla glauca* Lindley.

Distribution: About a dozen species of arid regions extending from mid-Africa to western and central Asia; only the type species is known from India.

1. **Pycnocycla glauca** Lindley in Royle, Ill. Bot. Himal.

Mount. 232. pl. 51. 1835; C. B. Clarke in Hook. f.,
Fl. Brit. Ind. 2: 694. 1879.

Chamaesciadium glaucum (Lindley) Hiroe, Umbell. World, 995. 1979.

Plants branched from a woody rootstock, 15–20 cm tall, the stems terete, striate, and subglaucous. Leaves pinnately dissected, the divisions remote, filiform, glabrous, 1–6 cm long. Umbels terminal on pubescent peduncles up to 9 cm long. Involucral bracts several, linear-lanceolate, 3–6 mm long; bracteoles 6–8, hairy, ca 2 mm long. Rays 8–12, up to only 4 mm long at maturity. Pedicels of the 6–8 staminate flowers ca 1 cm long. Fruit oblong-lanceolate, ca 8 mm long, 2 mm broad, villous. Vittae 3 or 4 in intervals, 4–6 on commissure. Commissural face sulcate. May–Nov.

Type locality: "Banks of Sutlej," Punjab.

Distribution: Ethiopia and Yemen to India, where it occurs in drier areas of N. W. India, Madhya Pradesh, and Bihar.

Representative specimens: Bihar: Tongo, Ranchi Distr., *Bressers 128* (CAL). Madhya Pradesh: Jubbalpore, *Beddome* (CAL). "N. W. India," *Royle* (CAL, K).

GROUP II

1. Fruit densely longly hairy or bristly, or provided with prominent prickles or tubercles.
  2. Calyx teeth obsolete or at most evident; carpophore entire or bifid only at apex or divided to the middle.
    3. Fruit compressed dorsally, its secondary ribs bearing barbed or glochidiate prickles joined into a wing. .... 7. *Daucus.*
    3. Fruit terete or slightly compressed laterally, its secondary ribs bearing distinct hairs, bristles, prickles, or tubercles.
      4. Bracts and bracteoles neither conspicuous nor scarious. .... 6. *Torilis.*
      4. Bracts and bracteoles conspicuously scarious or scarious-margined.
        5. Plants coarse and prickly; styles short; fruit with glochidiate-prickly ribs. .... 8. *Turgenia.*
        5. Plants delicate, glabrous or pubescent but not prickly; styles long; fruit with hirtellous or bristly ribs. .... 10. *Psammogeton.*
  2. Calyx teeth prominent, subulate, unequal; carpophore bipartite; fruit strongly aromatic. .... 9. *Cuminum.*
1. Fruit essentially glabrous, or at least lacking conspicuous hairs, bristles, prickles, or tubercles.
  6. Annual; ultimate divisions of cauline leaves very narrow; marginal petals usually radiant; stylopodium conical; styles slender; fruit broadly ovoid to subglobose with a hard pericarp. .... 11. *Coriandrum.*
  6. Perennial or biennial; leaflets broad, ovate to orbicular; marginal petals not radiant; stylopodium depressed; styles short; fruit ovoid to subquadrate, with a hollow, spongy pericarp. .... 12. *Chamaesium.*

6. **TORILIS** Adans.
Fam. Pl. 2: 99. 1763.

Caucalis sensu C. B. Clarke in Hook. f., Fl. Brit. Ind. 2: 718. 1879, *pro parte*, non Linn. 1753.

Erect or ascending, herbaceous, caulescent, branching, pubescent or hispid annuals, from slender taproots. Leaves petiolate, membranaceous, pinnately or ternate-pinnately divided or decompound, the ultimate divisions small and narrow; petiole sheathing. Inflorescence of lax or compact, terminal or lateral, pedunculate or rarely sessile compound umbels. Involucre of small bracts, or wanting. Rays spreading-ascending to obsolete. Involucel of several narrow bracteoles. Flowers white or rose tinged; petals obovate, with a narrower inflexed apex, the outer sometimes radiant; calyx teeth evident to obsolete; stylopodium thick, conical. Fruit ovoid to linear, compressed laterally, bristly, prickly, or tuberculate, the mericarps compressed dorsally in transection; primary ribs filiform, setulose or bristly, the lateral on the commissure; secondary ribs occupying entire interval and covered with glochidiate prickles or tubercles. Vittae solitary under secondary ribs, 2 on commissure. Commissural face concave to sulcate. Carpophore bifid at apex or to middle.

Type species: *Tordylium anthriscus* Linn. [*Torilis japonica* (Houtt.) DC.].

Distribution: About a dozen species, primarily of the Mediterranean region, several of them cosmopolitan weeds.

Note. Hamal & Koul, in their interesting paper on *Torilis* cytotaxonomy (1988), speak of obtaining material of *Torilis leptophylla*, *T. japonica*, *T. arvensis* (Huds.) Link, and *T. stocksiana* (Boiss.) Drude from "various sites in Kashmir Himalayas". The weedy *T. nodosa* (Linn.) Gaertner has been reported from Dehra Dun and Kashmir. We have seen no specimens of the last three taxa from our area, but included them in the key.

KEY TO SPECIES

1. Umbels open, the rays spreading, usually longly pedunculate, terminal and/or lateral; mericarps subequal.
  2. Petals glabrous; stigmas sessile; fruit 6–7 mm long. .... 2. *T. leptophylla*.
  2. Petals setose dorsally; stigmas on styles ca. 1 mm long; fruit 2.5–5 mm long.
    3. Stem erect; rays 2–12; fruit ovoid or ellipsoid.
      4. Involucral bracts 4–6; fruit bristles incurved-ascending, shorter than fruit width. .... 1. *T. japonica*.
      4. Involucral bracts 1 or 0; fruit bristles ± straight, about equaling fruit width. *T. arvensis*.
    3. Stems usually low or procumbent; rays 2 or 3; fruit linear-cylindric. .... *T. stocksiana*.
1. Umbels capitate, sessile or shortly pedunculate, leaf opposed; mericarps distinctly dimorphic. .... *T. nodosa*.

1. **Torilis japonica** (Houtt.) DC., Prodr. 4: 219. 1830.

Caucalis japonica Houtt., Nat. Hist. II. 8: 42. 1777; *Torilis anthriscus* β *japonica* (Houtt.) Boiss., Bull. Herb. Boiss. II. 3: 856. 1903.
Tordylium anthriscus Linn., Sp. Pl. 1: 240. 1753; *Caucalis anthriscus* (Linn.) Hudson, Fl. Angl. 99. 1762; C. B. Clarke in Hook. f., Fl. Brit. Ind. 2: 718. 1879, non *Torilis anthriscus* Gaertner 1788. Based on "In Europae septentrionalis arvis ruderatis," Linnean Herbarium.
Caucalis elata D. Don, Prodr. Fl. Nepal. 183. 1825; *Torilis elata* (D. Don) DC., Prodr. 4: 220. 1830. Based on *Buchanan-Hamilton* (Wallich, Cat. 564).

Plants 30–100 cm tall, the branches ascending, appressed-hispid throughout. Leaves triangular-ovate, ternate-pinnate or pinnate, the leaflets lanceolate to oblong, 0.5–4 cm long, 2–15 mm broad, acute or acuminate, regularly laciniate-dentate, or incised, or divided; petiole 2.5 cm long. Umbels terminal and lateral ca. 10-flowered, the peduncle 2–20 cm long. Involucre of 4–12 linear, hispid bracts up to 1 cm long. Rays 4–12, unequal, 0.7–3 cm long. Involucel of 3–7 linear to filiform bracteoles, ca. 4 mm long. Outer petals slightly radiant, all hirsute dorsally; calyx teeth triangular-lanceolate, ca. 0.5 mm long. Pedicels 1–4 mm long. Fruit ovoid, 1.5–4 mm long, 2–3 mm broad, beset with incurved, ascending, uncinate bristles or glochidiate prickles. Commissural face sulcate. May–Oct.

Type locality: Japan.

Distribution: Eurasian, widely adventive in India and elsewhere.

Representative specimens: Jammu & Kashmir: Baisaran, *T. A. Rao 558* (BSD). Himachal Pradesh: Manali, *Bhattacharyya 40,321* (BSD). Uttar Pradesh: Chakrata, *Bhattacharyya 16,058* (BSD). Nepal: Nepal, *Wallich 564* (CAL); Joka, *Wallich 564A* (CAL). Sikkim: Hee, *Clarke 13,036C* (CAL). Bhutan: Gomuchu Hill, *A. S. Rao 43,403* (CAL). Arunachal Pradesh: Lohit, Songodam, *Rao 10,396* (ASSAM). Nagaland, *Bor 2673* (DD). Mizoram, *Dutta 33,932* (ASSAM).

Note. Although *Torilis elata* (D. Don) DC. is said to possess more divided leaves and curved fruit prickles, it appears to be conspecific. (P.K.M.)

2. **Torilis leptophylla** (Linn.) Reichenb. f., Ic. Fl. Germ. 2: 83, pl. 169. 1867.

Caucalis leptophylla Linn., Sp. Pl. 242. 1753; C. B. Clarke in Hook. f., Fl. Brit. Ind. 2: 719. 1879; *Anthriscus leptophylla* (Linn.) Kozo-Polj., Bull. Soc. Nat. Mosc. II. 29: 151. 1916.

Plants slender, 20–45 cm tall, branched from base, appressed-hispid throughout. Leaves triangular-ovate to ovate-lanceolate, 1–2-pinnate, the leaflets ovate to narrowly oblong-lanceolate, 5–15 mm long, 2–10 mm broad, acute or acuminate, regularly incised and divided. Umbels arising opposite leaves, the peduncles 1–8 cm long. Involucre wanting, the involucel of several subulate bracteoles. Rays 2–4, unequal, 0.5–2 cm long. Calyx teeth prominent. Pedicels 1–4 mm long. Fruit oblong-ovoid, 3–5 mm long, 2–3 mm broad, the ribs slender and beset with straight or uncinate bristles as long as fruit width. Commissural face deeply sulcate. Mar.

Type locality: "Galla," no collector indicated.

Distribution: Mediterranean region, to northwestern India.

Representative specimens: Jammu & Kashmir: Kashmir, *Thomson* (CAL). Punjab: Punjab, *Clarke 104* (CAL). Himachal Pradesh: Kangra, *Burkill* (CAL); Rainuka Lake, *Rau 11,355* (BSD).

7. DAUCUS Linn.

Sp. Pl. 242. 1753; C. B. Clarke in Hook. f., Fl. Brit. Ind. 2: 718. 1879.

Erect, herbaceous, caulescent, pubescent annuals or biennials from taproots. Leaves petiolate, membranaceous, pinnately decompound, the ultimate divisions small and narrow; petiole sheathing. Inflorescence of lax, terminal or lateral, pedunculate, compound umbels. Involucre of numerous dissected or entire bracts, rarely wanting. Rays few to many, spreading. Involucel of numerous toothed or entire bracteoles, rarely wanting. Flowers mostly white (or central flower or flowers of each umbel colored); petals obcordate, unequally cleft, with a narrower inflexed apex, the outer petals commonly radiant; calyx teeth obsolete to evident; stylopodium conical. Fruit oblong to ovoid, compressed dorsally, bristly or prickly, the mericarps flattened dorsally in transection; primary ribs filiform, bristly, secondary ribs with barbed or glochidiate prickles joined into a wing. Vittae solitary under secondary ribs, 2 on commissure. Commissural face shallowly concave. Carpophore entire or shortly bifid.

Type species: *Daucus carota* Linn.

Distribution: A semicosmopolitan genus of ca two dozen species, especially well developed in the Mediterranean region, but lacking indigenous species in southern and eastern Asia. Only the type species is represented in India, through introduction and cultivation.

1. **Daucus carota** Linn., Sp. Pl. 242. 1753; C. B. Clarke in Hook. f., Fl. Brit. Ind. 2: 718. 1879.

Plants normally biennial, 30–120 cm tall, papillate-hispid to glabrate, from a stout fusiform taproot. Leaves oblong, 5–15 cm long, 2–7 cm broad, pinnately decompound with acute, mucronate, linear to oval ultimate divisions 2–12 mm long, 0.5–2 mm broad, usually hispid especially on nerves and margins; petiole 3–10 cm long; divisions of cauline leaves often elongated. Umbels lax in anthesis, strongly contracted in fruit. Involucre of conspicuous, usually pinnately divided, scarious-margined, reflexed bracts 3–30 mm long. Rays numerous, unequal, 3–7.5 cm long. Umbellets 20–30-flowered, often with colored sterile central flowers. Involucel of linear, mostly entire, ± scarious, ciliate bracteoles. Pedicels 3–10 mm long. Fruit ovoid, 3–4 mm long, ca 2 mm broad. Mar.–May.

Type locality: "In Europae campis exaridis," collector not known.

Distribution: Eurasia, but widely adventive and cultivated throughout warmer parts of the earth, including India.

Representative specimens: Jammu & Kashmir: Kashmir, *Stoliczka* (CAL). West Bengal: Raneeganj, *Raya* (CAL). Tamil Nadu: Coimbatore, *Saroja 8731* (MH).

8. **TURGENIA** Hoffm.
Gen. Umbell. 59. 1814

Caucalis subgen. *Turgenia* Drude in Engl. & Prantl, Pflanzenfam. 38: 157. 1898.
Caucalis sensu C. B. Clarke in Hook. f., Fl. Brit. Ind. 2: 719. 1879, *pro parte*, non Linn. 1753.

Erect, herbaceous, caulescent, prickly hispid annuals from taproots. Leaves petiolate, membranaceous, oblong to oval, 1-2-pinnate, with rather large, oblong to oval leaflets; petiole sheathing. Inflorescence of few-rayed, lax, terminal or lateral compound umbels. Involucre of several linear to lanceolate, acute or acuminate, scarious bracts, or lacking; the bracteoles similar. Umbellets few flowered, the flowers white or rose tinged, the central flower staminate; petals obovate with a narrower inflexed apex, the outer petals radiant; calyx teeth evident; stylopodium conical. Fruit oblong to ovoid, compressed laterally, glochidiate-prickly or tuberculate, the mericarps subterete in transection, both primary and secondary ribs provided with 2 or 3 rows of stout, curved or straight glochidiate prickles, the lateral ribs on the commissure. Vittae solitary under secondary ribs, 2 on commissure. Commissural face involute. Carpophore bifid.

Type species: *Caucalis latifolia* Linn. [*Turgenia latifolia* (Linn.) Hoffm.]

Distribution: A single variable Eurasian species.

1. **Turgenia latifolia** (Linn.) Hoffm., Gen. Umbell. 59. 1814.

Tordylium latifolium Linn., Sp. Pl. 240. 1753; *Caucalis latifolia* (Linn.) Linn., Syst. Nat. ed. 12. 2: 205. 1767; C. B. Clarke in Hook. f., Fl. Brit. Ind. 2: 719. 1879.

Plants rather rigid, divaricately branched, 15–60 cm tall. Leaves 4–16 cm long, 2–9 cm broad, the leaflets 1–4.5 cm long, 3–15 mm broad, coarsely dentate to lobed, the upper decurrent, the lower subsessile; petiole 2–6 cm long, scarious-margined at base; uppermost leaves with inflated sheaths. Peduncles stout, 2–15 cm long. Involucral bracts 2 or 3, 5–8 mm long. Rays 2 or 3, stout, 2–5 cm long. Involucel of 4 or 6 bracteoles 3–5 mm long. Fruit ca 10 mm long, ca 5 mm broad, the primary and secondary ribs aculeate-prickly and equally prominent. May–July.

Type locality: "Gallia, Italia," collector not known.

Distribution: Southern Europe and western and southern Asia; at 1,000–3,000 m in Kashmir.

Representative specimens: Jammu & Kashmir: Kashmir, *Falconer 498* (CAL); Pangi, *Nair 22,455* (BSD).

9. CUMINUM Linn.

Sp. Pl. 254. 1753; C. B. Clarke in Hook. f., Fl. Brit. Ind. 2: 717. 1879.

Slender, erect, glabrous annuals from slender taproots. Leaves petiolate, membranaceous, ternately or pinnately dissected with filliform ultimate divisions; petiole sheathing. Inflorescence of lax, terminal or lateral, pedunculate compound umbels. Involucre of several linear, entire or ternate reflexed bracts, the bracteoles similar. Rays few, spreading-ascending, subequal or unequal. Flowers white or rose tinged; petals oblong with a narrower inflexed apex, the outer radiant; calyx teeth subulate, unequal; stylopodium conical, attenuate into the short, rigid styles. Fruit oblong, narrowed at both apex and base, slightly compressed laterally, setose and bristly, the mericarps subterete in transection; primary ribs filiform, minutely setulose; secondary ribs usually more prominent and bristly. Vittae solitary under secondary ribs, 2 on commissure. Commissural face slightly concave. Carpophore bipartite.

Type species: *Cuminum cyminum* Linn.

Distribution: A small genus of northern Africa to southwestern Asia.

1. **Cuminum cyminum** Linn., Sp. Pl. 254. 1753; C. B. Clarke in Hook. f., Fl. Brit. Ind. 2: 718. 1879.

Plants 15–40 cm tall. Leaves ovate, 5–10 cm long, the filiform ultimate divisions entire, 1–5 cm long; petiole 1–2 cm long. Peduncles 3–10 cm long. Involucral bracts 5–8, unequal, 1–5 cm long. Rays 2–6, 1–2 cm long. Umbellets up to 8-flowered. Involucel of usually 5 entire or pinnatisect bracteoles 5–10 mm long. Calyx teeth ca 0.5 mm long. Pedicels ca 5 mm long. Fruit 5–7 mm long, 2.5–3 mm broad. Jan.–Mar.

Type locality: "In Aegypto, Aethiopia," collector not known.

Distribution: Probably indigenous to northern Africa, but widely cultivated and sometimes adventive in warmer regions.

Representative specimens: Punjab: Gurgaon, *Khan* (CAL). Uttar Pradesh: Saharanpur Garden, *Duthie* (CAL).

10. **PSAMMOGETON** Edgew.

Proc. Linn. Soc. 1: 253. 1846; C. B. Clarke in Hook. f., Fl. Brit. Ind. 2: 719. 1879.

Cuminum subgen. *Psammogeton* Kozo-Polj., Bull. Soc. Nat. Mosc, II. 29: 208. 1916.

Slender, erect, caulescent, branching, pubescent or glabrous annuals from slender taproots. Leaves petiolate, membranaceous, pinnately dissected, the ultimate divisions narrow; petiole sheathing. Inflorescence of lax few-rayed, terminal or leaf-opposed compound umbels. Involucre and involucel of several, narrow, entire or parted, scarious bracts or bracteoles respectively. Rays spreading-ascending. Flowers white to pink or purplish; petals obovate, bilobed, with a narrower inflexed apex, the outer sometimes radiant; calyx teeth evident to obsolete; stylopodium low conical, much shorter than the elongate styles. Fruit oblong to ovoid, compressed laterally, subconstricted at commissure, densely hairy or hispid, the mericarps terete in transverse section; primary ribs armed with a single row of long, glochidiate hairs; secondary ribs bearing shorter biseriate setose or glochidiate hairs. Vittae large, solitary under secondary ribs, 2 on commissure. Commissural face plane to shallowly concave. Carpophore bifid at apex.

Type species: *Athamanta canescens* DC. [*Psammogeton canescens* (DC.) Vatke]

Distribution: A small genus of central Asia.

1. **Psammogeton canescens** (DC.) Vatke, App. ad Ind. Sem. Hort. Berol. 3. 1876.

Athamantha canescens DC., Prodr. 4: 155. 1830.

Pimpinella crinita Boiss., Ann. Sci. Nat. Bot. 3: 131. 1844; *Psammogeton crinitum* (Boiss.) Boiss., Fl. Orient. 2: 1078. 1872. Based on *Auchery 4587*, Persia.

Psammogeton biternatum Edgew., Trans. Linn. Soc. 20: 57. 1851; C. B. Clarke in Hook. f., Fl. Brit. Ind. 2: 719. 1879. Based on *Edgeworth 167*, "In arenosis Ind. Prov. Bor. Occ. Ambala. Balawali."

Psammogeton biternatum var. *villosa* C. B. Clarke in Hook. f., Fl. Brit. Ind. 2; 719. 1879. Based on *Stocks*, Scinde.

Plants 5–45 cm tall, softly pubescent or tomentose, branching from base. Leaves oblong to ovate, 2–8 cm long, 2.5–5 cm broad, ternate-pinnate or bipinnate, the leaflets lanceolate, oblong, or cuneate, entire to pinnatisect, 5–10 mm long, up to 2 mm broad; petiole scarious-margined, half as long as lamina. Involucre of 5–8 lanceolate, scarious-margined bracts 3–8 mm long, the bracteoles similar, 2–6 mm long. Rays 4–12, spreading-ascending, subequal, 0.5–2.5 cm long. Flowers white or rose coloured. Fruit narrowly oblong to ovoid, ca 2 mm long, 0.5 mm broad, densely hairy with slender bristles longer than fruit width. Apr.–Jul.

Type locality: "Inter Aleppo et Baghdad," *Olivier & Bruguiere.*

Distribution: Central Asia to northwestern India, up to 1,000 m.

Representative specimens: Punjab: Punjab, *Clarke 136* (CAL). Uttar Pradesh: Dehra Dun, Rispan, *Babu 35,029* (BSD).

Note. Wagenitz (1956) treats *P. canescens* as a polytypic species consisting of four subspecies. Apparently only *P. canescens* subsp. *biternatum* (Edgew.) Wagenitz occurs in India, the other three being restricted to Afghanistan, Iran, and Pakistan.

11. **CORIANDRUM** Linn.

Sp. Pl. 256. 1753; C. B. Clarke in Hook. f., Fl. Brit. Ind. 2: 717. 1879.

Slender, erect, caulescent, branching, glabrous annuals from slender taproots. Leaves petiolate, membranaceous, pinnately dissected; petiole sheathing. Inflorescence of lax, few-rayed, terminal or lateral compound umbels. Involucre usually wanting. Rays spreading-ascending, unequal. Involucel of a few filiform bracteoles. Flowers white to purple; petals obovate, emarginate, with a narrower inflexed apex, the outer usually radiant; calyx teeth prominent, unequal; stylopodium conical, the styles slender. Fruit broadly ovoid to subglobose, terete, with a hard pericarp, glabrous, the mericarps hemispherical, compressed dorsally, not separating readily at maturity, both primary and secondary ribs filiform. Vittae obscure. Commissural face concave. Carpophore bipartite.

Type species: *Coriandrum sativum* Linn.

Distribution: A Mediterranean genus, apparently of two species.

1. **Coriandrum sativum** Linn., Sp. Pl. 256. 1753; C. B. Clarke in Hook. f., Fl. Brit. Ind. 2: 717. 1879.

Plants 20–70 cm tall. Lower leaves ovate, 3–15 cm long, 2–10 cm broad, simple, ternately or pinnately lobed or pinnate, with large flabelliform, cuneate, toothed or incised leaflets 10–20 mm long, 5–10 mm broad; petiole 2–15 cm long; upper cauline leaves ternately or pinnately decompound with linear to filiform, entire, obtuse, ultimate divisions 2–15 mm long, 5 mm broad. Peduncles 2–10 cm long. Involucre wanting, or of a small solitary bract. Rays 3–8, 1–2.5 cm long. Umbels 5–8-flowered. Involucel dimidiate, the bracteoles 3–5, linear, ca 5 mm long. Pedicels 3–5 mm long. Fruit 1.5–5 mm in diameter. Dec.–Feb.

Type locality: "In Italiae agris," collector not known.

Distribution: Indigenous to Mediterranean area, widely cultivated throughout the world.

Representative specimens: Jammu & Kashmir: Pangee, *Stoliczka* (CAL). Punjab: Punjab, *Clarke 157* (CAL). Uttar Pradesh: Rispana, *Babu 33,872* (BSD). Bihar: Chotanagpur, *Wood* (CAL). West Bengal: Bengal, *Hooker & Thomson* (CAL); Sukna, *Ribu & Rhomoo* (CAL). Tripura: Agartala, *Debbarman 634* (CAL). Tamil Nadu: R. S. Puram, *Saroja* (MH). Bangladesh: Purniya, *Heyne* (Wallich, Cat. 594C) (CAL).

12. CHAMAESIUM H. Wolff

Notizbl. Bot. Gart. Berlin-Dahlem 9: 275. 1925.

Low to strictly erect, simple or few-branched, perennial or biennial herbs, glabrous to sparsely pubescent, from fusiform taproots. Leaves petiolate, membranaceous to thick, pinnate with 5–12 oblong to orbicular, entire to lobed, palmately or pinnately nerved leaflets; petiole sheathing, sometimes inflated. Inflorescence of a lax, terminal and sometimes also lateral, compound umbels, the many-flowered individual umbellets often appearing as simple umbels if the main stem is strongly condensed. Involucre and involucel of 1–3-pinnatifid bracts or bracteoles respectively, or the latter sometimes entire. Rays strictly ascending to spreading-ascending, very unequal. Flowers greenish-white; petals spatulate to obovate, with a shortly incurved apex; calyx teeth evident to obsolete; stylopodium much flattened and depressed, broader than summit of fruit, the styles short, divergent or erect. Fruit oblong to ovoid or subquadrate, somewhat narrowed to sharply constricted at apex, rounded or cordate at base, glabrous, the mericarps terete or slightly compressed dorsally in transection; primary and secondary ribs subequal, hollow, elevated, smooth, obtuse or acute, or greatly extended into conspicuous undulate and even lobed wings. Vittae small or large, 5 in each mericarp, alternating with the ribs, an additional "pseudovitta" at junction of each primary with a secondary rib, 2 on commissure. Commissural face concave. Carpophore bipartite.

Type species: *Chamaesium paradoxum* H. Wolff.

Distribution: A small genus of the Eastern Himalaya, extending from Yunnan and Tibet to Sikkim and central Nepal.

Note. Farille (1985) makes the interesting observation that the apparent 10 individual ribs of each mericarp are actually 5 anatomically double or paired ribs, which form a "pseudo-interval" between the members of each pair, and that is underlain by a "pseudovitta."

KEY TO SPECIES

1. Fruit about as broad as long, not sharply constricted at apex, the ribs low, prominent but unwinged.
   2. Plants low, acaulescent or very shortly caulescent; leaflets ovate to orbicular. .... 1. *C. novem-jugum*.
   2. Plants slender, strict, caulescent; leaflets elliptic to cuneate-obovate. .... 2. *C. viridiflorum*.
1. Fruit much longer than broad, sharply constricted at apex, the ribs strongly undulately winged.
   3. Leaves membranaceous, glabrous; petals broad, oboval; calyx teeth evident, persistent. .... 3. *C. malleanum*.
   3. Leaves thick, crystalline-scarious, sparsely appressed-pubescent; petals narrower, spatulate; calyx teeth 0. .... 4. *C. shrestaeanum*.

1. **Chamaesium novem-jugum** (C. B. Clarke) Norman,
Jour. Bot. 76: 231. 1938.

Trachydium novem-jugum C. B. Clarke in Hook. f., Fl. Brit. Ind. 2: 672. 1879.

Plants low, 3–10 cm tall, perennial from a base covered with fibrous leaf remains and surmounting a stout taproot. Leaves 1.5–5 cm long, including petiole, with 4–6 pairs of ovate to orbicular, crenate leaflets 8–12 mm long, 6–10 mm broad; petiole 1.5–3 cm long, broadly scarious-dilated. Terminal umbel pedunculate to sessile and often appearing as a crown of simple umbels borne on stout and very unequal ascending rays. Involucre of several foliaceous, pinnate bracts, with narrower divisions. Rays 5–15(–20), 2–8 cm long. Umbellets 10–20-flowered, the mature pedicels 4–6 mm long. Involucel of 5–10 oblong or spatulate, entire, lobed, or subpinnatifid bracteoles 3–10 mm long, about equaling the umbellet. Petals obovate; calyx teeth obsolete; stylopodium low conical, swollen and as broad as fruit width, the styles very short. Fruit 2.5–3 mm long, 1.5–2.5 mm broad, the ribs low, spongy. July–Sept.

Type locality: "SIKKIM, alt. 12,000–15,000 ft., Lama Kangra, Tungu and Yeumtong," *Hooker* (*Hooker & Thomson*, Herb. Ind. Orient., *Trachydium 5*) (K, holotype; CAL, GH, NY).

Distribution: Himalaya, from Bhutan to Nepal, 4,000–5,600 m; Tibet.

Representative specimens: Sikkim: Llouk, *Smith & Cave 1785* (CAL); Kapoop, *Smith 3412* (CAL); Chumolari, *Rohmoo 470* (AAH, BM).

2. **Chamaesium viridiflorum** (Franchet) H. Wolff ex Shan,
Sinensia 8: 87. 1937.

Trachydium viridiflorum Franchet, Bull. Soc. Philom. Paris VIII. 6: 111. 1894.
Trachydium affine W. Smith, Rec. Bot. Surv. India 4: 374. 1913. Based on *W. W. Smith 4109*, "Sikkim Himalayas, Ningbil, Chola Range, 4500–4650 m."

Plants slender, erect, sparsely branching above, glabrous, 10–25 cm tall from a fibrous base surmounting a fusiform root. Leaves 2 or 3, 5–10 cm long, including petiole, oblong, pinnate with 5–7 elliptic to cuneate-obovate, serrate or lobed leaflets 5–10 mm long, 2–5 mm broad; petiole slender, oblong-sheathing at base, about equaling limb; upper cauline leaves with linear-oblong leaflets and inflated sheaths. Umbels terminal and sometimes also lateral, longly pedunculate, 1–2.5 cm long. Involucre of several (2–4) linear to spatulate, entire to pinnatifid bracts. Rays 5–15, in terminal umbels, 2–3 cm long, unequal. Involucel 0. Umbellets 10–15-flowered, the pedicels 3 times as long as the flowers. Petals ovate, acute and nearly plane at apex, rounded at base; calyx teeth evident, broad; stylopodium low conical, massive, the styles short, reflexed. Fruit suborbicular, ca 1.5 mm in diameter, the ribs low, obtuse. Commissural surface slightly concave. Aug.–Sep.

Type locality: "Yun-nan, in rupibus adumbratis prope cacumina montis Ma-eul-chan, alt. 3500 m," 6 Aug. 1889, *Delavay 4092*, 10 Sept. 1889, *3900* (P, lectotype).

Distribution: Yunnan to Sikkim Himalaya, 3,500–4,000 m.

Representative specimen: Sikkim: Ningbil, Chola Range, *Smith 4115* (CAL).

3. **Chamaesium malleanum** Farille & Malla in Farille, Cauwet-Marc & Malla, Candollea 40(2): 536. 1985.

Plants slender, aespitose, the stem simple, perennial, 30–40 cm tall. Basal leaves unknown, the lower cauline 10–25 cm long including petiole, oblong to oval, pinnate with 5–7 oboval leaflets 1–2 cm in diameter, irregularly and coarsely dentate or lobed distally, cuneate at base; petiole slender, equaling or exceeding lamina, scarcely dilated. Umbel terminal, solitary, longly pedunculate. Involucre of several foliaceous, pinnatifid and apically 3–5-lobed bracts 2–3 cm long. Rays usually 7–15, very unequal, 1–8 cm long, ascending. Umbellets 15–20-flowered, the pedicels capillary, unequal, 2–10 mm long. Involucel of several toothed or lobed bracteoles 3–7 mm long. Petals oboval, indistinctly clawed; calyx teeth ovate, persistent; stylopodium low conical, swollen and as broad as fruit width, the styles short. Fruit narrowly oblong, 4–5 mm long, 1–1.5 mm broad, sharply constricted at apex, the mericarps subterete, the ribs extended into unequal, somewhat undulate wings. Vittae large, alternating with the wings, 2 on commissure, a conspicuous "pseudovitta" at junction of each primary with a secondary rib. Commissural surface lightly concave. Sept.

Type locality: "C. Népal: massif de l'Annapurna, versant nord du col de Namun (Langtang himal), vers 3800 m," 24 Sept. 1981, *Farille 81,537* (G, JAY, isotypes).

Distribution: Known only from type collection.

4. **Chamaesium shrestaeanum** Farille & Malla in Farille, Cauwet-Marc & Malla, Candollea 40(2): 537. 1985.

Plants low, stout, erect, the stem solitary, simple, monocarpic, 10–30 cm tall, perennial from a fibrous base. Leaves thick, oblong, pinnate, with 4–8 pairs of narrow leaflets 2–2.5 mm long, 3–6 mm broad, with a crystalline-scarious surface and sparsely appressed-pubescent, the veins impressed; petiole slender, scarcely dilated. Umbel terminal, solitary, the peduncle stout, of variable height. Involucre of several foliaceous, petiolate, pinnatifid bracts 2–3 cm long. Rays 10–15, stout, angled, very unequal, 1–10 cm long, ascending but diverging at summit. Umbellets 10–20-flowered, the pedicels angled, very unequal, 1–7 mm long. Involucel of several foliaceous bracteoles, like the bracts, entire to few lobed, 5–10 mm long. Petals spatulate, clawed; calyx teeth 0; stylopodium low conical, swollen and nearly as broad as fruit width, its margin lobed, the styles short. Fruit oblong-oval, 3–5 mm long, 1.5–2 mm broad, sharply constricted at apex, the mericarps subterete, the ribs extended into unequal, undulate and lobed wings. Vittae large, alternating with the wings, 2 on the commissure, a conspicuous "pseudovitta" at juncture of each primary with a secondary rib. Commissural surface grooved. Oct.

Type locality: "E. Népal: vallée de Rolwalling, en amont de Beding, á Na," 24 Sept. 1981, *Farille 81,776* (G, JAY, isotypes).

Distribution: Known only from the type collection.

GROUP III

1. Leaves pinnately, ternate-pinnately, or ternately divided to decompound.
 2. Flowers white (in our species) or rarely purplish; involucel usually present (or if obscure, umbellet with a single fertile flower).
  3. Beak of fruit much longer than body; leaf divisions filiform. .....13. *Scandix.*
  3. Beak of fruit much shorter than body, or lacking; leaf divisions linear-oblong to lanceolate or ovate.
   4. Fruit prominently caudate at base (in our species), linear-clavate, 10–25 mm long; pedicels 10–30 mm long. .... 14. *Osmorhiza.*
   4. Fruit rounded at base, not caudate, linear to fusiform or oblong-ovoid, 3–10 mm long; pedicels 1–10 mm long.
    5. Stem attenuate at base and attached to a globose tuber. .... 17. *Neoconopodium.*
    5. Stem not attenuate at base and usually not tuberous.
     6. Umbellets with both perfect and staminate flowers; fruit glabrous, the ribs distinct and well defined; vittae distinct, solitary in intervals, 2 on commissure. .... 15. *Chaerophyllum.*
     6. Umbellets with all perfect flowers; fruit hispid or tuberculate, the ribs obscure or obsolete; vittae obscure. ....16. *Anthriscus.*
 2. Flowers bright yellow; involucel lacking; plants with licorice odor. .... 18. *Foeniculum.*
1. Leaves all simple and entire. .... 19. *Bupleurum.*

13. **SCANDIX** Linn.

Sp. Pl. 256. 1753; C. B. Clarke in Hook. f., Fl. Brit. Ind. 2: 692. 1879.

Low, slender, spreading or erect, caulescent, branching, usually pubescent annual herbs, from slender taproots. Leaves petiolate, membranaceous, pinnately decompound, the ultimate divisions small, linear or filiform; petiole sheathing. Inflorescence of lax compound, or some simple umbels. Involucre wanting, or of one or more foliaceous bracts. Rays few, rarely only 1, spreading-ascending. Involucel of several lobed or dissected bracteoles. Flowers white; petals oblong to obovate, emarginate, with a narrower inflexed apex, the outer somewhat radiant; calyx teeth minute or obsolete; stylopodium short conical, the styles very short to elongate. Pedicels slender to obsolete, unequal. Fruit narrowly oblong-linear, compressed laterally, terminating in an elongated, linear, subterete or laterally or dorsally compressed, hispid beak, the mericarps subterete or subcompressed dorsally in transection, the ribs prominent, filiform, obtuse, usually bristly. Vittae solitary in intervals, 2 on commissure. Commissural face concave to sulcate. Carpophore entire to deeply bifid.

Type species: *Scandix pecten-veneris* Linn.

Distribution: A genus of ca 15 species of the Mediterranean region, only the type species reaching northwestern India.

1. **Scandix pecten-veneris** Linn. Sp. Pl. 256. 1753; C. B. Clarke in Hook. f., Fl. Brit. Ind. 2: 692. 1879.

Plants 10–25 cm tall, hispid, usually branched from base. Leaves ovate-oblong, 2–10 cm long, ca 5 cm broad, pinnately decompound, the ultimate divisions linear, ca 2 mm long, 0.5 mm broad; petiole 1–5 cm long. Rays 1–4, 1–3 cm long. Umbellets about 10-flowered. Involucel of usually 2 ovate or oblong ciliate bracteoles, 3–6 mm long, 1.5–2 mm broad, lobed above. Stylopodium depressed, the styles 1/3 length of petals. Pedicels from 8 mm long to obsolete. Fruit 6–12 mm long, 1–2 mm broad, the beak flattened laterally, ciliate, 15–35 mm long, the mericarps subcompressed dorsally in transection, the ribs filiform. Commissural face sulcate. Carpophore entire or shortly bifid. Mar.–Sep.

Type locality: "Inter Germaniae & Europae australiores segetes," collector not known.

Distribution: Eurasia, extending into northwestern India up to altitudes of 2,000 m; a widespread weed.

Representative specimens: Jammu & Kashmir: Kashmir, *Thomson* (CAL), *Falconer 499* (CAL). Himachal Pradesh: Kangra, *A.R.E.P.'s Tour 15,472* (CAL).

14. **OSMORHIZA** Raf.

Jour. de Phys. 89: 257. 1819; C. B. Clarke in Hook. f., Fl. Brit. Ind. 2: 790. 1879.

Uraspermum Nutt., Gen. 1: 192. 1818, non *Urospermum* Scop. 1777; *Scandix* subgen. *Uraspermum* (Nutt.) Kozo-Polj., Bull. Sci. Nat. Mosc. II. 29: 142. 1916.
Washingtonia Raf., Amer. Month. Mag. 2: 176. 1818, *nomen nudum*; J. Coulter & Rose, Contr. U. S. Nat. Herb. 7: 60. 1900.

Slender to rather stout, erect, caulescent, branching, perennial herbs, pubescent to glabrate, from thick, fascicled roots. Leaves petiolate, membranaceous, biternate or ternate-pinnate, the leaflets lanceolate to orbicular, serrate to pinnatifid; petiole sheathing. Inflorescence of lax, lateral and terminal, compound umbels. Involucre usually wanting, or rarely of 1 to several lanceolate bracts. Rays few, slender, ascending and diverging, unequal. Involucel of several narrow foliaceous bracteoles, or wanting. Flowers polygamous, white, greenish-yellow, or purple; petals spatulate to obovate, with a narrower inflexed apex; calyx teeth obsolete; stylopodium conical, the styles slender to obsolete. Pedicels spreading to divaricate. Fruit linear-oblong, clavate, compressed laterally, shortly beaked to obtuse at apex, caudate to rounded at base, bristly hispid to glabrate, the mericarps subterete or slightly subcompressed dorsally in transection, the ribs filiform, acute, bristly. Vittae obscure or wanting. Commissural face sulcate or concave. Carpophore bifid less than half its length.

Type species: *Myrrhis claytonii* Michaux [*Osmorhiza claytonii* (Michaux) C. B. Clarke].

Distribution: A genus of about 10 species in North and South America and one in eastern and southeastern Asia, reaching the Himalaya.

1. **Osmorhiza aristata** (Thunb.) Rydb., Bot. Surv. Nebr. 3: 37. 1894.

Chaerophyllum aristatum Thunb., Fl. Jap. 119. 1784; *Myrrhis aristata* (Thunb.) Sprengel, Sp. Umbell. 133. 1818; *Uraspermum aristatum* (Thunb.) Kuntze, Rev. Gen. 1: 270. 1891; *Scandix aristata* (Thunb.) Kozo-Polj., Bull. Soc. Nat. Mosc. II. 29: 143. 1916.
Osmorhiza laxa Royle ex Lindley in Royle, Ill. Bot. Himal. Mount. 233, pl. 52, fig., 1. 1839; *Washingtonia laxa* (Lindley) Kozo-Polj. ex B. Fedtsch., Fl. Asiat. Russ. 15: 52. 1920; *Osmorhiza aristata* var. *laxa* (Lindley) Constance & Shan, Univ. Calif. Publ. Bot. 23: 130. 1948. Based on *Royle*, "HAB. Sirmore in the Himalayan Mountains."
Osmorhiza claytonii sensu C. B. Clarke in Hook. f., Fl. Brit. Ind. 2: 690. 1879, *pro parte*, non *Myrrhis claytonii* Michaux, 1803.

Plants slender, 30–100 cm tall, the stems glabrous. Leaves triangular-ovate to orbicular, 7–20 cm in diameter, biternate, hirtellous, the leaflets coarsely serrate or sparsely lobed at base, 1.5–9 cm long, 1–6 cm broad, obtuse to acuminate; petiole 5–26 cm long. Peduncles 3.5–25 cm long. Rays 3–6, spreading-ascending, 3.5–14 cm long. Involucel of 1 to several linear or lanceolate, ciliate or hirtellous, reflexed bracteoles 3–18 mm long. Pedicels spreading, 1–3 cm long. Fruit linear-clavate, 1–2.5 cm long, ca 2 mm broad, obtuse to abruptly acute at apex, caudate at base, the appendages 5–8 mm long, sparingly bristly on ribs, densely so on appendages. Carpophore bifid to middle. Apr.–July.

Type locality: Presumably based upon Japanese material, but neither locality nor collector indicated.

Distribution: Western Himalaya to Tibet, southwestern China, and Japan, 2,000–4,000 m.

Representative specimens: Jammu & Kashmir: Gulmarg, *Duthie 13,036* (DD). Punjab: w/o further locality, *Royle 93/46* (DD). Himachal Pradesh: between Kinanipani and Jakopass, *Duthie 1136* (DD). Uttar Pradesh: Bamsu Valley, *Duthie 19,819* (DD). Nepal: between Pisan and Tat Pani, *Nakao* (UC). Bhutan: "Boutan," *Griffith 2062* (BM). Arunachal Pradesh: Talung Dzong, Balipara Frontier, *Kingdon-Ward 11,398* (BM); Phutang, *Kingdon-Ward 11,445* (BM).

Note. Lowry & Jones (1985) do not recognize the distinctness of the described variety.

15. CHAEROPHYLLUM Linn.

Sp. Pl. 1: 258. 1753; C. B. Clarke in Hook. f., Fl. Brit. Ind. 2: 690. 1879.

Tall and slender to low and spreading, caulescent, branching, pubescent to glabrous biennial, or less commonly perennial or annual herbs, from tubers, rhizomes, or taproots. Leaves petiolate, membranaceous, pinnately or ternate-pinnately divided or decompound, the ultimate divisions rather small; petiole sheathing. Inflorescence of lax terminal or axillary compound umbels, occasionally some sessile. Involucre usually wanting, or occasionally of 1 to several linear bracts. Rays rather few, spreading-ascending. Involucel of few to numerous, often conspicuous, narrow or broader bracteoles, reflexed or spreading at maturity. Umbellets polygamous, the perfect and staminate flowers equally longly pedicellate, or the central one perfect and sessile or subsessile. Flowers white or yellowish or purplish; petals obcordate or obovate, emarginate, with a narrower inflexed apex, the outer often radiant; calyx teeth usually obsolete, rarely evident; stylopodium conical or depressed, the styles short or slender. Pedicels spreading, unequal. Fruit linear to oblong-ovoid or fusiform, compressed laterally and usually constricted at commissure, often narrowed toward apex and shortly beaked or beakless, often narrowed toward base, glabrous or pubescent, the mericarps usually terete, or sometimes pentagonal or subcompressed dorsally in transection, the ribs evident (rarely obscure), obtuse, broader or narrower than intervals, the lateral ribs ± commissural. Vittae solitary in intervals, 2 on commissure, rarely obscure. Commissural face sulcate or strongly concave. Carpophore stout, bifid or entire.

Type species: *Chaerophyllum temulentum* Linn.

Distribution: A circumboreal genus of an estimated 30-40 species, the great majority of them Eurasian. Five reach India, chiefly in the Himalaya, two being endemic, the others extending to Pakistan and sometimes Afghanistan.

KEY TO SPECIES

1. Calyx teeth evident, to 0.5 mm long; mature fruit all slender pedicellate .... 1. *C. orientale*.
1. Calyx teeth obsolete; central fruit sessile or subsessile, or mature fruit solitary in umbellet.
   2. Involucel usually obscure; fertile fruit usually solitary in umbellet; fruit ribs usually obscure .... 2. *C. cachemiricum*.
   2. Involucel conspicuous; fertile fruit 2-6 in at least some umbellets (often solitary in others); fruit ribs evident to prominent.
      3. Cauline leaflets lanceolate, closely and deeply regularly serrate to incised (falcate-serrate); fruit lanceolate-oblong ....3. *C. acuminatum*.
      3. Cauline leaflets usually smaller and more strongly or irregularly dissected; fruit linearly-oblong to fusiform.
         4. Stylopodium depressed into a broad disk surmounting a very short, thick, cylindrical beak; styles very short; involucel bracteoles usually glabrous. .... 4. *C. reflexum*.

4. Stylopodium low conical, the fruit essentially beakless; styles equaling or exceeding stylopodium; involucel bracteoles usually ciliate. .... 5. *C. villosum.*

1. Chaerophyllum orientale (C. B. Clarke) P. K. Mukherjee, Bull. Bot. Surv. India 24(1–4): 44. 1983.

Chaerophyllum reflexum Lindley var. *orientalis* C. B. Clarke, Jour. Linn. Soc. 25: 28, pl. 16. 1899.

Plants slender, erect or scandent, 30–120 cm tall, widely branched, sparsely appressed-pubescent or scaberulous. Leaves ovate, 4–12 cm long, 5–15 cm broad, ternate-pinnately decompound, the ultimate divisions ovate to lanceolate, dentate, incised, or pinnately lobed, 4–8 mm long, 3–6 mm broad, a little strigulose above; petiole 2–7 cm long. Peduncles terminal and lateral, 10–15 cm long. Involucre of a few minute linear bracts ca 5 mm long. Fertile rays 6–12, spreading, unequal, 2–3.5 cm long. Involucel of 4–6 inconspicuous linear bracteoles 2–3 mm long. Umbellets with 1–8 perfect and several staminate flowers, all pedicellate, the pedicels of perfect flowers longer than sterile, 6–10 mm long. Flowers white; petals oblanceolate, subequal; calyx teeth prominent, lanceolate, to 0.5 mm long; stylopodium stout, conical, about equaling the short, recurved styles. Fruit oblong-ovoid, compressed laterally and constricted at commissure, 4–6 mm long, 2–3 mm broad, narrowed into a short neck toward apex, the mericarps terete in transection, the ribs evident. Vittae solitary in intervals, 2 on commissure. Commissural face sulcate. Carpophore entire. Sept.–Nov.

Type locality: Jakpho, 2330 m, Naga hills, *Clarke 41,861*.

Distribution: Naga Hills, northeastern India.

Representative specimens: Bhutan: Shingbe La, Me La Chu, *Ludlow, Sheriff & Hicks 20,815* (BM). Nagaland: Kegwima Edge, Naga Hills, *Clarke 41,861A, B, E* (K), *D* (CAL); Kohima, Naga Hills, *Koelz 26,220* (UC); Esakan, Mt. Victoria, *Kingdon-Ward 22,619* (BM).

2. Chaerophyllum cachemiricum C. B. Clarke in Hook. f., Fl. Brit. Ind. 2: 691. 1879.

Plants 60–100 cm tall, pubescent with reflexed hairs, or rarely glabrous. Leaves ovate, ternate-bipinnate, the ultimate divisions lanceolate, 2–5 mm long, 1–2.5 mm broad, deeply incised to subpinnatifid; petiolar sheath white-margined. Peduncles terminal and lateral, stout. Involucre wanting. Rays 5–8, 8–14 mm long, glabrate. Involucel of several linear bracteoles. Umbellets about 12-flowered, only one (rarely 2) maturing, the pedicels unequal. Flowers not seen. Stylopodium low conical, shorter than the slender styles. Fruit oblong-ovoid, compressed laterally, narrowed toward apex and constricted at commissure, ca 6 mm long, 3 mm broad, the mericarps terete in transection, the ribs inconspicuous and slender, or rarely prominent. Vittae solitary in intervals, 2 on commissure. Commissural face deeply sulcate. Carpophore entire, or rarely bipartite. Sept.

KEY TO VARIETIES

1. Plants pubescent; bracteoles inconspicuous; fruit ribs usually slender; carpophore entire. .... 2a. var. *cachemiricum.*
1. Plants glabrous; bracteoles as long as flowering pedicels; fruit ribs prominent; carpophore bipartite. .... 2b. var. *dissectum.*

2a. var. **cachemiricum**

Plants pubescent. Bracteoles of the involucel inconspicuous, ca 2 mm long. Fruit ribs slender, filiform. Carpophore entire.

Type locality: "NORTH-WEST HIMALAYAS; Dalhousie, alt. 7000 ft.," *Clarke*.

Distribution: Known only from Dalhousie in Western Himalaya, at 2,000 m.

Representative specimens: Himachal Pradesh: Dalhousie, *Clarke 22,618A* (K), *B* (CAL), *23,270* (CAL).

2b. var. **dissectum** C. B. Clarke in Hook. f., Fl. Brit. Ind. 2: 692. 1879.

Chaerophyllum reflexum var. *dissectum* (C. B. Clarke) Hedge & Lamond, Notes. Roy. Bot. Gard. Edinburgh 38: 254. 1980.

Plants glabrous. Bracteoles 3–5, as long as flowering pedicels. Fruit blackish, the ribs prominent. Carpophore bipartite.

Type locality: "Kashmir, Sind Valley, alt. 12,000 ft.," *Clarke*.

Distribution: Known only from the type locality.

Representative specimens: Jammu & Kashmir: Sind Valley, *Clarke 31,009C* (K), *31,089A* (K).

Note. Clarke (l. c.) states: "Perhaps a distinct species, but the fruit closely agrees with that of *C. cachemiricum* except that it is a little larger." Hedge & Lamond (1980), who regard *C. cachemiricum* as synonymous with typical *C. reflexum* Lindley, retain var. *dissectum* as "a low-growing variety .... found mainly at high altitudes" and to be "recognized by its distinctive branching."

3. **Chaerophyllum acuminatum** Lindley in Royle, Ill. Bot. Himal. Mount. 232. 1835; C. B. Clarke in Hook. f., Fl. Brit. Ind. 2: 691. 1879.

Chaerophyllum reflexum var. *acuminatum* (Lindley) Hedge & Lamond, Notes. Roy. Bot. Gard. Edinburgh 38: 254. 1980.

Chaerophyllum reflexum var. *tuberosum* Hedge & Lamond, Notes. Roy. Bot. Gard. Edinburgh 38: 255. 1980. Based on *R. Stewart 22,051A*, "Kashmir: Burzil pass, 3660–3960 m."

Plants slender, 15–60 cm tall, few to many branched above or below, pubescent or hispid, rarely glabrate, annuals or biennials, from slender taproots. Leaves ovate, 3.5–30 cm long, 2.5–24 cm broad, 1-2-pinnate, the leaflets lanceolate, 5–10 cm long, 0.5–3 cm broad, strigose-hispid, closely and deeply regularly serrate to incised; petiole slender, 5–15 cm long. Peduncles terminal and axillary, 2–8 cm long. Involucre wanting, or of a few minute, linear, scarious bracts, ca 5 mm long. Rays 5–10, slender, unequal, 1.5–6 cm long. Involucel of 4–6 lanceolate-acuminate, scarious-margined, arachnoid-ciliate bracteoles 3–7 mm long. Umbellets with 1–6 perfect and several staminate flowers, pedicellate except central flower sessile or subsessile, the mature pedicels up to 7 mm long. Flowers white (rarely purple); petals obcordate, the outer radiant; stylopodium low conical, the styles a little longer than stylopodium, diverging. Fruit lanceolately oblong, compressed laterally, 5–8 mm long, 1.5–2.5 mm broad, constricted into a very short neck at apex, the mericarps terete in transection, the ribs prominent, obtuse. Vittae solitary in intervals, 2 on commissure. Commissural face concave. Carpophore stout, bifid at apex. June–Aug.

Type locality: "Hab. Choor," N. W. India, *Royle.*

Distribution: Western Himalaya from Kumaon to Kashmir, 2,000–4,000 m; Pakistan.

Representative specimens: Jammu & Kashmir: Gulmarg, *Aitchison 31* (CAL, UC); Kishtawar, *Thomson* (CAL, K); Baltistan, Dras Valley, *Duthie 11,680* (K); Burzil Pass, *R. R. Stewart 22,051a* (K). Himachal Pradesh: Lahul, *Jaeschke 102* (CAL); Choor, *Royle 93/44* (DD), *s. n.* (K). Uttar Pradesh: Ganges Valley, *Duthie 1125a* (DD); Gori Valley, Kimaon, *Inayat Khan 24,455* (DD).

Note. Hedge and Lamond (1980) comment that, although their vars. *acuminatum* and *tuberosum* are recognizable in their most characteristic states, many intermediates between them and "var. *reflexum*" exist.

4. **Chaerophyllum reflexum** Lindley in Royle, Ill. Bot. Himal. Mount. 232. 1835; C. B. Clarke in Hook. f., Fl. Brit. Ind. 2: 691. 1879.

Plants slender, erect, 15–100 cm tall, freely branched, sparsely hirsute to glabrous, from a taproot. Leaves ovate, 4–15 cm long, 3–11 cm broad, ternate-pinnate or bipinnate, the leaflets oblong to lanceolate, 0.5–3 cm long, 3–12 mm broad, very sparsely appressed-hirsute to glabrate, closely and regularly serrate to pinnatifid; petiole slender, 4–8 cm long. Peduncles terminal and lateral, 3–12 cm long. Involucre usually wanting. Rays 5–15, slender, unequal, 1–6 cm long. Involucel of 3–8 linear to lanceolate-acuminate, scarious-margined, mostly glabrous bracteoles 2–5 mm long. Umbellets with 1–7 perfect flowers and several staminate flowers, pedicellate except the central perfect flower sessile or subsessile, the mature pedicels up to 5 mm long. Flowers white; petals obovate, the outer radiant; stylopodium depressed to form a flat disk, the styles very short, diverging. Fruit linearly oblong to fusiform, compressed laterally, constricted at commissure, 6–10 mm long, 2–3 mm broad, narrowed toward apex and terminating in a very short, thick, cylindrical beak under stylopodium disk, the mericarps terete in transection, the ribs prominent, obtuse. Vittae solitary in intervals, 2 on commissure. Commissural face concave to sulcate. Carpophore bifid at apex. June–Oct.

Type locality: "Hab. Kedarkanta and Nagkanda," N. W. India, *Royle.*

Distribution: Western Himalaya, from Kumaon to Kashmir, 1,500–3,000 m, to Pakistan and Afghanistan.

Representative specimens: Jammu & Kashmir: Srinagar, *Gammie* (DD); Gulmarg, *Aitchison 11* (CAL). Himachal Pradesh: Manali, *Bhattacharyya 40,339* (BSD), *Mukherjee*

326 (CUH, UC). Uttar Pradesh: Dhauli Valley, Kumaon, *Duthie 5597* (CAL); Kedarkanta, *Royle 93/49* (DD), *s. n.* (K).

Note. Hedge & Lamond (1980) have combined *C. cachemiricum* and *C. acuminatum* with *C. reflexum*, which they construe as comprising 5 varieties, at least 3 of them occurring in India. As they remark (p. 255), "many intermediates exist and it is not possible to assign every specimen to a particular variety with confidence."

5. **Chaerophyllum villosum** DC., Prodr. 4: 225. 1830; C. B. Clarke in Hook. f., Fl. Brit. Ind. 2: 690. 1879.

Plants compact or slender, 15–100 cm tall, usually branched above or throughout, sparsely hispid to glabrous above, usually hirsute below, from slender, fusiform taproots. Leaves ovate, 10–20 cm long, 2–15 cm broad, pinnately or ternate-pinnately dissected or decompound, the leaflets small, lanceolate, 2–5 mm long, ca 2.5 mm broad, pinnately lobed or incised, sparsely hispid to glabrous; petiole 2–10 cm long. Peduncles terminal or axillary, 5–9 cm long. Involucre wanting, or occasionally of a linear bract. Rays 2–12, slender, reflexed-hirsute to glabrous, 1–5 cm long. Involucel of 3–6 linear-acuminate, scarious-margined, usually ciliate bracteoles 2–5 mm long. Umbellets of 2–6 fertile and 2 or 3 times as many staminate flowers, pedicellate except the central flower subsessile, the mature pedicels 3–8 mm long. Flowers white; petals obovate, the outer radiant; stylopodium low conical, the styles equaling or a little longer than stylopodium. Fruit linearly oblong, compressed laterally, constricted at commissure, 6–10 mm long, 1.5–2.5 mm broad, narrowed toward apex and very shortly constricted beneath stylopodium, the mericarps subterete in transection, the ribs prominent, obtuse. Vittae solitary in intervals, 2 on commissure. Commissural face deeply sulcate. Carpophore bifid at apex. May–Oct.

Type locality: "Nepaliae monte Chandaghiry," *Wallich*.

Distribution: Himalaya, from Bhutan to Kashmir, and Khasia Hills, 2,000–4,000 m; Nepal, Afghanistan, and Pakistan.

Representative specimens: Jammu & Kashmir: Kilanmarg, *Aitchison 13* (CAL, UC). Himachal Pradesh: Polang, Kangra Valley, *Vaid 24,235* (DD). Uttar Pradesh: Dombita Gadh, *Duthie 4062* (CAL). Nepal: Chandaghiry, *Wallich 558* (CAL). West Bengal: Manebhanjan, Darjeeling Distr., *Mukherjee 3707* (CUH). Sikkim: Chungthang, *Smith & Cave 908* (CAL); Bakhim, *Mukherjee 315* (CUH, UC). Bhutan: Wamrung, *A. S. Rao 43,338* (CAL). Meghalaya: Khasia, Mwaphlong, *Clarke 38,280A* (CAL); Khasia, *Hooker & Thomson* (CAL).

16. ANTHRISCUS Pers.

Pers. Syn. Pl. 1: 320. 1805, *nomen conserv.*, non Bernh. 1800;
C. B. Clarke in Hook. f., Fl. Brit. Ind. 2: 692. 1879.

Slender to stout, erect, caulescent, branching, hispid or hirsute to glabrous annual, biennial, or rarely perennial herbs, from slender taproots or fascicled fibrous roots. Leaves petiolate, membranaceous, ternate-pinnately or pinnately divided or decompound, the leaflets lanceolate to ovate, dentate to pinnatifid; petiole sheathing. Inflorescence of lax compound umbels. Involucre usually wanting. Rays rather few, spreading-ascending or spreading. Involucel of several, narrow, entire, usually reflexed bracteoles. Flowers white; petals oblong to obovate, with a narrower inflexed apex, the outer radiant or all subequal; calyx teeth obsolete; stylopodium conical, the styles short. Fruit linear to ovoid, terminating in a slender beak, compressed laterally, constricted at commissure, the mericarps subterete in transection, the ribs obsolete, smooth, tuberculate, or uncinate-prickly. Vittae obscure to obsolete. Commissural face sulcate. Carpophore entire or bifid.

Type species: *Scandix anthriscus* Linn. [*Anthriscus caucalis* M. Bieb.].

Distribution: A small Eurasian-African genus of about a dozen species, one of which reaches western Himalaya and a second is apparently adventive in India as it is in temperate North America.

KEY TO SPECIES

1. Plants 1–3 m tall; leaf divisions rather coarse; fruit lanceolate-oblong, 5–10 mm long. . . . . 1. *A. nemorosus.*
1. Plants usually less than 1 m tall; leaf divisions very fine; fruit oblong-ovoid, 3–5 mm long. . . . . 2. *A. caucalis.*

1. Anthriscus nemorosus (M. Bieb.) Sprengel, Pl. Umbell.

Prodr. 27. 1813; C. B. Clarke in Hook. f., Fl. Brit. Ind. 2: 692. 1879.

Chaerophyllum nemorosum M. Bieb., Fl. Taur. 1: 232. 1808.
Anthriscus nemorosus var. *glabriusculus* Nasir, Biologia 3: 61. 1957. Based on *Stewart & Nasir 22,724*, "Kashmir, near Kamri Bungalow, 9–10,000 ft."

Plants stout, branching annuals 1–3 m tall, hirtellous to glabrate. Leaves ovate, 20–40 cm in diameter, bipinnate, the leaflets lanceolate to ovate, 1–6 cm long, 1–2 cm broad, pinnately toothed to pinnatifid; petiole 5–20 cm long. Peduncles terminal and lateral, 3–8 cm long. Fertile rays 4–12, slender, spreading-ascending, 2–5 cm long. Involucel of 3–5 lanceolate to ovate-acuminate, entire or ciliate, usually reflexed bracteoles 2–5 (–8) mm long. Umbellets with 3–8 perfect and several staminate flowers, all pedicellate, the mature pedicels filiform, hispidulous at summit, 5–10 mm long. Flowers white; petals oblanceolate to obovate, the outer radiant; stylopodium conical, the styles a little longer than stylopodium. Fruit lanceolate-oblong, 8–10 mm long, 2.5–3 mm broad, tapering into a slender beak at apex, black and shining, covered with short, curved, appressed bristles or rarely glabrate. Carpophore bifid at apex. May–July.

Type locality: "In sylvis ad fluuim Terak sub arboribus frequens," Caucasus, no collector indicated.

Distribution: Eurasia; western Himalaya to Pakistan.

Representative specimens: Jammu & Kashmir: Jhelum Valley, *Keshavanand 964* (DD); Kishanganga Valley, *Duthie 12,549* (DD). Himachal Pradesh: Manali, *S. Singh 47* (DD). Punjab: w/o further locality, *Gamble 26,765* (CAL). Uttar Pradesh: Mussoorie, *Royle 93153* (DD).

Note. Although Hiroe & Constance (1958, p. 26) combined *Anthriscus nemorosus* with *A. sylvestris* (Linn.) Hoffm. and the two taxa are undoubtedly close, only one of some 35 East Asiatic specimens at UC show any trace of fruit bristles, suggesting that Chinese-Japanese and Indian material may not be identical. (L. C.)

2. Anthriscus caucalis M. Bieb., Fl. Taur. 1: 230. 1808.

Scandix anthriscus Linn., Sp. Pl. 257. 1753; *Chaerophyllum anthriscus* (Linn.) Crantz, Class Umbell. 76. 1767; *Myrrhis anthriscus* (Linn.) Lagasca, Amen. Nat. 98. 1821; *Anthriscus anthriscus* (Linn.) Karsten, Deuts. Fl. 857. 1882; *Myrrhodes anthriscus* (Linn.) Kuntze, Rev. Gen. 1: 268. 1891; *Chaerefolium anthriscus* (Linn.) Schinz & Thell., Viert. Nat. Ges. Zürich 53: 554. 1909.

Anthriscus vulgaris Pers., Syn. Pl. 1: 320. 1805, non Bernh. 1800; *Cerefolium vulgare* Bubani, Fl. Pyren. 2: 411. 1900.

Caucalis scandix Scop., Fl. Carn. ed. 2. 1: 191. 1772; *Anthriscus scandix* (Scop.) Asch., Fl. Brand. 1: 260. 1860, non M. Bieb. 1808.

Anthriscus scandicina Mansf., Feddes Repert. 46: 309. 1939.

Plants slender, much branched, hispid or hirsute, annual, 5–100 cm tall. Leaves oblong-ovate to triangular-ovate, 5–15 cm long, 3–20 cm broad, pinnately decompound, the ultimate divisions linearly oblong, 1–5 mm long; petiole 1.5–8 cm long. Peduncles lateral, leaf opposed, 3–20 mm long. Fertile rays 3–6, slender, spreading, 0.5–3 cm long. Involucel of 4 or 5 linearly lanceolate, ciliate bracteoles 1–3 mm long. Umbellets with 3–7 perfect and 0 staminate flowers, all pedicellate, the mature pedicels stout, hispidulous at summit, 1–9 mm long. Flowers white; petals oblanceolate to narrowly spatulate, subequal; stylopodium conical, the styles very short, shorter than stylopodium. Fruit oblong-ovoid, 3–5 mm long, 2–3 mm broad, abruptly narrowing into a slender beak at apex, black and shiny, covered with short uncinate bristles. Carpophore bifid at apex. Apr.-Jun.

Type locality: "In Europae aggeribus terrenis," no collector indicated.

Distribution: Eurasia; widely naturalized in temperate regions of Northern Hemisphere.

Specimen examined: Uttar Pradesh: Dehra Dun, Rajpur, *Babu 35,225* (BSD).

17. NEOCONOPODIUM Pimenov & Kljuykov
Feddes Repert. 98: 377. 1987.

Conopodium subgenus (?) *Neoconopodium* Kozo-Polj., Bull. Soc. Nat. Mosc. II. 29: 206. 1916.

Slender and fragile, weakly erect, simple or few branched, glabrous to pubescent, perennial herbs with a single stem, attenuate at base, arising from a globose tuber. Leaves petiolate, membranaceous, triangular-ovate, biternate to bipinnate with oblong to ovate, entire or lobed leaflets; petiole slender, narrowly sheathing; cauline leaves few, reduced upward. Inflorescence of terminal and lateral compound umbels, the peduncles elongate. Involucre lacking. Rays few, very slender, spreading-ascending to divaricate. Involucel of several lanceolate to ovate semiscarious bracteoles. Umbellets rather few flowered, some flowers staminate; petals obovate with a small, narrower inflexed apex, often slightly radiant; calyx teeth 0; stylopodium low conical, the styles slender, reflexed. Pedicels slender, reflexed. Fruit linearly-oblong, attenuate at apex into a well-developed beak, rounded at base, constricted at commissure, the mericarps subterete in transection, glabrous or pilose, the ribs subequal, prominent, filiform. Vittae obscure. Commissural face sulcate. Carpophore bifid at apex.

Type species: *Butinia capnoides* Decne. [*Neoconopodium capnoides* (Decne.) Pimenov & Kljuykov].

Distribution: A newly described genus of two species, extending from northwestern India to Pakistan and Afghanistan, at moderate elevations.

1. **Neoconopodium capnoides** (Decne.) Pimenov & Kljuykov, Feddes Repert. 98: 377. 1987.

Butinia capnoides Decne. in Jacquem. Voy. Bot. Inde 70, pl. 80. 1844; *Chaerophyllum capnoides* (Decne.) C. B. Clarke in Hook. f., Fl. Brit. Ind. 2: 691. 1879; *Conopodium capnoides* (Decne.) Kozo-Polj., Bull. Soc. Nat. Mosc. II. 29: 206. 1916.

Plants slender, weak, simple or few-branched, glabrous, 1.5–8 cm tall from a globose tuber 1 cm or more in diameter. Leaves broadly triangular-ovate, 10–20 cm in diameter, biternate to bipinnate, with oblong to ovate-lanceolate, entire-margined, mucronulate, petiolulate leaflets 5–30 mm long, 3–10 mm broad, glabrous to minutely scaberulous beneath; petiole slender, narrowly sheathing, 3–6 cm long. Peduncles terminal and lateral, 3–15 cm long. Involucre 0. Rays 4–8, filiform, 3–15 cm long. Umbellets with 2–5 perfect and several staminate flowers on pedicels 5–13 mm long. Involucel of 3–7 lanceolate to ovate, semiscarious bracteoles ca. 2.5 mm long. Flowers white; petals obovate with a small inflexed apex, often slightly radiant; stylopodium low conical, the styles slender, ca. 1 mm long. Fruit linearly oblong, 8–10 mm long, ca. 2 mm broad, attenuate into a beak at apex, constricted at commissure, the mericarps terete in transection, the ribs prominent. Vittae obscure.

Type locality: Kashmir: Ouri and Nouchaira, banks of the Djhelone, *Jacquemont 308* (K).

Distribution: Western Himalaya (Garhwal to Kashmir), up to 2,500 m; Pakistan.

Representative specimens: Jammu & Kashmir: Banahal, *Thomson* (CAL). Uttar Pradesh: Matkangra, W. Chakrata, *Gamble* (DD); Tehri Garhwal, *Duthie* (DD).

18. **FOENICULUM** Adans.

Fam. Pl. 2: 101. 1763; C. B. Clarke in Hook. f., Fl. Brit. Ind. 2: 695. 1879.

Seseli subgen. *Foeniculum* (Adans.) Kozo-Polj., Bull. Soc. Nat. Mosc. II. 29: 183. 1916.
Ozodia Wight & Arn., Prodr. Fl. Ind. Orient. 1: 374. 1834.

Slender to stout, erect, caulescent, branching, glabrous and glaucous perennial or biennial, strongly anise-scented herbs, from taproots. Leaves petiolate, membranaceous, pinnately decompound, the ultimate divisions filiform, spreading; petiole sheathing, sometimes fleshy. Inflorescence of lax terminal and lateral compound umbels, some lateral umbels often staminate only. Involucre and involucel lacking. Rays few to numerous, spreading-ascending. Flowers yellow; petals oblong, with a slightly narrower inflexed apex; calyx teeth obsolete; stylopodium conical, the styles very short. Pedicels usually numerous, spreading. Fruit oblong or ovoid-oblong, slightly compressed laterally to subterete, not constricted at commissure, the mericarps pentagonal and a little compressed dorsally in transection, the ribs very prominent, acute or obtuse, subequal or the lateral a little broader than the dorsal. Vittae solitary in intervals, 2 on commissure. Commissural face plane to concave. Carpophore bipartite.

Type species: *Anethum foeniculum* Linn. [*Foeniculum vulgare* Miller].

Distribution: A Mediterranean genus, apparently consisting of a single variable species, widely cultivated and adventive in warmer parts of the world.

1. **Foeniculum vulgare** Miller, Gard. Dict. ed. 8. *Foeniculum*, No. 1. 1768; C. B. Clarke in Hook. f., Fl. Brit. Ind. 2: 695. 1879.

Anethum foeniculum Linn., Sp. Pl. 263. 1753; *Ligusticum foeniculum* (Linn.) Crantz, Class. Umbell. 82. 1767; *Meum foeniculum* (Linn.) Sprengel, Umbell. Prodr. 32. 1813; *Foeniculum foeniculum* (Linn.) Karsten, Deuts. Fl. 837. 1882; *Selinum foeniculeum* (Linn.) E.H. Krause in Sturm, Fl. Deuts. ed. 2. 12: 115. 1904; *Seseli foeniculum* (Linn.) Kozo-Polj., Bull. Soc. Nat. Mosc. II. 29: 183. 1916.
Foeniculum officinale All., Fl. Ped. 2: 25. 1785.
Anethum panmorium Roxb. in Fleming, Cat. Ind. Med. Pl. 6. 1810; *Foeniculum panmorium* (Roxb.) DC., Prodr. 4: 142. 1830. Based on plants cultivated in Bengal, no collector indicated.
Ozodia foeniculacea Wight & Arn., Prodr. Fl. Ind. Orient. 375. 1834. Based on *Wight 1210*.

Plants slender to stout, 60–250 cm tall, often forming stout clumps. Leaves triangular-ovate, 4–40 cm in diameter, the ultimate divisions 5–50 mm long; petiole 7–14 cm long, wholly and broadly sheathing. Peduncles 1.5–6.5 cm long. Rays 5–40, spreading to ascending, subequal to unequal, 1–6.5 cm long. Umbellets mostly 12–25-flowered, the mature pedicels 2–10 mm long. Fruit ovoid-oblong to oblong, 4–10.5 mm long, 1.5–2.5 mm broad, the ribs very conspicuous, subequal. Dec.–Feb.

Type locality: "In Narbonne, Aremoriae, Maderae rupibus cretaceis," no collector indicated.

Distribution: Mediterranean, widely adventive in warmer parts of the world.

Representative specimens: Punjab: Banda, *Bell* (CAL). West Bengal: Uttarpara, Hooghly Distr., *Mukherjee* (CUH). Tripura: Agartala, *Debbarman 1069* (CAL). Madhya

Pradesh: Mandla, *Joseph 12,290* (MH). Tamil Nadu: R. S. Puram, *Saroja 9869* (MH). Karnataka: Bangalore, *Cameron* (CAL).

19. BUPLEURUM Linn.

Sp. Pl. 236. 1753; C. B. Clarke in Hook. f., Fl. Brit. Ind. 2: 674. 1879.

Low to slender, decumbent to erect, usually caulescent, branching, glabrous and often glaucous, perennial (sometimes suffrutescent or woody based) to annual herbs, from taproots. Leaves all entire, the basal distinctly petiolate or not, membranaceous or chartaceous, parallel veined, the cauline often sessile to perfoliate; petiole if present, sheathing. Inflorescence of lax terminal and usually some lateral compound umbels. Involucre of foliaceous bracts, or rarely lacking. Rays few (rarely only 1) to many, spreading-ascending to divaricate. Involucel usually of foliaceous, often connate bracteoles, frequently exceeding umbellet. Flowers yellow to purplish-black; petals oblong to orbicular, acute and slightly incurved at apex; calyx teeth obsolete; stylopodium conical to depressed, the styles short. Pedicels spreading. Fruit oblong to ellipsoid, compressed laterally, constricted at commissure, the mericarps subterete in transection, glabrous to tuberculate, the ribs usually prominent, acute to distinctly winged, less commonly obtuse. Vittae 1–4 in intervals, 2–6 on commissure. Commissural face plane to slightly concave. Carpophore entire, bifid, or bipartite.

Type species: *Bupleurum rotundifolium* Linn.

Distribution: A very distinctive genus of 100–150 species concentrated in Eurasia (one North American, one South African), with a strong development in the Himalaya. Of the 21 species known to occur in India, 14 occur in the Himalaya, 4 in Assam (two endemic), and 5 in peninsular India (two of these also in Sri Lanka).*

KEY TO SPECIES

1. Plants simple or usually branched above or from base, but neither dichotomously branched nor with filiform branches; umbels with few to many (2–15) rays: flowers 4–50 per umbellet; mericarp surface smooth, except for ribs.
    2. Leaves broadly ovate or ovate-lanceolate, 2–5 cm broad, the basal and lower cauline conspicuously slender, petiolate. .... 1. *B. lanceolatum.*
    2. Leaves linear, lanceolate, oblong, or rarely obovate, 3–20 mm broad, or if rarely broader, then without a distinct petiole.
        3. Low, compact, high altitude Himalayan perennials, with several to numerous, mostly decumbent branches arising from base of a much shortened main stem, 10–20 (35) cm tall.
            4. Involucral bracts linear to linearly lanceolate, 3-nerved; rays filiform. ... . 16. *B. gracillimum.*
            4. Involucral bracts lanceolate to ovate or oval, 5–11-nerved; rays slender to stout.

---

*In addition to the taxa included here, Nasir (1955 & 1972) recorded the following from Kashmir: *B. atroviolaceum* (O.E. Schulz) Nasir, *B. canaliculatum* Diels, *B. imaicolum* (H. Wolff) J. Kerner, *B. longicaule* DC. var. *ramosum* Nasir, *B. rotundifolium* Linn. and *B. tenue* Buch.-Ham. ex Don var. *attenuatum* Nasir. They are excluded here because we could not see specimens for them.

5. Flowers yellow or brown; stylopodium conical; fruit 2.5–4 mm long.
6. Upper cauline leaves acuminate; fruit ribs thick and prominent.
7. Leaves with white margin; pedicels about as long as fruit. ....15. *B. aitchisonii.*
7. Leaves lacking white margin; pedicels much shorter than fruit. . ...2. *B. dalhousieanum.*
6. Upper leaves obtuse; fruit ribs slender. .... 7. *B. sikkimense.*
5. Flowers purple; stylopodium depressed; fruit ca. 5 mm long. ....6. *B. longicaule.*
3. Slender perennials or annuals with one to few erect stems, often simple below and usually branching above, mostly 30–300 cm tall.
8. Involucel and often also involucre conspicuous, foliaceous, some bracteoles broad, often imbricate laterally.
9. Flowers deep purple; Himalayan.
10. Stems usually several from base; bracteoles 5, oblong-oval to orbicular, obtuse; fruit ribs prominent but scarcely winged. ....6. *B. longicaule.*
10. Stems usually simple below; bracteoles 6–10, ovate-lanceolate to oval, acuminate or acute; fruit ribs broadly thin-winged. ....3. *B. rupestre.*
9. Flowers yellow; plants of Himalaya, Assam, or peninsular India.
11. Rays filiform; fruit small, 2–2.5 mm long. .....8. *B. hamiltonii.*
11. Rays stout to slender, but not filiform; fruit larger, 3.5–8 mm long.
12. Plants stout, often shrubby at base, up to 3 m tall; peninsular India.
13. Involucel bracteoles 6–12 mm long, about equaling umbellet; fruit 6–8 mm long, the ribs prominent, thick and subwinged; vittae solitary in intervals, 2 on commissure. ....12. *B. plantaginifolium.*
13. Involucel bracteoles 2–6 mm long, much shorter than umbellet; fruit 3–3.5 mm long, the ribs filiform, unwinged; vittae 3 or 4 in intervals, 4 or 5 on commissure. ....20. *B. andhricum.*
12. Plants slender, not shrubby, rarely more than 1 m tall; Assam or Himalaya.
14. Stem copiously branched; leaves without white margin; involucral bracts 15–20-veined; stylopodium conical; Assam. ....10. *B. kurzii.*
14. Stem sparsely branched; leaves with white margin; involucral bracts 4–13-veined, or 0; stylopodium low conical; Himalaya.
15. Basal and lower cauline leaves 9–20-veined; upper leaves cordate and broadly amplexicaul; fruit ribs prominent, winged or subwinged. ....4. *B. jucundum.*

15. Basal and lower leaves 5- or 7-veined; upper leaves rounded at base and narrowly amplexicaul; fruit ribs filiform, obtuse, unwinged. .... 11. *B. candollei.*

8. Involucel and involucre inconspicuous, scarcely foliaceous, the bracteoles narrow and rarely imbricate laterally, or involucre and involucel 0.

16. Upper cauline leaves cordate-amplexicaul; fruit ribs winged or subwinged; Himalaya.

17. Basal leaves 10–15 mm long, acute or obtuse, white-margined; veins 9–20. .... 4. *B. jucundum.*

17. Basal leaves 3–8 mm broad, attenuate, without white margin; veins 5 or 7. .... 5. *B. thomsonii.*

16. Upper cauline leaves narrowed at base, not cordate; fruit ribs prominent or inconspicuous, but not winged.

18. Woody based perennials; foliage white-margined; involucre obsolete; Himalaya.

19. Bracteoles distinctly shorter than flowering umbellet; fruit ribs not prominent. .... 17. *B. falcatum.*

19. Bracteoles about equaling flowering umbellet; fruit ribs prominent. .... 18. *B. maddenii.*

18. Strictly herbaceous perennials or annuals; foliage lacking white margin, or involucre evident if white margin present; Assam or peninsular India.

20. Rays filiform; bracts and bracteoles longly acuminate; leaves often punctate, not at all coriaceous; fruit oblong; Assam. ....
.... 14. *B. khasianum.*

20. Rays slender but scarcely filiform; bracts and bracteoles acute or abruptly acuminate; leaves not punctate, or if punctate also slightly coriaceous; fruit ovoid or ovoid-elliptic; peninsular India.

21. Lower leaves 3–10 mm broad, neither distichous nor crowded; fruit 2–3 mm long, the ribs prominent; stylopodium depressed or obsolete.

22. Lower leaves acute or obtuse, usually white-margined and ± punctate; involucel equaling or surpassing fruiting umbellet; vittae mostly solitary in intervals. ....
.... 9. *B. mucronatum.*

22. Lower leaves shortly acuminate, neither white-margined nor punctate; involucel shorter than fruiting umbellet; vittae 3 in intervals. .... 19. *B. wightii.*

21. Lower leaves 2–3 mm broad, the lower cauline distichous and crowded; fruit 1.5–2 mm long, the ribs obscure; stylopodium conical. .... 13. *B. distichophyllum.*

1. Plants profusely dichotomously branching with filiform, sparsely leafy branches; umbels with a single ray, thus appearing simple; flowers usually only 2 or 3 per umbellet; mericarp surface rugose. .... 21. *B. subuniflorum.*

1. **Bupleurum lanceolatum** DC., Prodr. 4: 132. 1830; C. B. Clarke in Hook. f., Fl. Brit. Ind. 2: 674. 1879.

Bupleurum lanceolatum Wallich, Cat. 551. 1829, *nomen nudum.*

Plants slender, perennial, erect, simple or branching above, 50–150 cm tall, from a slender, woody taproot. Basal and lower cauline leaves conspicuously petiolate, ovate-lanceolate to broadly ovate, 3.5–10 cm long, 2–5 cm broad, shortly acuminate, rounded or tapering at base, the veins usually 5 or 7; petiole slender, 1–6 cm long; cauline leaves diminishing upward, the uppermost subsessile. Inflorescence terminal and lateral from branches arising in upper leaf axils, the peduncles very slender, 1–5 cm long. Involucre usually lacking. Rays 3–8, filiform, spreading-ascending, unequal, 5–20 mm long. Umbellets 5–12-flowered, the mature pedicels 1–3 mm long. Involucel of 3–5 linear to lanceolate, usually acute bracteoles 0.5–2 mm long, most 1- or 3-veined, rarely lacking. Flowers lemon-yellow; petals broadly oval; stylopodium depressed, forming a flattened disk. Fruit oblong to ovoid, 3–5 mm long, 2–3.5 mm broad, the ribs filiform, unwinged. Vittae large, 3 in intervals, 4 on commissure. Commissural face plane or slightly concave. Carpophore entire. June–Oct.

Type locality: Nepal, *Wallich 551.*

Distribution: Himalaya, from Nepal to Kumaon, Kashmir, and Pakistan, 1,500–4,000 m.

Representative specimens: Jammu & Kashmir: Rutton Pir, *Clarke 28,430B* (CAL); Lidder Valley, *Duthie 13,468* (CAL); Kashmir, *Royle 93/18* (DD). Himachal Pradesh: Wangtu, Simla Distr., *Brandis 3722* (DD); Dalhousie, *Clarke 23,162* (CAL). Uttar Pradesh: Mussoorie, *Mackinnon* (CAL); Naini Tal *Duthie 4050* (DD). Nepal: Nepal, *Wallich 551* (CAL).

2. **Bupleurum dalhousieanum** (C. B. Clarke) Kozo-Polj., Act. Hort. Petrop. 30: 165. 1915.

Bupleurum longicaule var. *dalhousieana* C. B. Clarke in Hook. f., Fl. Brit. Ind. 2: 677. 1879.

Plants low, perennial, decumbent, usually with several stems arising from base and branching above, 10–15 cm tall. Basal and lower cauline leaves sessile, linear, 2–3 cm long, ca. 3 mm broad, acuminate, the veins 7–11; upper cauline leaves ovate-lanceolate to ovate, the veins up to 13, narrowed and amplexicaul at base. Inflorescence of terminal and some axillary lateral umbels, the peduncles stout, 1–3 cm long. Involucre of 1–4 oval to ovate, unequal, acute or obtuse, mucronate, 5–13-veined bracts up to 1 cm long and 5 mm broad. Rays 2–4, slender, spreading-ascending, unequal, 5–20 mm long. Umbellets several flowered, the mature pedicels much shorter than fruit. Involucel of usually 5 ovate-lanceolate to obovate, obtuse or acute, mucronate, 5–7-veined bracteoles ca. 5 mm long, longer than flowering umbellet. Flowers yellow; stylopodium conical, dark brown. Fruit oblong-ovoid, ca. 3 mm long, 2 mm broad, the mericarps pentagonal in transection, the ribs, prominent, thick, very narrowly winged. Vittae 1–3 in intervals, 2 on commissure. Commissural face plane. Carpophore bifid to bipartite. June–Sept.

Type locality: "Dalhousie and Dhurmsala, 8,000-12,000 ft.," *Clarke.*

Distribution: Central and western Himalaya, 2,000–4,000 m.

Representative specimens: Himachal Pradesh: Dalhousie, *Clarke 22,121, 23,016,*

23,260 (K), *22,987C* (CAL); Lahul, *Brandis 3727* (DD); Laka, *Clarke 24,536* (K). Uttar Pradesh: Kedarnath, *Nair 35,909* (BSD); northern Kumaon, *La Touche* (DD). Nepal: Langtang, *Mukherjee 278* (CUH, UC).

3. **Bupleurum rupestre** Edgew., Trans. Linn. Soc. 20: 52. 1851.

Bupleurum himalayense Klotzsch in Klotzsch & Garcke, Bot. Ergebn. Reise Prinz Waldemar 146. 1862; *B. longicaule* var. *himalayense* (Klotzsch) C. B. Clarke in Hook. f., Fl. Brit. Ind. 2: 677. 1879. Based on "Himalaya," neither precise locality nor collector indicated.

Bupleurum longicaule var. *himalayense* f. *obtusiusculum* Nasir, Univ. Calif. Publ. Bot. 27: 431. 1955. Based on *Stewart & Stewart 18,017*, Rajdhiangu Pass.

Plants slender, perennial, erect, simple or branching from base, 20–50 cm tall, from a taproot. Basal and lower cauline leaves linear to linear-lanceolate, 6–15 cm long, 2 5 mm broad, acuminate to obtuse, narrowed and subpetiolate at base, narrowly white-margined, the veins usually 3–7; cauline leaves lanceolate to ovate-lanceolate, acuminate, dilated at base, sessile and broadly amplexicaul, 11–19-veined. Inflorescence terminal or terminal and lateral, the peduncles slender, 3–10 cm long. Involucre of 1–4 lanceolate to ovate, acuminate, unequal, 7–11-veined bracts, 5–30 mm long, 3–10 mm broad. Rays 2–8, slender, spreading-ascending, unequal, 0.5–6 cm long. Umbellets 15–50-flowered, the mature pedicels 1–4 mm long. Involucel of 6–10 ovate-lanceolate to oval, acuminate or acute, imbricate, basally connate, spreading bracteoles usually 5–10 mm long, 4–5 mm broad, 5- or 7-veined, conspicuously exceeding flowers and fruit. Flowers deep purple; petals oval to obovate; stylopodium low conical, deep purple. Fruit oblong, 2.5–3 mm long, 2 mm broad, the mericarps slightly compressed dorsally, smooth, the ribs equally thin winged. Vittae usually 3 in intervals, 4 on commissure. Commissural face plane. Carpophore bipartite. Aug.–Oct.

Type locality: "Chur," *Edgeworth.*

Distribution: Himalaya, from Nepal to Kashmir, 2,000–4,500 m.

Representative specimens: Jammu & Kashmir: Kashmir, *Clarke 31,033* (CAL); Zozila, *Stoliczka* (CAL). Himachal Pradesh: Lakha, *Edgeworth 4010* (CAL). Uttar Pradesh: Hemkund, *Bhattacharyya 24,256* (BSD); Ralam Glacier, *Pant & Naithani 39,683* (BSD); Kedarnath, 25 Oct. 1982, *Mukherjee* (CUH, UC). Nepal: Mugu Khola, *Polunin, Sykes & Williams 5346* (CAL); w/o locality. *Jacquemont 1202* (K), *1756* (K).

Note. From an examination of both specimens and literature, I am convinced that the original Wallich collections examined by de Candolle in publishing *B. longicaule* were of two different taxa. That from Nepal (represented in CAL by *Wallich 557A*), with acute but unwinged fruit ribs, best matches the protologue of that species and markedly resembles *B. longicaule* var. *stricta* Clarke. That from Kumaon (represented in CAL by *Wallich 557B*), on the other hand, has immature fruits that apparently are winged, as they are in other collections from the same area. This should be referred to *B. rupestre (B. himalayense). Bupleurum longicaule* and *B. rupestre* are treated here as separate species, although Nasir (1955, 1972) regarded them as conspecific. (P. K. M.)

4. **Bupleurum jucundum** Kurz, Jour. Bot. 5: 240. 1867;
C. B. Clarke in Hook. f., Fl. Brit. Ind. 2: 675. 1879.

Plants slender, perennial, erect, branching above, 30–70 cm tall. Basal and lower cauline leaves linearly oblong or lanceolate to obovate, shortly petiolate to sessile, up to 10 or 12 cm long, 1-1.5 cm broad, acute or obtuse, mucronate, narrowed at base, conspicuously white-margined, the veins 9–20; petiole 2–3 mm long, or obsolete; upper cauline leaves broadly lanceolate to oblong-ovate or oval, sessile, cordate-amplexicaul, 15–25-veined. Inflorescence terminal and lateral, the peduncles up to 10 cm long. Involucre of 1-3 oblong or ovate-lanceolate to ovate, acute or obtuse, 5–13-veined bracts, or sometimes lacking. Rays 5–10, stout, unequal, 4–25 mm long. Umbellets 10–20-flowered, the mature pedicels 2–3 mm long. Involucel of 3–5 linear or oblong to ovate-lanceolate, acute bracteoles 2–3 mm long, equaling or shorter than umbellet, 3- or 5-veined, or lacking. Flowers yellow; stylopodium low conical. Fruit oblong to oval, 3.5–5 mm long, 1.5 mm broad, the mericarps subterete in transection, glabrous, the ribs prominent, acute, subwinged to narrowly winged. Vittae 3 in intervals, 4 on commissure. Commissural face plane. Carpophore bipartite.

KEY TO VARIETIES

1. Lower leaves obovate, the upper ovate, deeply cordate-amplexicaul; involucre and involucel prominent; fruit ribs equally subwinged. .... a. var. *jucundum.*
1. Lower leaves linearly oblong or lanceolate, the upper ovate-lanceolate, shallowly cordate-amplexicaul; involucre and involucel inconspicuous; fruit ribs subequally winged. .... b. var. *cachemiricum.*

4a. var. **jucundum**

Plants up to 60 cm tall, corymbosely branched above. Lower cauline leaves obovate, very shortly petiolate, the upper cauline broad, deeply cordate-amplexicaul. Involucre of 1–3 bracts, usually conspicuous; involucel of 4 or 5 evident bracteoles. Fruit ribs acute, subwinged. July–Sept.

Type locality: Lahul, *Jaeschke.*

Distribution: Himalaya of Kashmir, perhaps to Pakistan, 2,000–4,000 m.

Representative specimens: Jammu & Kashmir: Banihal Ridge, Pir Pass, *T. A. Rao 7677* (BSD); Tilail Valley, *Duthie* (DD); Lahul, *Jaeschke* (CAL).

4b. var. **cachemiricum** C. B. Clarke in Hook. f., Fl. Brit. Ind. 2: 675. 1879.

Plants 30–45 cm tall, branched. Lower cauline leaves linearly oblong or lanceolate, sessile, amplexicaul, the upper cauline rather narrow, shallowly cordate-amplexicaul. Involucre obsolete or of a single bract; involucel of up to 4 inconspicuous bracteoles, or frequently absent. Fruit ribs narrowly winged. July–Sept.

Type locality: "Kashmir, alt. 7500–10,000 ft.", *Clarke.*

Distribution: Himalaya of Kashmir and Himachal Pradesh, perhaps to Pakistan, 2,300–3,650 m.

Representative specimens: Jammu & Kashmir: Srinagar, *Clarke 29,112D* (CAL). Himachal Pradesh: Sangla, *Nair 34,266* (BSD).

5. **Bupleurum thomsonii** C. B. Clarke in Hook. f., Fl. Brit. Ind. 2: 675. 1879.

Plants slender, perennial, erect, with 1 or more stems from base, branching above, 10–100 cm tall. Basal and lower cauline leaves linear to lanceolate, 7–14 cm long, 3–8 mm broad, attenuate at apex, tapering and subpetiolate at base, the veins usually 5 or 7; upper cauline leaves lanceolate to ovate, acuminate to attenuate, broadly dilated and deeply cordate-amplexicaul at base, up to 45-veined. Inflorescence terminal, or with some lateral umbels in upper leaf axils, the peduncles 2–8 cm long. Involucre obsolete, or occasionally of 1–3 narrow bracts. Rays 3–10, spreading-ascending, unequal, 1.5–3 cm long. Umbellets ca. 20-flowered, the mature pedicels 2–4 mm long. Involucel of 5–10 linear or linearly lanceolate, acuminate, basally connate bracteoles 5–7 mm long, 3-veined, as long as or longer than umbellet. Flowers yellow; petals obovate; stylopodium depressed to form a flattened, dark-coloured disk. Fruit oblong-ellipsoid, 3–7 mm long, 1.5–2 mm broad, the mericarps subterete in transection, glabrous, the ribs broadly and equally winged. Vittae 3 in intervals, 4 on commissure. Commissural face plane. Carpophore bipartite. June–Sept.

Type locality: "KASHMIR; Sonamurg, alt. 9250 ft.," *Clarke*. "NORTH-WEST HIMALAYA, alt. 7000–9000 ft.," *Thomson*.

Distribution: Himalaya of Kashmir and Himachal Pradesh to Pakistan.

Representative specimens: Jammu & Kashmir: Sind Valley, *Duthie 11,567* (CAL, DD); Kishanganga Valley, *Keshavanand 1488* (DD). Himachal Pradesh: Lahul, Gunglong, *Bor 9922* (DD).

6. **Bupleurum longicaule** DC., Prodr. 4: 131. 1830; C. B. Clarke in Hook. f., Fl. Brit. Ind. 2: 677. 1879.

Bupleurum longicaule Wallich, Cat. 557. 1829, *nomen nudum*.
Bupleurum longicaule var. *stricta* C. B. Clarke in Hook. f., Fl. Brit. Ind. 2: 677. 1879. Based on *Clarke*, "Sikkim; Jongri, alt. 12,000–13,000 ft."

Plants rather stout, low, perennial, branching from base with several simple erect stems 20–35 cm tall. Basal and lower cauline leaves linear, up to 6 cm long, ca. 3 mm broad, tapering toward base and amplexicaul, the veins 5 or 7; upper cauline leaves sparse, lanceolate, narrowed at apex, broadly amplexicaul, 5- or 7-veined. Inflorescence terminal, the strict peduncles 5–15 cm long. Involucre of 2 or 3 ovate-lanceolate, unequal, 7–9-veined bracts 1–5 cm long, reflexed in fruit. Rays 3–5, stout, ascending, unequal. Umbellets 8–15-flowered, the mature pedicels strict, 1–2 mm long. Involucel of usually 5 oblong-oval to orbicular, obtuse, mucronate bracteoles 1.5–2 cm long, equaling or longer than flowering umbellet, reflexed at maturity. Flowers purple; stylopodium depressed, purplish. Fruit oblong-ellipsoid, ca. 5 mm long, 2 mm broad, the mericarps slightly compressed dorsally, glabrous, the ribs prominent but unwinged. Vittae 3 in intervals, 4 on commissure. Commissural face slightly concave. Carpophore entire. Aug.–Oct.

Type locality: Gossainthan, Nepal, *Wallich 557*.

Distribution: Eastern Himalaya, from Yunnan and Tibet to Sikkim and Nepal, 3,000–5,000 m.

Representative specimens: Nepal: Gossainthan, *Wallich 557A* (CAL). Sikkim: Dzongri, *Mukherjee 318* (CUH, UC); Jongri to Aloklong, *Anderson 607* (CAL); Jongri, *Clarke 25,805, 25,944, 26,100* (K). Arunachal Pradesh: Shyamrang, Kameng Distr., *Hajra 64,875* (ASSAM).

7. **Bupleurum sikkimense** P. K. Mukherjee, Actes II Sympos. Internat. Ombell. Perpignan, 54. 1978.

Plants low, perennial, decumbent and freely branching at base, with numerous slender erect or ascending stems 8–20 cm tall, from a slender taproot. Basal and lower cauline leaves sessile, linearly oblong to oblanceolate, 2.5–5 cm long, ca. 2 mm broad, tapering toward base, white-margined, the veins 5 or 7; upper cauline leaves sparse, lanceolate, obtuse, narrowing toward the sessile, amplexicaul base, white-margined, 7-veined. Inflorescence terminal, the peduncles up to 3 cm long. Involucre of 1–3 ovate-lanceolate, obtuse, unequal, 7- or 9-veined bracts 3–10 mm long, 5 mm broad. Rays 2–5, slender, unequal, 5–20 cm long. Umbellets 5–7-flowered, the mature pedicels ca. 1 mm long. Involucel of usually 5 ovate or oval, 3-veined bracteoles ca. 3 mm long, exceeding flowering umbellet. Flowers dark brown; stylopodium conical, dark brown. Fruit oblong, ca. 2.5 mm long, 1.5 mm broad, the mericarps subterete, glabrous, the ribs slender, unwinged. Vittae 3 in intervals, 4 on commissure. Commissural face plane. Carpophore? June–July.

Type locality: Sikkim Himalaya, Dowthung, 21 June 1878, *Dungboo s. n.* (CAL, holotype).

Distribution: Known only from the Sikkim Himalaya, 3,300–4,000 m.

Representative specimen: Sikkim. Laghep, *W. W. Smith 3073* (CAL).

8. **Bupleurum hamiltonii** Balakr., Jour. Bombay Nat. Hist. Soc. 63. 328. 1966.

Bupleurum tenue Buchanan-Hamilton ex D. Don, Prodr. Fl. Nepal. 182. 1825; C. B. Clarke in Hook. f., Fl. Brit. Ind. 2: 677. 1879, non Salisb. 1796.

Plants annual or perennial, with 1 to several slender stems ascending or erect, 15–100 cm tall, branching from base or above, arising from a slender taproot, the foliage sometimes evidently punctate and usually scaberulous. Basal and cauline leaves sessile to shortly petiolate, linear to oblanceolate, 2.5–10 cm long, 3–12 mm broad, acute or obtuse, mucronate, slightly narrowed and amplexicaul at base, the veins 5–9; petiole slender, 5–10 mm long, or usually lacking; upper cauline leaves lanceolate to oblong, acute or obtuse, reduced upward, sessile and amplexicaul, 7–11-nerved. Inflorescence open, with numerous terminal and lateral umbels, the peduncles very slender, 1–4 cm long. Involucre of 1–4 lanceolate to oval, unequal, 3- or 5-veined bracts 2–8 mm long. Rays 3–8, filiform, spreading-ascending, unequal, 3–20 mm long. Umbellets ca. 10-flowered, the mature pedicels 1–2 mm long. Involucel of 5 or 6 linear-lanceolate to ovate or oval, acute or acuminate bracteoles 1–4 mm long, about equaling flowering umbellet, 3- or 5-veined. Flowers yellow; petals oval to obovate; stylopodium depressed to form a yellow or darker-coloured flattened disk. Fruit oblong-ovoid, 2–2.5 mm long, 1–1.5 mm broad, the mericarps subterete in transection, glabrous, the ribs prominent, acute to narrowly winged. Vittae solitary in intervals, 2 on commissure. Commissural face plane. Carpophore bipartite.

Type locality: Nepalia, *Buchanan-Hamilton s. n.*

Distribution: China and southeastern Asia; Himalaya, from Arunachal Pradesh and Nepal to Kashmir and Pakistan, 1,500–3,000 m.

Representative specimens: Jammu & Kashmir: Patni Top/Sannar, *T. A. Rao 7450* (BSD). Himachal Pradesh: Rampur, *Nair 29,850* (BSD). Punjab: Pathankot, *Clarke 21,973* (CAL). Uttar Pradesh: Raneekhet, *Duthie 4054* (CAL); Dehra Dun, *Babu 33,482* (BSD); Kali Valley, *Duthie 5610* (CAL). Nepal: Nepal, *Buchanan-Hamilton* (CAL), *Wallich 553A* (CAL), *554A* (CAL); Gossainthan, *Wallich 555B*. Arunachal Pradesh: Twang, *Mukherjee* (CUH).

9. **Bupleurum mucronatum** Wight & Arn., Prodr. Fl. Ind. Orient. 370. 1834; C. B. Clarke in Hook. f., Fl. Brit. Ind. 2: 676. 1879, non Brouss. ex Sprengel in Roem. & Schult. 1820, *nomen in synon.*

Bupleurum ramosissimum Wight & Arn., Prodr. Fl. Ind. Orient. 370. 1834; *B. falcatum* var. *ramosissimum* (Wight & Arn.) Dalzell & Gibson, Bombay Fl. 108. 1861; *B. mucronatum* var. *ramosissimum* (Wight & Arn.) C. B. Clarke in Hook. f., Fl. Brit. Ind. 2: 676. 1879. Based on *Wight 1198* (K).

Bupleurum wightii P. K. Mukherjee, Indian Forester 95: 567. 1969, non Kozo-Polj. 1913.

Plants slender, perennial, erect, branching above, 20–150 cm tall, from a slender taproot, the foliage sometimes slightly coriaceous and punctate. Basal and lower cauline leaves sessile to petiolate, narrowly linearly oblong to oblanceolate, 6–16 cm long, 3–10 mm broad, acute, mucronate, white-margined, tapering at base, the veins 3–7; petiole slender, up to 3 cm long, or obsolete; upper cauline leaves reduced upward, narrowed and amplexicaul at base, 5- or 7-nerved. Inflorescence open, with several terminal and lateral umbels, the peduncles slender, 1–3 cm long. Involucre of 3–5 lanceolate, acuminate, unequal, 3- or 5-veined bracts 3–10 mm long, 0.5–2 mm broad. Rays 3–5 (-12?), very slender, spreading-ascending, unequal, 5–35 mm long. Umbellets 4–10-flowered, the mature pedicels 1–2 mm long. Involucel of usually 5 lanceolate, acuminate bracteoles 1–5 mm long, about equaling umbellet, 3-veined. Flowers yellow (or purplish); petals obovate; stylopodium depressed to form a flattened disk. Fruit ovoid-elliptic, 2–3.5 mm long, 1.5–2.5 mm broad, the mericarps subterete-pentagonal in transection, glabrous, the ribs prominent, acute, but scarcely winged. Vittae solitary (rarely 2) in intervals, 2 on commissure. Commissural face slightly concave. Carpophore bipartite. July–Oct.

Type locality: w/o locality, *Wight 1197* (K).

Distribution: Bihar to peninsular and southwestern India, and Sri Lanka.

Representative specimens: Bihar: Sargooja, *Clarke 34,000A* (CAL). Orissa: Kalahandi, *Mooney 3190* (DD). Madhya Pradesh: Panchmarhi, *Duthie 10,405* (CAL, DD); Bastar, *Mooney 1487* (DD). Andhra Pradesh: Chintapalli, Vishakapatnam Distr., *Subbarao 28,228* (MH). Tamil Nadu: Coonor, *Gamble 122,282* (DD); Pulney, *Wight 1167* (CAL). Kerala: Lockhert Gap, Kottayam Distr., *Sebastine 17,494* (MH). Karnataka: Attapadi Hills, *Fischer 2320* (CAL). Maharashtra: Bombay Presidency, *Dalzell & Gibson* (DD).

Note. The existence of *B. mucronatum* Brouss. ex Sprengel in Roem. & Schult., Syst.

Veg. 6: 377. 1820, published as a synonym of *Tenorea fruticescens* (Linn.) Sprengel, according to Article 34 of the International Rules of Botanical Nomenclature, cannot be used to invalidate *B. mucronatum* Wight & Arn. (L.C.)

10. **Bupleurum kurzii** P. K. Mukherjee, Actes II Sympos. Internat. Ombell. Perpignan, 54. 1978.

Plants erect, copiously branched, the stems brown, the foliage sometimes punctate. Basal and lower cauline leaves oblanceolate, 5–10 cm long, up to 8 mm broad, obtuse, mucronate, narrowed and amplexicaul at base, the veins 7 or 9, secondary veins prominent; upper cauline leaves similar but reduced. Inflorescence open, with numerous terminal and lateral umbels, the peduncles slender, divaricate, 1–3 cm long. Involucre of 2–4 ovate or oval, unequal, 15–20-veined foliaceous bracts 3–10 mm long. Rays 2–6, slender, spreading, unequal, 5–20 mm long. Umbellets ca. 10-flowered. Involucel of 5 ovate, obtuse, mucronate, imbricate bracteoles, about equaling umbellet, 5- or 7-veined. Flowers yellow (?); stylopodium conical, dark brown. Fruit elliptic-oblong, 4–4.5 mm long, 2–2.5 mm broad, the mericarps terete in transection, the ribs prominent, acute but unwinged. Vittae 2 or 3 in intervals, 4 on commissure. Carpophore bipartite. Oct.–Nov.

Type locality: Tenasserim, Burma, 7000 ft., *Kurz s. n.* (CAL, holotype).

Distribution: Burma to Assam, 2,000–2,700 m.

Representative specimens: Nagaland: Kegwima Edge, *Clarke 41,848B* (CAL). Manipur. On way to Seeriphani, *Watt 5934* (CAL).

Note. Allied to *B. hamiltonii* Balakr., but differing in the 2- or 3-vittate valleculae and distinguishable from *B. falcatum* by the foliaceous involucre and involucel and punctate leaves. Clarke mistakenly cited the type collection (without having seen it) under his *B. tenue* var. *khasiana*. (P. K. M.)

11. **Bupleurum candollei** DC., Prodr. 4: 131. 1830; C. B. Clarke in Hook. f., Fl. Brit. Ind. 2: 674. 1879.

Bupleurum candollii Wallich, Cat. 552. 1829, *nomen nudum.*

Bupleurum candollei f. *acutifolium* H. Wolff and f. *obtusifolium* H. Wolff in Engl., Pflanzenreich 43 ($4^{\underline{228}}$): 150. 1910, no types indicated.

Plants slender, perennial, erect, branching above, 30–100 cm tall, from a stout, woody taproot. Basal and lower cauline leaves linearly oblong to lanceolate, 5–12 cm long, 5–20 mm broad, acute, mucronate, white-margined, narrowed at base into a short petiole, or sessile, the veins 5 or 7, secondary veins prominent; upper cauline leaves oblong or ovate to oval, obtuse to acute, rounded and sessile or subsessile, amplexicaul at base, mostly 7- to 13-veined. Inflorescence of terminal and lateral umbels, the peduncles slender, 2–5(–9) cm long. Rays 4–11, slender, unequal, spreading-ascending, 1–3.5 cm long. Involucre of 1–4 ovate-lanceolate to oval, unequal, acute or obtuse, 5- or 7-veined bracts 1–3.5 cm long, 5–20 mm broad. Umbellets ca. 20-flowered, the mature pedicels 1.5–2 mm long. Involucel of 4 or 5 lanceolate to oval, acute or obtuse foliaceous bracteoles 2–6 mm long, 1–3 mm broad, about equaling flowering umbellet, 3- or 5-veined. Flowers light yellow; petals obovate; stylopodium low conical. Fruit oblong-ovoid, 3.5–5 mm long, 2–3 mm broad, a little narrowed toward apex, rounded at base, the mericarps subterete in transection, the ribs filiform, obtuse, unwinged. Vittae usually 3 in intervals, 4 on commissure. Commissural face slightly concave. Carpophore bipartite. July–Oct.

Type locality: Gossainthan, Nepal, *Wallich 552.*

Distribution: Himalaya, from Arunachal Pradesh and Sikkim to Kashmir and Pakistan, 2,000–4,500 m.

Representative specimens: Jammu & Kashmir: Kishanganga Valley, *Keshavanand 1571* (DD). Himachal Pradesh: Durmsala, *Clarke 24,103* (CAL); Choor, Nagtibba, *Royle 93/17* (DD). Uttar Pradesh: Mussoorie, *Royle 49/15* (DD); Naini Tal, *T. A. Rao 4990* (BSD); Rambara, 23 Oct. 1982, *Mukherjee* (CUH, UC). Nepal: Gossainthan, *Wallich 552* (CAL). West Bengal: Darjeeling, *Clarke 13,422D, 13,422E, 13,422S* (CAL); Sandakphu, *Gamble 85F* (CAL). Sikkim: Zemu Valley, *Smith & Cave 2799* (CAL). Bhutan: Thinleygaon, *Hajra 393* (CAL). Arunachal Pradesh: Zimithang, Kameng Distr., *Hajra 68,343* (ASSAM).

12. **Bupleurum plantaginifolium** Wight., Ic. Pl. Ind. Orient. 1: 281. 1840; C. B. Clarke in Hook. f., Fl. Brit. Ind. 2: 674. 1879.

Plants stout, perennial, shrubby, corymbosely branching above, glabrous to scaberulous, 0.6–3 m tall. Basal and lower cauline leaves oblong to broadly oblanceolate, 10–25 cm long, 12–25(–60?) mm broad, obtuse, mucronate, gradually narrowed toward base and then dilated and amplexicaul, the veins 11–17; upper cauline leaves crowded, broadly oblong to oblanceolate, shortly narrowed and subsessile at base, 10–30-nerved, secondary veins prominent. Inflorescence of terminal and lateral umbels, the peduncles elongate. Involucre of 3–5 oblanceolate to obovate, obtuse, subequal, foliaceous, 7–11-veined bracts 1–2.5 cm long, ca. 5 mm broad. Rays (3–)6–12, spreading, unequal, 1–3 cm long. Umbellets 4–14-flowered, the mature pedicels ca. 3 mm long. Involucel of 4–6 oblong-oval, obtuse, mucronate, 5- or 7-veined bracteoles 6–12 mm long, about equaling umbellet. Flowers yellow; petals elliptic; stylopodium depressed, forming a flattened disk. Fruit oblong-ovoid, 6–8 mm long, 4 mm broad, subquadrate, the mericarps pentagonal in transection, glabrous, the ribs prominent, thick and corky, elevated and subwinged. Vittae large, solitary in intervals, 2 on commissure. Seed grooved beneath valleculae, the commissural face plane or slightly concave. Carpophore bipartite. July–Oct.

Type locality: Elk Hill, "Nilgherries," *Munro.*

Distribution: Peninsular India, 2,500–3,000 m.

Representative specimens: Tamil Nadu: Dodabetta, *Gamble 16,985* (CAL); Ooty, *Clarke 11,151* (CAL); Peninsular India Oriental., *Wight 1169* (CAL).

13. **Bupleurum distichophyllum** Wight & Arn., Prodr. Fl. Ind. Orient. 370. 1834; C. B. Clarke in Hook. f., Fl. Brit. Ind. 2: 677. 1879.

Plants slender, perennial, erect, flexuous and branching above, 8–40 cm tall. Basal leaves petiolate, early deciduous, oblong-obovate, ca. 2 cm long, 3 mm broad, the lower cauline leaves distichous, coriaceous, crowded and overlapping, linear or linear-subulate, 2.5–6 cm long, 2–3 mm broad, long-acuminate, subcuspidate, broad at base, the veins 5 or 7; upper cauline leaves sparse, reduced upward, subamplexicaul. Inflorescence of terminal and lateral umbels, the peduncles 1–3 cm long. Involucre of 5 or 6 linearly lanceolate, acuminate, subequal bracts 5–10 mm long, the veins 3 or 5. Rays 5–8, slender, subequal, ca. 2 cm long. Umbellets 15–20-flowered, the mature pedicels ca. 2 mm long. Involucel of ca. 5 linearly lanceolate, acuminate, 3-nerved bracteoles 3–5 mm long, exceeding umbellet. Flowers yellow (?); petals obovate; stylopodium conical. Fruit ovoid, 1.5–2 mm long, the mericarps glabrous, the ribs prominent, acute, unwinged. Vittae solitary in intervals, 2 on commissure. Carpophore bipartite. June–Oct.

Type locality: Peninsular India, *Wight 1196*.

Distribution: Peninsular India, 2,000–25,000 m.

Representative specimens: Tamil Nadu: Ootacamund, *Gamble 12,448* (CAL); Kodaikanal, *Bourne 154* (CAL); Devicolam, *Meebold 13,152* (CAL); Peninsular India Orient., *Wight 1168, 1170* (CAL).

14. **Bupleurum khasianum** (C. B. Clarke) P. K. Mukherjee, Bull. Bot. Surv. India 24(1–4): 43. 1983.

Bupleurum tenue var. *khasiana* C. B. Clarke in Hook. f., Fl. Brit. Ind. 2: 678. 1879; *B. hamiltonii* var. *khasiana* (C. B. Clarke) Balakr., Jour. Bombay Nat. Hist. Soc. 63: 328. 1968.

Plants very slender, perennial, erect, branching from base or above, 0.9–1.5 m tall from a slender taproot, the foliage sometimes punctate. Basal and lower cauline leaves linearly lanceolate, 5–8 cm long, 3–6 mm broad, acute, mucronate, narrowed at base either into a short petiole or subamplexicaul, the veins 3 or 5 (7 ?); upper cauline leaves rather sparse, similar, sessile, reduced upward. Inflorescence of terminal and lateral umbels, the peduncles slender, 2–4 cm long. Involucre of 2–4 linearly lanceolate, longly acuminate, subequal, 3- or 5-nerved bracts 7–13 mm long, 1–2 mm broad. Rays 2–5, filiform, spreading-ascending, unequal, 1–3 cm long. Umbellets 4–10-flowered, the mature pedicels 1–1.5 mm long. Involucel of 4 or 5 linearly lanceolate, longly acuminate, 3-nerved bracteoles 1–4 mm long, longer than umbellet. Flowers lemon-yellow; petals orbicular, stylopodium low conical. Fruit oblong, ca. 4 mm long, 1–1.5 mm broad, the mericarps subterete in transection, glabrous, the ribs prominent, acute but unwinged. Vittae solitary in intervals, 2 on commissure. Commissural face not seen. Carpophore bipartite. Aug.–Nov.

Type locality: “Khasia Mts., alt. 4000–6000 ft.,” Assam, no collector indicated.

Distribution: Khasia Hills, Meghalaya, 1,500–2,000 m.

Representative specimens: Meghalaya: Khasia, Nongram, *Clarke 16,708* (CAL); Makadokadok, *Clarke 40,385A, 45,265B* (CAL); Khasia, *Griffith 2608* (CAL).

15. **Bupleurum aitchisonii** (Boiss.) H. Wolff in Engl., Pflanzenreich 43 ($4\frac{228}{}$): 142. 1910.

Bupleurum olympicum var. *aitchisonii* Boiss., Fl. Orient. Suppl. 252. 1888.
Bupleurum aitchisonii var. *caudatum* H. Wolff in Engl., Pflanzenreich 43 ($4\frac{228}{}$): 142. 1910. Based on *Jaeschke 80*, “Himalaya: Lahul.”

Plants low, decumbent and densely branching at base of the short stem, the branches ascending or erect, 10–15 cm long or tall. Basal and lower cauline leaves crowded, linear to linearly lanceolate, 2–4 cm long, 2–4 mm broad, acuminate, narrowed at base but rarely petiolate, the veins 5–9; upper cauline leaves sparse, linear to ovate-lanceolate, white-margined, broadly amplexicaul at base, 7-11-veined. Inflorescence of terminal umbels, the peduncles usually ca. 2 cm long. Involucre of 1-3 lanceolate to oval, shortly acuminate, 5-9-veined, unequal bracts 2-7 mm long. Rays 2-5, unequal, 0.5-2 cm long. Umbellets 10-15-flowered, the mature pedicels 1-4 mm long. Involucel of 5-7 lanceolate to oval, acute, 3-veined bracteoles 2-6 mm long, about equaling umbellet. Flowers yellow; petals obovate, emarginate; stylopodium conical, yellow or purple. Fruit ovoid-oblong, 2-4 mm long, ca. 1.5 mm broad, the mericarps subterete in transection, glabrous, the ribs prominent, thick, unwinged. Vittae 2 or 3 in intervals, 4-6 on commissure. Commissural face concave. Carpophore bipartite. June-Oct.

Type locality: Kurram Valley, Afghanistan, *Aitchison 929*.

Distribution: Western Himalaya, to Afghanistan and Pakistan, 4,000-5,500 m.

Representative specimens: Himachal Pradesh: Lahul, Rotang, *Bor 9811* (CAL); Dudu Glacier, *Duthie 1149* (DD). Uttar Pradesh: Kedarnath, 25 Oct. 1982, *Mukherjee* (CUH, UC).

16. **Bupleurum gracillimum** Klotzsch in Klotzsch & Garcke, Bot. Ergebn. Reise Prinz Waldemar 148. 1862.

Bupleurum falcatum ssp. *eufalcatum* var. *gracillimum* (Klotzsch) H. Wolff in Engl., Pflanzenreich 43 ($4\frac{228}{}$): 132. 1910.

Bupleurum falcatum var. *nigrocarpa* Jacquem. ex C. B. Clarke in Hook. f., Fl. Brit. Ind. 2: 676. 1879.

Plants low, spreading, densely branching at base of the short stem, the branches decumbent or ascending, 10-35 cm long or tall. Basal and lower cauline leaves crowded, linear to linearly lanceolate, 3-6 cm long, 1.5-3 mm broad, acuminate, narrowed to an amplexicaul base; upper cauline leaves sparse, linear to ovate-lanceolate, reduced upward, longly acuminate, dilated and broadly amplexicaul at base, the veins 3-7. Inflorescence of mostly terminal and some lateral axillary umbels, the peduncles 2-9 cm long. Involucre of 2-5 linear to linearly lanceolate, unequal, 3-nerved bracts 2-10 mm long. Rays usually 3 (2-5), filiform, unequal, 1-3(-6?) cm long. Umbellets 5-15-flowered, the mature pedicels 1-1.5 mm long. Involucel of 5-7 linear to linearly lanceolate, 3-veined bracteoles 1-4 mm long, about equaling flowering umbellet. Flowers yellow; petals obovate; stylopodium low conical, with a dark-coloured disk. Fruit narrowly oblong, 3-5 mm long, ca. 1.5 mm broad, the mericarps terete in transection, glabrous, the ribs prominent, acute, subwinged. Vittae 3 in intervals, 4-6 on commissure. Carpophore bipartite. July-Aug.

Type locality: "N. W. Himalayas," *Waldemar*.

Distribution: Himalaya, Bhutan to Pakistan, 3,000-5,000 m.

Representative specimens: Jammu & Kashmir: Burzilla Pass, *Clarke 29,894* (CAL). Himachal Pradesh: Lahul, Keylang, *Bor 16,457* (DD). Nepal: Katkl-Langu, Magu Khola Confluence, *Polunin, Sykes & Williams 5266* (CAL). Bhutan: Bhutan, *White* (CAL); w/o locality, *Jacquemont 1492* (K).

17. **Bupleurum falcatum** Linn., Sp. Pl. 237. 1753; C. B. Clarke in Hook. f., Fl. Brit. Ind. 2: 676. 1879.

Plants stout or slender, perennial (rarely annual?), 30-100 cm tall, erect, the stems one to many, simple or branching from a woody base, the leaves ± white-cartilaginous-margined, the cauline often falcate. Basal and lower cauline leaves linear, lanceolate or oblong to obovate, 4–25 cm long, 2.5–10 mm broad, acute, narrowed to a petiole or sessile and amplexicaul at base, the veins 5–9, secondary veins prominent to obscure; upper cauline leaves linearly lanceolate to lanceolate or oblong, abruptly acute to acuminate, mucronate, reduced upward, narrowed and amplexicaul at base, the veins 3–9. Inflorescence terminal and lateral, often much branched, the peduncles slender, 1–10 cm long. Involucre of 1–5 linearly-subulate to lanceolate, acuminate or acute, unequal, 3-(5-) nerved bracts 0.5–5 (–10?) mm long, or sometimes lacking. Rays 3–15, slender or filiform, spreading or spreading-ascending, equal or unequal, 1–2 cm long. Umbellets 5–15-flowered, the mature pedicels 1–3 mm long. Involucel of 5–8 linearly lanceolate, acute or acuminate, 3–(–5-) veined bracteoles 1–3 mm long, or sometimes lacking, much shorter than umbellet. Flowers yellow; petals oval to obovate; stylopodium very low conical, forming a thick yellow disk. Fruit oblong to oblong-ovoid, 2.5–6 mm long, 1–2 mm broad, narrowing slightly toward apex, the mericarps subterete in transection, the ribs slender and inconspicuous to stout and prominent, unwinged to ± winged. Vittae 3 in intervals, 4 on commissure. Commissural face plane or slightly concave. Carpophore bipartite.

Type locality: "Habitat in Misniae, Vallesiae sepibus," no collector indicated.

Distribution: Eurasia.

Note. "A very variable species which requires investigation before infra-specific taxa can be satisfactorily delimited" (Tutin, 1968). Indian material appears best treated as comprising two varieties and two *formae*, in close agreement with the revision of northwestern Himalayan *Bupleurum* by Nasir (1955), except that vars. *linearifolium* and *marginatum* are combined in the present work. Clearly, the Indian forms of this and other species of the genus are in need of further study.

KEY TO INDIAN INFRASPECIFIC TAXA

1. Leaves broadly white-cartilaginous-margined, linearly lanceolate, acuminate; secondary veins obscure; fruit ribs slender, inconspicuous. .... a. var. *marginatum.*
    2. Stems thick, simple and solitary at base, 40–100 cm tall; lower leaves 5–10 mm broad. .... f. *marginatum.*
    2. Stems slender, usually branching and bushy at base, 30–45 cm tall; lower leaves mostly 2.5–4 mm broad. .... f. *stenophyllum.*
1. Leaves narrowly or obsoletely white-cartilaginous-margined, lanceolate, abruptly acute; secondary veins prominent; fruit ribs stout, prominent. ... b. var. *hoffmeisteri.*

17a. var. **marginatum** (DC.) C. B. Clarke in Hook. f., Fl. Brit. Ind. 2: 676. 1879.

Bupleurum marginatum Wallich, Cat. 566. 1829, *nomen nudum*; *B. marginatum* DC., Prodr. 4: 132. 1830; *B. falcatum* ssp. *marginatum* (DC.) H. Wolff in Engl., Pflanzenreich 43 ($4^{\underline{228}}$): 133. 1910.

Lower leaves linear to oblong or spatulate, 6–25 cm long, 3–10 m broad, the margin broad, the secondary veins obscure; upper leaves lanceolate to oblong, broadest below but narrowed at the amplexicaul base; rays 3–15, 10–20 mm long, equal; involucral bracts 1–5, linearly lanceolate, 1–10 mm long, rarely obsolete; involucel bracteoles ca. 5, 1–2.5 mm long; stylopodium flat; fruit ribs prominent. June–Oct.

Type locality: "ad Kamoon et in Nepalia," *Wallich 566.*

Distribution: Khasia Hills, Assam, throughout the Himalaya to Kashmir and Pakistan, 1,500–3,500 m.

f. **marginatum**

Representative specimens: Jammu & Kashmir: Kashmir, *Thomson* (CAL); Lidder Valley, *T. A. Rao 542* (BSD). Himachal Pradesh: Pangi, *Nair 22,555* (BSD); Lahul, *Bor 9702* (CAL). Uttar Pradesh: Chakrata, *Bhattacharyya 14,888* (BSD); Mussoorie, *Brandis 2368* (DD); Gori Valley, *Inayat Khan 24,451* (DD). Nepal: Gossainthan, *Wallich 566A* (CAL). Meghalaya: Shillong, *Clarke 38,450B & D, 44,267D* (CAL).

f. **stenophyllum** (H. Wolff) P. K. Mukherjee,
Edinburgh Jour. Bot. 48(1): 43. 1991.

Bupleurum falcatum ssp. *marginatum* var. *stenophyllum* H. Wolff in Engl., Pflanzenreich 43 ($4\frac{228}{}$): 133. 1910.

Bupleurum marginatum var. *linearifolium* Thomson ex H. Wolff, l.c., *nomen in synon.*; *B. falcatum* var. *linearifolium sensu auct., non B. linearifolium* DC. 1830.

Type locality: "Himal. Bor. Occ.," *Thomson s. n.*

Distribution: Himalaya, from Uttar Pradesh to Himachal Pradesh.

Representative specimens: Himachal Pradesh: Tapri, Kinnaur Distr., *Nair 30,058* (BSD). Uttar Pradesh: Ramni-Sutot, Garhwal, *Rau 10,159* (BSD).

17b. var. **hoffmeisteri** (Klotzsch) C. B. Clarke in Hook. f.,
Fl. Brit. Ind. 2: 676. 1879.

Bupleurum hoffmeisteri Klotzsch in Klotzsch & Garcke, Bot. Ergebn. Reise Prinz Waldemar, 147. 1862.

Lower leaves lanceolate, falcate, 4–10 cm long, 3–5 mm broad, the margin narrow, the secondary veins ± prominent; upper leaves lanceolate, narrowed at the amplexicaul base; rays 8–12, 8–20 mm long, unequal; involucral bracts 2–5, lanceolate, ca. 3 mm long, frequently lacking; involucel bracteoles 5–8, 2–2.5 mm long; stylopodium depressed; fruit ribs slender. June–Sept.

Type locality: Himalaya, *Hoffmeister*.

Distribution: Northwestern Himalaya, Himachal Pradesh and Kashmir to Pakistan, 1,800–4,000 m.

Representative specimens: Jammu & Kashmir: Banihal Range, *T. A. Rao 7602* (BSD). Himachal Pradesh: Sangla, Bashahr, *Nair 33,084* (BSD).

18. **Bupleurum maddenii** C. B. Clarke in Hook. f.,
Fl. Brit. Ind. 2: 678. 1879.

Plants slender, perennial, branching above, the numerous branches erect, 30–60 cm tall. Basal and lower leaves linear, 1.5–5 cm long, 1–1.5 mm broad, acuminate, narrowed and slightly amplexicaul at base, white-margined, the veins 7 or 9, secondary veins distinct, resinous; upper cauline leaves shorter, linear but broader, acuminate, 7–11-nerved, slightly amplexicaul. Inflorescence terminal and lateral, the umbels numerous, the peduncles 5–50 mm long. Involucre of 2 or 3 linear to lanceolate, unequal, 3- or 5-veined bracts up to 4 mm long, or lacking. Rays 5–8, slender, unequal, 1–1.5 cm long. Umbellets ca. 20-flowered, the mature pedicels 2–7 mm long. Involucel of 3–5 linear to lanceolate, acute, 3-nerved bracteoles about equal to flowering umbellet, but only 1/2 length of fruiting umbellet. Flowers subsessile (?); stylopodium low conical, forming a flattened disk. Fruit ovoid-oblong, 2–3 mm long, 1–1.5 mm broad, the mericarps glabrous and glaucous, the ribs prominent, narrow, unwinged. Vittae 2 or 3(?) in intervals.

Type locality: "Kunawur, alt. 6000–8000 ft.," *Thomson* (*Hooker & Thomson*, Herb. Ind. Orient., *Bupleurum 8*).

Distribution: Western Himalaya, 2,000–3,000 m.

Representative specimens: Himachal Pradesh: Kunwar, *Thomson* (CAL, K); Simla, *Lady Dalhousie* (K); Mussoorie, *Jameson 284* (K), *Madden* (K).

19. **Bupleurum wightii** Kozo-Polj., Epitome Bupleurorum Rossiae, 166. 1913.

Bupleurum virgatum Wight & Arn., Prodr. Fl. Ind. Orient. 370. 1834; *B. mucronatum* var. *virgata* (Wight & Arn.) C. B. Clarke in Hook. f., Fl. Brit. Ind. 2: 676. 1879, *non B. virgatum* Cav. 1827.

Plants slender, erect, annual or perennial, 30–60 cm tall, the stem usually simple below. Basal and lower cauline leaves sessile, linearly oblong, 4–7 cm long, 3–5 mm broad, sharply acuminate; amplexicaul at base; upper cauline leaves sparse, shorter. Umbels terminal and lateral, the peduncles to 3 cm long. Involucre of 4 or 5 linearly lanceolate, sharply acuminate, 3-nerved bracts ca. 7 mm long. Rays ca. 6, unequal, spreading-ascending, 1–2 cm long. Umbellets ca. 10-flowered, the mature pedicels 1–1.5 mm long. Involucel of 4–6 linearly lanceolate, 3-nerved bracteoles 1.5–2 mm long, about equal to flowering umbellet, but shorter than fruiting umbellet. Flowers yellow; petals broad; stylopodium obsolete. Fruit ovoid, ca. 3 mm long, 2 mm broad, the mericarps subterete in transection, glabrous, the ribs prominent, acute but unwinged. Vittae 3 in intervals, 2 on commissure. Commissural face plane. Carpophore bipartite. Nov.–Feb.

Type locality: "Ceylon," no collector indicated.

Distribution: Sri Lanka; peninsular India, 1,000–1,700 m.

Representative specimens: Madhya Pradesh: Panchmarhi, *Duthie 10,405* (CAL). Tamil Nadu: Sengatheri, Tinnavelli Distr., *Joseph 15,884* (MH). Kerala: Attapadi Hills, *Fischer 2320* (MH). Karnataka: Bababudan Hills, *Talbot 3183* (CAL).

20. **Bupleurum andhricum** Nayar & R. N. Banerjee, Bull. Bot. Surv. India 12: 255. 1970.

Plants stout, perennial, erect, branching above. Basal and lower cauline leaves oblong-lanceolate, 7–17 cm long, 13–25 mm broad, obtuse at apex, narrowed and amplexicaul at base, the veins 7–11; upper cauline leaves oval or lanceolate-oval, sessile, usually 5-nerved. Inflorescence terminal and lateral, the peduncles 6–25 mm long. Involucre of 3 or 4 oval to obovate, obtuse, 5–7-veined bracts 5–15 mm long. Rays 3–8, slender, 1.5–3 cm long. Umbellets 4–6-flowered, the mature pedicels 1–3 mm long. Involucel of 3–4 obovate, mucronate, 3–5-nerved bracteoles 2–6 mm long, much shorter than umbellet. Flowers pale yellow; calyx teeth evident; petals obovate or suborbicular; stylopodium low conical, forming a flattened fleshy disk. Fruit oblong-oval, 3–3.5 mm long, 2–2.5 mm broad, the mericarps subterete in transection, glabrous, the ribs filiform, unwinged. Vittae 3 or 4 in intervals, 4 or 5 on commissure. Commissural surface slightly concave. Aug.–Dec.

Type locality: Andhra Pradesh: Godavari District, Palakonda, 1 Oct. 1920, *Narayanaswami 370A* (CAL, holotype).

Distribution: Peninsular India, 1,150–1,300 m.

Representative specimens: Orissa: Lakshmipur, Manasmara Hill, *Raju 2366* (CAL). Andhra Pradesh: Palakonda, Godavari Distr., *Narayanaswami 370A–C* (CAL).

Note. *Bupleurum falcatum* var. *biharensis* Haines, Bot. Bihar & Orissa, 406. 1925, based on *Haines 4737*, "High mountains of Chota Nagpur. Netarhat!", Bihar, may be this species but the prominently white-margined cauline leaves, inconspicuous involucre, narrow bracteoles, and slender pedicels mentioned in the original description seem to be at variance. No specimen of this could be seen.

21. **Bupleurum subuniflorum** Boiss. & Heldr. in Boiss., Diagn. ser. 1, 10: 28. 1849.

Bupleurum setaceum sensu C. B. Clarke in Hook. f., Fl. Brit. Ind. 2: 678. 1879, non Fenzl 1866.

Plants very slender, annual, erect, with profuse dichotomous branching and filiform, quadrangular, sparsely leafy branches, 20–50 cm tall from a slender taproot. Basal and lower cauline leaves linearly lanceolate to linear, 1–5 cm long, 1–4 mm broad, acute, narrowed and amplexicaul at base, the veins 3 or 5; upper cauline leaves strongly reduced upward, linearly setaceous or acicular, obscurely 3-nerved. Inflorescence very open, of terminal and lateral simple or 1-rayed umbels, the peduncles or rays filiform, 5–20 mm long. Involucre of a solitary, acicular, 3-veined, appressed bract 2–5 mm long. Umbellets mostly 2- or 3-flowered, the mature pedicels only 0.4–0.8 mm long. Involucel of 4–6 acicular, obscurely 3-veined bracteoles 2–3 mm long, longer than flowering umbellet. Flowers yellow; petals obovate; stylopodium depressed. Fruit oval, 1.5–2 mm long, 1.5 mm broad, the mericarps subterete in transection, rugose, the ribs evident, filiform, unwinged. Vittae solitary in intervals, 2 on commissure. Commissural face concave. Carpophore bipartite. July–Aug.

Type locality: Pamphylia, Turkey, *Heldreich 1845*.

Distribution: Southwestern Asia to the Himalaya of Kashmir, Afghanistan, and Pakistan, 1,500–2,000 m.

Representative specimens: Jammu & Kashmir: below Tragbol, *R. Stewart 19,445* (DD); Jhelum Valley, *R. Stewart 9981* (CAL).

GROUP IV

1. Leaves all simple and entire. .... 19. *Bupleurum.*
1. Leaves pinnately, ternately, or ternate-pinnately divided to decompound, or if some (rarely all) leaves simple, their margins variously serrate, toothed, or incised.
    2. Plants arising from a single globose, ovoid, or fusiform tuber or corm.
        3. Petals broad, with a narrower inflexed apex.
            4. Leaf divisions narrow, oblong or lanceolate to filiform; calyx obsolete.
                5. Fruit ribs evident to prominent; vittae large, solitary in intervals. .... 20. *Bunium.*
                5. Fruit ribs obscure to lacking; vittae small, several in intervals, or obliterated in fruit. .... 21. *Elaeosticta.*
            4. Leaf divisions broad, ovate; calyx evident. ....22. *Karnataka.*
        3. Petals very narrow to broad, plane or shallowly concave and without an inflexed apex, obtuse, acute, or longly caudate.
            6. All flowers regular and the petals subequal; petals acuminate to longly caudate apically ....43. *Acronema.*
            6. Peripheral flowers irregular, the distal petals radiant; all petals obtuse or rounded at apex ....44. *Sinocarum.*
    2. Plants arising from taproots, fascicled roots, or rhizomes.
        7. Aquatic or marsh plants arising from fascicled fibrous or tuberous roots or creeping rootstocks.
            8. Involucre conspicuous; carpophore halves each adhering to a mericarp and falling with it.
                9. Fruit ribs prominent, corky, subwinged; vittae solitary in intervals, 2 on commissure. ....23. *Sium.*
                9. Fruit ribs inconspicuous in thick corky pericarp; vittae numerous, closely surrounding seed cavity. ....24. *Berula.*
            8. Involucre lacking or obsolete; carpophore bipartite and becoming free of mericarps or carpophore lacking.
                10. Calyx teeth forming a prominent corona on fruit; mericarps compressed dorsally; carpophore lacking. ....25. *Oenanthe.*
                10. Calyx teeth evident but not conspicuous in fruit; mericarps subterete; carpophore present, bipartite. ....28. *Cicuta.*
        7. Terrestrial plants from taproots or rootstocks.
            11. Flowers bright yellow; ultimate leaf divisions filiform. . . 18. *Foeniculum.*
            11. Flowers white, pink, red, or purple, or if greenish-yellow or yellowish-white, the fruit didymous or subdidymous or villous and the ribs inconspicuous or obscured, or plants essentially stemless; ultimate leaf divisions broader.
                12. Forest plants, arising from definite horizontal rootstocks.
                    13. Leaves apparently 5-parted (digitate); flowers radiant; mericarps compressed dorsally; carpophore lacking. ....31. *Vanasushava.*

13. Leaves ternate or ternately, pinnately, or ternate-pinnately compound; flowers regular, mericarps subterete or compressed laterally; carpophore present (not seen in 34. *Haplosphaera*).
    14. Umbellets 2–4-flowered, one or more subsessile; petals acute and sometimes slightly incurved, but without a narrower inflexed apex; calyx prominent. .... 32. *Pternopetalum.*
    14. Umbellets rather numerous flowered, all pedicellate; petals with a narrower inflexed apex; calyx minute or lacking.
        15. Umbels clearly compound, the rays elongate; involucel usually lacking; flowers white (rarely reddish); fruit ribs inconspicuous; vittae obscure. .... 33. *Aegopodium.*
        15. Umbels appear simple, the rays short to abortive; involucel present; flowers purplish-brown; fruit ribs prominent, subwinged; vittae evident. .... 34. *Haplosphaera.*

12. Plants mostly of open habitats, arising from taproots.
    16. Plants glabrous or essentially so.
        17. Involucre of conspicuous pinnatifid bracts (cf. also 49. *Seseli*).
            18. Plants low, acaulescent or nearly so.
                19. Rays strikingly unequal, 1–20 cm long, the umbellets appearing like simple umbels; calyx teeth evident.
                    20. Fruit ribs neither hollow, inflated, nor rugulose; carpophore bipartite.
                        21. Flowers regular; fruit ribs prominent but unwinged; seed face plane.
                            22. Stems prominently fibrous at base; mericarps compressed dorsally; fruit ribs obtuse, corky thickened. .... 35. *Kedarnatha.*
                            22. Stems lacking fibers at base; mericarps pentagonal or terete; fruit ribs slender, unthickened .... 36. *Schulzia.*
                        21. Flowers (at least the peripheral) slightly irregular; fruit ribs narrowly winged; seed face deeply concave. .... 37. *Keraymonia.*
                    20. Fruit ribs hollow, inflated, and rugulose; carpophore entire. .... 38. *Trachydium.*
                19. Rays subequal, the umbellets clearly part of a compound umbel; calyx teeth obsolete. .... 36. *Schulzia.*
            18. Plants slender, distinctly caulescent.
                23. Perennial; cauline leaves few; rays 6–15, unequal; umbellets 10–15-flowered; fruit ribs thick, undulate. .... 39. *Physospermopsis.*
                23. Annual or biennial; cauline leaves numerous; rays 20–60, subequal; umbellets 30–50-flowered; fruit ribs thin, straight. .... 28. *Ammi.*

17. Involucre lacking entirely, or else of entire bracts (one or more less commonly divided but scarcely pinnatifid).
24. Involucel present and persistent.
25. Fruit ribs undulate; vittae usually obliterated in ripe fruit, a ring of secretory tissue surrounding seed. ....27. *Conium.*
25. Fruit ribs straight; vittae large or small but distinct, no secretory ring present.
26. Involucre present, the bracts nearly as long as rays; fruit ribs triangular, somewhat corky. ....49. *Seseli.*
26. Involucre lacking, or if present the bracts much shorter than rays; fruit ribs filiform to obsolete, scarcely corky thickened.
27. Rays and pedicels quadrangular; commissural face sulcate. ....40. *Vicatia.*
27. Rays and pedicels ± terete; commissural face plane or planoconcave.
28. Involucre lacking, or if present the plants perennial; vittae 2 or more in intervals, or if (rarely) solitary, the carpophore entire.
29. Roots digitately branched; involucel present; calyx teeth prominent ....41. *Meeboldia.*
29. Taprooted; involucel usually 0; calyx teeth minute or 0. ....46. *Pimpinella.*
28. Involucre present; vittae solitary in intervals, or if vittae 2 or more, plants annual. ....
.........................48. *Trachyspermum.*
24. Involucel lacking, or deciduous.
30. Fruit oblong, 3–5 mm long. ....42. *Carum.*
30. Fruit broadly ovoid or elliptical to globose, 1–3 mm in diameter.
31. Petals with a narrower inflexed apex; styles slender, usually recurved.
32. Carpophore slender, bifid or bipartite; vittae 2 or more in intervals. ....46. *Pimpinella.*
32. Carpophore stout, entire or shortly bifid; vittae solitary in intervals. ....29. *Apium.*
31. Petals without an inflexed apex; styles short to abortive.
33. Plants low, profusely branched, leafy throughout; peduncles short or abortive; rays 1–5; petals clawless.
.........................30. *Ciclospermum.*
33. Plants erect, few branched or simple, subscapose; peduncles elongate; rays 5–25; petals clawed. ....
.................................45. *Tongoloa.*

16. Plants pubescent throughout, or at least in inflorescence; ovaries and usually also the fruit pubescent, papillose, or conspicuously roughened.

34. Fruit constricted at commissure, often didymous or subdidymous, the mericarps usually subterete-pentagonal.

35. Plants perennial, sometimes woody based, or biennial, neither cultivated nor adventive.

36. Fruit vittae solitary in intervals, 2 on commissure.

37. Fruit oblong to oblong-elliptic or oblong-ovoid, the ribs obtuse and largely hidden by pubescence. ....47. *Eriocycla.*

37. Fruit ovoid to ovoid-globose, the ribs acute, prominent to inconspicuous (cf. also *Eriocycla stewartii*). ....48. *Trachyspermum.*

36. Fruit vittae 2–4 in intervals, 2–4 on commissure (cf. also *Eriocycla stewartii*). ....46. *Pimpinella.*

35. Plants annual; cultivated for their spicy seeds, often adventive. ....48. *Trachyspermum.*

34. Fruit broadest at commissure, the mericarps compressed dorsally. ....49. *Seseli.*

20. BUNIUM Linn.
Sp. Pl. 1: 243. 1753.

Carum subgen. *Bunium* Reichenb., Consp. 143. 1828; *Carum* sect. *Bunium* (Reichenb.) Benth. in Benth. & Hook., Gen. Pl. 1: 891. 1867; C. B. Clarke in Hook. f., Fl. Brit. Ind. 2: 681. 1879.
Bulbocastanum Adans., Fam. 2: 97. 1765; *Carum* sect. *Bulbocastanum* (Adans.) DC., Prodr. 4: 115. 1830.

Slender and erect or low and spreading, sparsely branching or simple, glabrous perennial herbs from a globose tuber. Leaves petiolate, membranaceous, pinnately decompound, the ultimate divisions linear to filiform; petiole slender, sheathing; cauline leaves reduced upward, the upper sessile. Inflorescence of lax, mostly terminal, compound umbels, the peduncles elongate. Involucre of several linear, sometimes apically cleft bracts, or bracts obscure or lacking. Rays rather few, subequal to unequal, spreading-ascending or spreading. Involucel of several short, inconspicuous, narrow bracteoles, or lacking. Umbellets several to many flowered, the flowers pedicellate. Flowers white or pinkish; petals obovate, emarginate, with a very short, narrowed, inflexed apex, subequal; calyx obsolete; stylopodium conical or flattened to form a thick disk, the styles slender, often reflexed. Pedicels slender, sometimes thickening at maturity. Fruit oblong-cylindrical to oblong-ovoid, rounded at both ends, the mericarps subterete or pentagonal to a little compressed dorsally in transection, glabrous, the ribs filiform or obtuse, ± prominent, narrower than intervals. Vittae usually solitary in intervals, 2 on commissure. Commissural face planoconcave. Carpophore bipartite.

Type species: *Bunium bulbocastanum* Linn.

Distribution: A genus of ca. 40 spp., Mediterranean to Central Asia. Of the two Indian species, one is widespread, the other endemic to peninsular India.

KEY TO SPECIES

1. Leaf segments linearly filiform; stylopodium low conical; fruit ribs filiform; Himalaya. .... 1. *B. persicum*.
1. Leaf segments narrow ± rigid; stylopodium obsolete; fruit ribs prominent, acute; peninsular India. .... 2. *B. nothum*.

1. Bunium persicum (Boiss.) B. Fedtsch.,
Rastit. Turkestan 612. 1915.

Carum persicum Boiss., Ann. Sci. Nat. Bot. III. 1: 138. 1844.
Carum bulbocastanum sensu C. B. Clarke in Hook. f., Fl. Brit. Ind. 2: 681. 1879, non Koch 1824.

Plants slender, erect, branched above, 15–70 cm tall. Leaves triangular-ovate, 3–8 cm in diameter, the basal leaves ternate-bipinnate with short, lanceolate, mucronate lobes up to 1.5 mm broad, all other leaves pinnately decompound, the ultimate divisions linearly filiform, 2–20 mm long, usually less than 0.5 mm broad; petiole slender, 1–2 cm long; upper cauline leaves sessile. Peduncles slender, 3–13 cm long. Involucre of 1–5 setaceous, basally scarious-margined bracts 2–10 mm long, or lacking. Rays 5–17, slender, spreading-ascending, unequal, 1.5–6 cm long. Umbellets usually 15–20-flowered, the mature pedicels capillary, unequal, spreading-ascending, 3–8 mm long. Involucel of 1–5 inconspicuous, linear-lanceolate bracteoles 1–4 mm long, or lacking. Stylopodium low conical to depressed, much shorter than reflexed styles. Fruit oblong, 3–4 mm long, 1–2 mm broad, the mericarps pentagonal, slightly compressed dorsally in transection, glabrous, the ribs low, filiform. Vittae large, solitary in intervals, 2 on commissure. Commissural face slightly concave. May–Oct.

Type locality: Persia: Fasa & Chyraz, *Auchery 4553 & 4555*; Persia, *Bushe*.

Distribution: Central Asia, to western Himalaya of India, Pakistan, and Afghanistan, 1,500–3,000 m.

Representative specimens: Jammu & Kashmir: Lolab Forest, Jhelum Valley, *Keshavanand 715* (DD). Himachal Pradesh: Lahul, *Stoliczka 101* (CAL); Lahul, Triloknath, *Brandis 3723* (DD).

2. **Bunium nothum** (C. B. Clarke) P. K. Mukherjee, Bull. Bot. Surv. India 24(1–4): 43. 1983.

Carum nothum C. B. Clarke in Hook. f., Fl. Brit. Ind. 2: 681. 1879.
Bunium nilghirense H. Wolff in Feddes Repert. 27: 334. 1930. Based on *Gamble 13,442*, Nilgiri, above Sirpaca.

Plants slender, erect, branched, 8–20 cm tall. Leaves broadly triangular to rhomboid, 5–8 cm long, pinnately decompound, the ultimate divisions linear, 3-parted or -lobed, 5–10 mm long, 0.5–1 mm broad. Peduncles elongate. Involucre of 3–6 linearly lanceolate, acute or obtuse bracts 7–10 mm long. Rays 4–7, subequal, spreading-ascending, 2–3.5 cm long. Umbellets 5–10-flowered, the mature pedicels ca. 2 mm long. Involucel of ca. 10 lanceolate to oblanceolate, acute bracteoles ca. 5 mm long, about equal to fruiting umbellet. Stylopodium depressed, obsolete. Fruit oblong-oval, ca. 4 mm long, 2.5 mm broad, the mericarps pentagonal, subterete in transection, glabrous, the ribs prominent, acute. Vittae large, solitary in intervals, 2 on commissure. Commissural face plane. Nov.

Type locality: "CEYLON?" [Probably Nilgiri Hills, Peninsular India], no collector indicated (K, holotype).

Distribution: Peninsular India, 2,500 m; Sri Lanka (?).

Representative specimen: Tamil Nadu: Sirpaca, Nilgiris Distr., *Gamble 13,442* (CAL, DD, K).

Note. No authentic material of *Carum nothum* C. B. Clarke exists in India. Norman annotated an isotype (DD) of *Bunium nilghirense* H. Wolff as the former, and Kew staff have verified the conspecificity of the two taxa. Subsequent inspection of material at Kew confirmed this judgment.

21. ELAEOSTICTA Fenzl
Flora 28: 458. 1843; C. B. Clarke in Hook. f., Flor. Brit. Ind. 2: 673. 1879.

Scaligeria sect. *Elaeosticta* (Fenzl) Boiss., Fl. Orient. 2: 877. 1872; *Scaligeria* subgen. *Elaeosticta* (Fenzl) Korovin, Fl. USSR 16: 207. 1950.

Slender, erect, usually branching, essentially glabrous and often glaucous perennial or biennial herbs from a globose or ovoid tuber. Leaves petiolate, membranaceous, the basal often early deciduous, pinnately or ternate-pinnately decompound, the ultimate divisions oblong or lanceolate to filiform; petiole slender, sheathing; cauline leaves few, reduced upward, the uppermost often sessile. Inflorescence of lax terminal and lateral compound umbels, the peduncles elongate. Involucre and involucel of several, oblong-oval to linearly lanceolate, often scarious or scarious-margined bracts or bracteoles, or involucre rarely lacking. Rays few to numerous, spreading-ascending to divaricate, usually unequal. Umbellets few to many flowered, the flowers pedicellate. Flowers white, rarely purplish; petals oval or obovate, with a narrower inflexed apex, subequal; calyx obsolete; stylopodium usually flattened to form a thick disk, less frequently shortly conical or conical, the styles short, curved. Pedicels slender, usually unequal. Fruit globose to ovoid-oblong, compressed laterally and often didymous, rarely linearly cylindrical, mostly rounded at both ends, the mericarps subterete to slightly compressed dorsally in transection, glabrous to scabrous, the ribs obscure or lacking. Vittae usually 2–5 in intervals, 4–6(–12) on commissure, sometimes obliterated in fruit. Commissural face plane or concave, sometimes furrowed. Carpophore bipartite or merely bifid.

Type species: *Elaeosticta meifolia* Fenzl.

Distribution: Kljuykov, Pimenov & Tikhomirov (1976c) have concluded that *Elaeosticta* Fenzl, treated as one of three subgenera of *Scaligeria* DC. by Korovin (1950, etc.), is not closely allied to the Mediterranean *Scaligeria*, *sensu stricto*, and should therefore be regarded as an independent genus of some two dozen species of Central Asia.

1. Elaeosticta aitchisonii (H. Wolff) Kljuykov, Pimenov & Tikhomirov, Bjull. Moskovsk. Obsc. Isp. Prir. Otd. Biol. 6: 93. 1976.

Elaeosticta meifolia sensu C. B. Clarke in Hook. f., Fl. Brit. Ind. 2: 673. 1879, non Fenzl 1843.
Scaligeria aitchisonii H. Wolff, Feddes Repert. 21: 109. 1925.
Scaligeria indica H. Wolff, Feddes Repert. 27: 126. 1929. Based on *Keshavanand 704*, Muzaffarabad, Kashmir, 2,500 ft.

Plants slender, erect, 1 or more stems from base, branched, 20–60 cm tall from a globose tuber. Leaves early deciduous, ovate-lanceolate, 5–12 cm long, 3–10 cm broad, pinnately decompound, the ultimate divisions linear to lanceolate, 5–20 mm long, 0.5–5 mm broad; petiole slender, 2–8 cm long, dilated and scarious-sheathing at base; cauline leaves strongly reduced upward, the uppermost sessile. Peduncles slender, 4–12 cm long. Involucre of 5 or 6 lanceolate, scarious, 3-veined bracts 2–5 mm long. Rays filiform, 5–15, 1–3 cm long, slightly unequal. Umbellets mostly 10–15-flowered, the mature pedicels very unequal, spreading-ascending, 1–5 mm long. Involucel of 6–8 linearly lanceolate to lanceolate, acute, scarious bracteoles 1.5–4 mm long. Stylopodium low conical, about equal to the short, recurved styles. Fruit broadly ovoid, 2–3 mm long, 1.5–2.5 mm broad, didymous, the mericarps subterete in transection, glabrous, the ribs obscure. Vittae minute, 2 or 3 in intervals, mostly 4 on commissure. Commissural face narrowly sulcate. Carpophore bipartite. Apr.–Aug.

Type locality: Punjab, *Aitchison.*

Distribution: Western Himalaya, Kashmir to Punjab and Pakistan, 1,500–2,500 m.

Representative specimens: Jammu & Kashmir: Sadhuganga, *Keshavanand 738* (DD); Muzaffarabad, *Keshavanand 704* (DD, isotype of *Scaligeria indica* H. Wolff); Nasim Bagh, Srinigar, *R. Stewart 23,024* (UC). Punjab: w/o further locality, *Aitchison* (CAL).

22. **KARNATAKA** P. K. Mukherjee & Constance
Brittonia 38(2): 145. 1986.

Slender, erect, caulescent, branching, from a swollen and tuberous caudex. Leaves petiolate, membranaceous, 2–3-pinnate, the leaflets broad, crenate-dentate, minutely scaberulous. Inflorescence consisting of several terminal and lateral compound umbels, the peduncles rather slender. Involucre and involucel of several narrow bracts and bracteoles. Rays few, spreading-ascending. Pedicels thick. Flowers polygamous: marginal flowers perfect, the central staminate, the petals white, obovate, with a narrower inflexed apex; calyx teeth evident, equal, linear; stylopodium low conical, the styles slender. Fruit narrowly ovoid, subterete, glabrous, the mericarps compressed dorsally, the ribs low, filiform. Vittae 1 or 2 in intervals, usually 6 on commissure. Commissural face shallowly planoconcave. Carpophore bipartite.

Type species: *Schultzia ? benthamii* C. B. Clarke [*Karnataka benthamii* (C. B. Clarke) P. K. Mukherjee & Constance].

Distribution: Known solely from the type species, native to southern Peninsular India.

1. **Karnataka benthamii** (C. B. Clarke) P. K. Mukherjee & Constance, Brittonia 38(2): 145. 1986.

Schultzia ? benthamii C. B. Clarke in Hook. f., Fl. Brit. Ind. 2: 697. 1879.

Plants slender, erect, diffusely branching, scaberulous, 30–60 cm tall. Leaves ovate, 5–10 cm in diameter, 2- or 3-pinnate, the leaflets ovate, 3–13 mm in diameter, crenate-dentate, sparsely scaberulous; petiole slender, 3–6 cm long, very narrowly sheathing; cauline leaves gradually reduced upward. Peduncles rather slender, 3–6 cm long. Involucre of 2–5 linearly lanceolate, white-margined, entire bracts 1–5 mm long. Rays 3–5, slender, unequal, 1.5–3 cm long. Involucel of 3–6 linearly lanceolate, entire or irregularly dentate bracteoles 2–6 mm long. Umbellets 8–12-flowered, the 1–3 perfect ones peripheral, the others staminate, the mature pedicels unequal, 3–5 mm long. Calyx teeth evident, subequal, persistent; stylopodium low conical, about equal to or exceeded by the recurved styles. Fruit narrowly ovoid, ca. 4 mm long, 2–3 mm broad, glabrous, the ribs slender.

Type locality: "Terr. Canara et Confin," *Metz (Hohenacker*, Pl. Ind. Orient. *657a. Pimpinella ? involucrata* Wight et Arn.), India (BM, holotype; K, LE, P, isotypes).

Distribution: Known only from the type collection.

23. **SIUM** Linn.

Sp. Pl. 257. 1753; C. B. Clarke in Hook. f., Fl. Brit. Ind. 2: 683. 1879.

Sium sect. *Eusium* Endl., Gen. Pl. 722. 1838; *Sium* subgen. *Eusium* Drude in Engl. & Prantl, Pflanzenfam. 38: 197. 1898.
Sisarum Miller, Gard. Dict. ed. 4. 1754; *Sium* sect. *Sisarum* (Miller) DC., Prodr. 4: 124. 1830; *Sium* subgen. *Sisarum* (Miller) Thell. in Hegi, Ill. Fl. Mittel-Europa 52: 1217. 1926.
Drepanophyllum Hoffm., Gen. Umbell. 109. 1814, *pro parte*.

Stout or slender, erect, branching, glabrous perennial herbs from fascicled fibrous or tuberous roots. Leaves petiolate, membranaceous, simply pinnate (or rarely simple), the leaflets linear to ovate, entire to serrate or incised; petiole often fistulose, sheathing at base; cauline leaves reduced upward. Inflorescence of lax, terminal or axillary compound umbels, the peduncles rather slender. Involucre and involucel of conspicuous, foliaceous, narrow, often reflexed bracts and bracteoles, the latter entire. Rays rather few, spreading-ascending, subequal to unequal. Umbellets several to many flowered, the flowers pedicellate. Flowers white; petals obovate, emarginate, with a narrower inflexed apex, sometimes unequal; calyx teeth evident to obsolete; stylopodium conical to depressed, the styles short to slender, spreading or recurved. Pedicels slender. Fruit oblong-ovoid to suborbicular, rounded at both ends, slightly compressed laterally, somewhat constricted at commissure, the mericarps subterete to pentagonal in transection, glabrous, the ribs prominent, filiform to narrowly corky winged, subequal. Vittae 1–3 in intervals, 2–4 on commissure. Commissural face plane. Carpophore bipartite, the halves adnate to mericarps or free, or carpophore entire.

Type species: *Sium latifolium* Linn.

Distribution: A largely circumboreal genus of about a dozen marsh plants, one of which reaches the western Himalaya, and another is endemic to sub-Saharan Africa.

1. **Sium latijugum** C. B. Clarke in Hook. f., Fl. Brit. Ind. 2: 683. 1879.

Plants stout, erect, the stem fistulose, 30–60 cm tall, from fascicled fibrous roots. Leaves oblong, 8–30 cm long, 6–16 cm broad, the leaflets ovate to linearly lanceolate, 20–80 cm long, 8–25 mm broad, acute or acuminate, sharply serrate; petiole fistulose, septate, 8–20 cm long; upper cauline leaves mostly ternate, sessile. Peduncles fistulose, 2–15 cm long. Involucre of 5–8 linear to lanceolate, entire, acute, ultimately reflexed bracts 3–8 mm long. Rays 8–15, slender, unequal, 1–3 cm long. Umbellets 10–20-flowered, the mature pedicels slender, spreading-ascending, 2–8 mm long. Involucel of 5–8 lanceolate bracteoles 1–2.5 mm long. Calyx obsolete; stylopodium depressed conical, shorter than the recurved styles. Fruit oblong-ovoid, 3–4 mm long, 2–2.5 mm broad, the mericarps subterete in transection, the ribs very prominent, corky, subwinged, much broader than intervals. Vittae large, solitary in intervals, 2 on commissure. Carpophore bipartite, each half adnate to a mericarp. June–Aug.

Type locality: "KASHMIR and BALTISTHAN, in ditches, alt. 5000–9000 ft.; very common," no collector indicated.

Distribution: Western Himalaya from Himachal Pradesh to Pakistan, 1,500–3,000 m.

Representative specimens: Jammu & Kashmir: Pahlgam, *Clarke 3116C* (CAL); w/o further locality, *Thomson* (CAL). Himachal Pradesh: Lahul, *Jaeschke 110* (CAL); Manali, Vashista, *Bhattacharyya 40,331* (BSD); Katrain, 1 Sep. 1981, *Mukherjee* (CUH, UC).

24. BERULA Koch
In Röhling, Deuts. Fl. 2: 25. 1826.

Sium subgen. *Berula* (Koch) Drude in Engl. & Prantl, Pflanzenfam. 38: 197. 1898.
Siella Pimenov, Bot. Zhurn. 63: 1746. 1978.

Robust or slender, erect or ascending, branching, glabrous perennial herbs from a creeping rootstock, rooting at nodes and bearing fascicled fibrous roots. Leaves petiolate, membranaceous, simply pinnate, the leaflets entire to lobed or dissected (especially in submerged leaves); petiole slender, sheathing at base; cauline leaves like the basal, reduced upward. Inflorescence of lax, lateral and terminal compound umbels, the peduncles short to elongate. Involucre and involucel of several conspicuous, narrow, usually entire, reflexed, herbaceous bracts and bracteoles. Rays several, spreading or spreading-ascending, unequal. Umbellets rather many flowered, the flowers pedicellate. Flowers white; petals obovate with a narrower inflexed apex; calyx teeth minute, subulate, and persistent, or calyx obsolete; stylopodium low conical, much shorter than to equal in length to ascending styles. Pedicels very slender, spreading. Fruit broadly ovoid to orbicular, a little narrowed toward apex, rounded at base, compressed laterally, a little constricted at commissure, the mericarps subterete to somewhat pentagonal in transection, glabrous, the ribs filiform, obscure in a thick corky pericarp; vittae numerous, contiguous, and closely surrounding seed cavity. Commissural face plane. Carpophore bipartite, each half adnate to a mericarp.

Type species: *Sium erectum* Hudson [*Berula erecta* (Hudson) Coville].

Distribution: An apparently monotypic (Cannon 1978) and essentially cosmopolitan genus of aquatic or marsh plants; part of its range may be due to introduction.

1. **Berula erecta** (Hudson) Coville, Contr. U. S. Nat. Herb. 4: 115. 1893.

Sium erectum Hudson, Fl. Angl. 103. 1762; *Siella erecta* (Hudson) Pimenov, Bot. Zhurn. 63: 1746. 1979.
Sium angustifolium Linn., Sp. Pl. ed. 2. 1672. 1763; *Berula angustifolia* (Linn.) Mert. & Koch in Röhling, Deuts. Fl. ed. 3. 2: 433. 1826. Based on "Habitat in Europae australis aquosis, nec extra aquas adscendit," no collector indicated.
Sium berula Gouan, Fl. Monsp. 218. 1765; *Apium berula* (Gouan) Caruel in Parl., Fl. Ital. 8: 463. 1889; *Selinum berula* (Gouan) E. H. Krause in Sturm., Fl. Deuts. ed. 2. 12: 33. 1904. Based on "HABITAT in rivulis juxta *Lattes*, & *Pascales*," no collector indicated.
Sium thunbergii DC., Prodr. 4: 125. 1830. Based on *Ecklon*, "ad cap Bonae-Spei in paludosis propè Seekuhwallei," South Africa.
Sium pusillum Nutt. ex Torr. & A. Gray, Fl. N. Amer. 1: 611. 1840; *Berula pusilla* (Nutt. ex Torr. & A. Gray) Fern., Rhodora 44: 189. 1942. Based on "Nuttall, Wappatoo Island, Oregon," non *Sium pusillum* Poiret 1810.
Sium ? incisum Torr. in Frémont, Rep. Calif. 90. 1845. Based on *Frémont*, "North fork of the Platte," North America.
Sium novae-mexicae Kozo-Polj., Bull. Soc. Nat. Mosc. II. 28: 173. 1915. Based on *Wooton 256*, "White Mountains, Lincoln County, New Mexico."

Plants 20–100 cm tall, with a fistulose, ridged stem. Leaves oblong, 10–50 cm long, 3–10 cm broad, with 4–12 pairs of rather remote, sessile, obtuse, oblong-lanceolate to oblong-oval, serrate, dentate, or lobed leaflets 1–8 cm long, 5–30 mm broad; petiole slender, 4–15 cm long, narrowly sheathing. Peduncles 2–30 cm long. Involucre of 6–10 linear or lanceolate, acute, scarious-margined, unequal bracts 5–15 mm long. Rays usually 10–20, 1–3 cm long. Umbellets 10–20-flowered, the mature pedicels capillary, unequal, 2–5 mm long. Involucel of 4–8 linear to lanceolate, acute, entire bracteoles 1–5 mm long. Fruit broadly ovoid to orbicular or even broader than long, 1.5–2 mm in diameter. Vittae small, numerous, deeply embedded in pericarp and continuous about seed cavity. June–Sept.

Type locality: England, no collector indicated.

Distribution: Circumboreal and African; Western Himalaya, extending to Pakistan and Afghanistan, 1,300–1,600 m.

Specimen examined: Jammu & Kashmir: near Drash, *H. Hamilton 17,878* (CAL).

25. OENANTHE Linn.

Sp. Pl. 254. 1753; C. B. Clarke in Hook. f., Fl. Brit. Ind. 2: 695. 1879.

Phellandrium Linn., Sp. Pl. 255. 1753.
Dasyloma DC., Prodr. 4: 140. 1830.

Slender or stout, decumbent to erect, usually branching, glabrous and often succulent perennial herbs from fascicled or tuberous roots or rootstocks, often rooting at lower nodes. Leaves petiolate, membranaceous, pinnate to pinnately decompound, the ultimate divisions fine or coarse, or leaves reduced to fistulose phyllodes without definite petioles; petiole conspicuously sheathing when present; cauline leaves like the basal, often wholly sheathing. Inflorescence of lax, mostly terminal, compound umbels, the peduncles rather stout, elongate, or opposite the leaves and often obsolete. Involucre usually lacking, or occasionally of one or more inconspicuous narrow bracts. Rays few to numerous, subequal to unequal, spreading. Involucel of several narrow bracteoles. Umbellets many flowered, the flowers pedicellate. Flowers white or pinkish; petals obovate, with a narrower inflexed apex, sometimes unequal; calyx teeth evident, narrow, acute, usually enlarging and persisting as a corona on fruit, or calyx rarely obsolete; stylopodium conical, often partially hidden by calyx, the styles erect or spreading, persistent or rarely deciduous. Fruit oblong-ovoid to quadrate-subglobose, terete or slightly compressed either laterally or dorsally, rounded or truncate at apex and base, broadest at commissure, the mericarps somewhat compressed dorsally in transection, glabrous, the ribs low, obtuse, often hidden and somewhat confluent in the corky pericarp, the marginal ribs larger. Vittae solitary in intervals, 2 on commissure. Commissural face plane. Carpophore lacking.

Type species: *Oenanthe fistulosa* Linn.

Distribution: A small but widespread, largely circumboreal genus of marsh and aquatic plants, the actual number of component species, depending on the taxonomic concept employed, perhaps 25–30.

KEY TO SPECIES

1. Leaves with a definite lamina, not reduced to fistulose phyllodes, with definite leaflets or ultimate leaf divisions, and a distinct petiole; calyx teeth forming a corona partly concealing stylopodium.
    2. Leaves 1–2-pinnate with linear to ovate leaflets 2–30 mm broad. ... 1. *O. javanica*.
    2. Leaves pinnately decompound, the ultimate divisions linear-filiform, less than 0.5 mm broad. .... 2. *O. thomsonii*.
1. Leaves reduced to fistulose phyllodes, with at most 1 or 2 distal pairs of minute linear lobes, and lacking a distinct petiole; calyx teeth attached well below and not concealing stylopodium. .... 3. *O. hookeri*.

1. Oenanthe javanica (Blume) DC., Prodr. 4: 138. 1830.

Sium javanicum Blume, Bijdr. Fl. Ned. Ind. 15: 1881. 1826; *Falcaria javanica* (Blume) DC., Prodr. 4: 110. 1830; *Dasyloma javanica* (Blume) Miq., Fl. Ind. Bat. 1(1): 741. 1856;

Oenanthe stolonifera var. *javanica* (Blume) Kuntze, Rev. Gen. 1: 269. 1891.
Phellandrium stoloniferum Roxb., Hort. Bengal. 21. 1814, *nomen nudum*; Fl. Ind., ed. 2. 2: 93. 1832; *Oenanthe stolonifera* (Roxb.) Wallich, Cat. 585. 1829, *nomen nudum*; DC., Prodr. 4: 138. 1830; C. B. Clarke in Hook. f., Fl. Brit. Ind. 2: 696. 1879. Based on *Roxburgh* (?), "in India orientali ad agros oryzaceous prov. Silhet Bengali orientale."
Seseli benghalensis Roxb., Hort. Bengal. 22. 1814, *nomen nudum*; Fl. Ind., ed. 2. 2: 94. 1832; *Biforis benghalensis* (Roxb.) Wallich, Cat. 588. 1829, *nomen nudum*; *Dasyloma benghalense* (Roxb.) DC., Prodr. 4: 140. 1830; *Oenanthe benghalensis* (Roxb.) Kurz, Jour. Asiat. Soc. Bengal. 2: 115. 1877; C. B. Clarke in Hook. f., Fl. Brit. Ind. 2: 696. 1879. Neither type locality nor collector indicated.
Biforis glauca Wallich, Cat. 587. 1829, *nomen nudum*; *Dasyloma glauca* DC., Prodr. 4: 140. 1830. Based on *Wallich 587*, "cult. in calcuttensis."
Oenanthe linearis Wallich, Cat. 586. 1829, *nomen nudum*; DC., Prodr. 4: 138. 1830; C. B. Clarke in Hook. f., Fl. Brit. Ind. 2: 696. 1879. Based on *Wallich*, Nepalia.
Dasyloma latifolia Lindley in Royle, Ill. Bot. Himalay. Mount. 232. 1839. Based on *Royle*, Choor.
Oenanthe corticata Edgew., Trans. Linn. Soc. 20: 53. 1851; *O. stolonifera* var. *corticata* (Edgew.) C. B. Clarke in Hook. f., Fl. Brit. Ind. 2: 696. 1879. Based on *Edgeworth*, Sirhind, Radur, India Boreali-Occidentali.
Oenanthe stolonifera var. *khasiana* C. B. Clarke in Hook. f., Fl. Brit. Ind. 2: 696. 1879. Based on *Hooker & Thomson*, Khasia Mts., near Moflong, alt. 6,000 ft.

Plants decumbent or ascending, branched below or above, 5–120 cm tall, usually rooting from lower nodes. Leaves oblong to ovate, 5–20 cm long, 6–17 cm broad, 1–2-pinnate, the leaflets ovate or ovate-lanceolate to linearly oblong, 3–60 mm long, 2–30 mm broad, mostly acute, serrate to incisely lobed; petiole 1–10 cm long; cauline leaves similar, often ternate. Peduncles stout, elongate and up to 20 cm long, to abortive, often a little scaberulous at summit. Involucre usually lacking. Rays 5–30, rather stout, spreading, 5–35 mm long. Umbellets 10–25-flowered, the mature pedicels slender to stout, unequal, spreading, 1–4 mm long. Involucel of 2–8 linear, acute bracteoles 2–6 mm long. Calyx teeth linearly subulate, 0.5–1 mm long; stylopodium conical, the styles slender, 1–3 mm long. Fruit oblong to oblong-ovoid, 2–3 mm long, 1.5–2 mm broad, yellowish-brown to purplish. Nov.-Dec. on plains, Apr.-June in hills.

Type locality: "In paludosis Javae ubi dicitur *Tespong*."

Distribution: As construed here, largely on the authority of Buwalda (1936, pp. 194–200) supplemented by comparison of Indian specimens, this is an extraordinarily variable and widespread species-complex, extending from Japan and China southward throughout Indonesia to New Guinea and Australia. In India it occurs in the Himalaya to Pakistan, in the Khasia Hills, and on the Gangetic Plains, at altitudes of up to 3,000 m.

Representative specimens: Himachal Pradesh: Choor, *Royle 93/41* (DD); Katrain, 1 Sep. 1981, *Mukherjee* (CUH, UC). Uttar Pradesh: Mussoorie, *Vicary* (DD); Ganges Valley, *Duthie 1124* (DD); Buland Shahi, *N. P. Singh 25,508* (BSD). Nepal: w/o further locality, *Wallich 506* (CAL). West Bengal: Botanical Garden, *Roxburgh* (CAL); Shibpur, *Prain* (CAL). Bhutan: w/o further locality, *Booth* (CAL). Arunachal Pradesh: Abor, Kobo, *Burkill 36,781* (CAL). Nagaland: Naga Hills, *Watt 7306* (CAL). Manipur: Kangpokpi, *Bor 17,786* (DD). Mizoram: Kohima, *Clarke 41,024A* (CAL). Meghalaya: Khasia, *Hooker & Thomson* (CAL). Assam: Gowhatty, *Simons* (DD). Madhya Pradesh: Near Goghar

River, Rewa Distr., *Sebastine 10,024* (MH). Bangladesh: Sylhet, *Clarke 3107* (CAL); Noakhali, *Clarke 6875* (CAL).

2. **Oenanthe thomsonii** C. B. Clarke in Hook. f., Fl. Brit. Ind. 2: 697. 1879.

Plants slender and diffuse, erect or ascending, simple or few branched, 15–90 cm tall, from slender taproots. Leaves triangular-ovate, 3–12 cm long, 3–10 cm broad, pinnately decompound, the ultimate divisions linearly-filiform, 0.5–1.5 mm long, less than 0.5 mm broad; petiole 2–6 cm long. Peduncles stout, opposite the leaves, 3–10 cm long. Involucre lacking. Rays 4–8, slender, spreading, unequal, 1–3 cm long. Umbellets 10–15-flowered, the mature pedicels slender, unequal, spreading, 1–4 mm long. Involucel of 5–9 inconspicuous linear bracteoles 1–4 mm long. Calyx teeth linearly-subulate, ca. 0.5 mm long; stylopodium conical, the styles slender, 1–3 mm long. Fruit oblong-ovoid to ovoid-globose, slightly compressed dorsally, 2–3 mm long, 2–2.5 mm broad, light brown. July–Oct.

Type locality: Khasia, 1330–2000 m, *Hooker & Thomson* (Herb. Ind. Orient., *Dasyloma* 2).

Distribution: China and Indochina to the Western Himalaya from Arunachal Pradesh to Nepal, and Khasia and Naga Hills, 1,500–2,000 m.

Representative specimens: Nepal: Memeny to Phedda, — —*859* (CAL, w/o collector). West Bengal: Darjeeling, *Anderson 633* (CAL). Sikkim: Jongri, *Anderson 621* (CAL). Arunachal Pradesh: Kameng F. D., *Sahni 5121* (DD). Nagaland: Naga Hills, *Bor 2677* (DD). Meghalaya: Khasia, *Hooker & Thomson* (CAL), *Griffith 1001* (CAL).

Note. This is probably the taxon referred to by Farille (1985, p. 551) as *Oenanthe* cf. *dielsii* de Boiss.

3. **Oenanthe hookeri** C. B. Clarke in Hook. f., Fl. Brit. Ind. 2: 697. 1879.

Plants slender, decumbent, sparsely branched, 30–60 cm tall, the ribbed, fistulose stem rooting at nodes. Leaves reduced to fistulose linear phyllodes, with or without 1 or 2 pairs of remote, minute, linear leaflets toward apex, a distinct petiole lacking. Peduncles opposite uppermost leaves, slender, 3–6 cm long. Involucre usually lacking. Rays 4–8, quadrangular, spreading, unequal, 5–15 mm long. Umbellets ca. 20-flowered, the mature pedicels stout, unequal, spreading, 1–3 mm long. Involucel of 8–10 linear bracteoles 1–3 mm long. Calyx teeth linearly-subulate, ca. 0.5 mm long, but set well below stylopodium; stylopodium obtusely conical, the styles short, ca. 1 mm long, spreading. Fruit oblong-ovoid, subterete, ca. 1.5 mm long, 1.2 mm broad. July–Aug.

Type locality: "Khasia Mts., alt. 5,000–6,000 ft.;" *Griffith, Hooker & Thomson, Clarke*.

Distribution: Known from the Khasia Hills, Bhutan, and Nepal, 1,500–2,000 m.

Representative specimens: Meghalaya: Khasia, *Hooker & Thomson* (CAL), *Garden collector* (CAL); Mawphlang, *Chand 8045A* (UC); Laitlynkot, *Koelz 23,241* (UC).

26. CICUTA Linn.
Sp. Pl. 255. 1753; C. B. Clarke in Hook. f., Fl. Brit. Ind. 2: 679. 1879.

Stout or slender, erect, caulescent, branching, glabrous perennial herbs from a swollen caudex bearing fascicled fibrous or tuberous roots. Leaves petiolate, membranaceous, 1-3-pinnate, the leaflets narrow, serrate to incised; petiole sheathing at base; cauline leaves similar to the basal, reduced upward. Inflorescence of lax terminal or lateral compound umbels, the peduncles usually elongate. Involucre lacking, or occasionally of one to several inconspicuous bracts. Rays numerous, slender, spreading-ascending, subequal to unequal. Involucel of several narrow bracteoles, rarely lacking. Umbellets many-flowered, the flowers pedicellate. Flowers white or greenish-white; petals obovate, emarginate, with a narrower inflexed apex; calyx teeth evident, acute; stylopodium low conical to depressed, the styles short. Pedicels slender, spreading. Fruit ovoid to globose, rounded at both ends or cordate at base, compressed laterally and sometimes constricted at commissure, the mericarps subterete in transection, glabrous, the ribs prominent, obtuse, corky, subequal. Vittae solitary in intervals, 2 on commissure. Commissural face plane to slightly concave. Carpophore bipartite, deciduous.

Type species: *Cicuta virosa* Linn.

Distribution: According to the latest revision (Mulligan, 1980), this is a genus of three American and one circumboreal species of marsh plants, the latter just reaching northwestern India.

1. Cicuta virosa Linn. Sp. Pl. 255. 1753; C. B. Clarke in Hook. f., Fl. Brit. Ind. 2: 679. 1879.

Plants stout, branched, 50–120 cm tall from a swollen, transversely septate, hollow caudex bearing fibrous or fleshy-fibrous roots. Leaves oblong to ovate, 15–30 cm long, 10–30 cm broad, 1-3-pinnate, the leaflets oblong-lanceolate, 3–8 cm long, 3–20 mm broad, acuminate, coarsely serrate to incised; petiole 15–45 long. Peduncles slender, 4–15 cm long. Involucre lacking, or of 1 or 2 inconspicuous bracts. Rays usually 10–20, 2–8 cm long. Umbellets many flowered, the mature pedicels slender, 2–20 mm long. Involucel of 6–8 linearly lanceolate to lanceolate, acute bracteoles 2–5 mm long. Stylopodium depressed or low conical. Fruit ovoid to globose, 1.5–3 mm in diameter, slightly constricted at commissure, the ribs low, corky, obtuse, subequal in surface display, narrower than intervals. Seed very oily, sulcate beneath vittae, the commissural face concave.

Type locality: "in paludibus Europe sterilibus," no collector indicated.

Distribution: Circumboreal marsh plant; in India, known only from Kashmir.

Representative specimen: Jammu & Kashmir: w/o further locality, *Falconer 488* (CAL).

27. CONIUM Linn.
Sp. Pl. 243. 1753.

Slender, erect, branching, glabrous biennial or perennial herbs from a stout taproot. Leaves petiolate, membranaceous, pinnately decompound, the ultimate divisions small, incised or pinnatifid into narrow lobes; petiole stout or slender, dilated and scarious sheathing at base; cauline leaves gradually reduced upward. Inflorescence of lax terminal and lateral compound umbels, the peduncles elongate to abortive. Involucre and involucel of several narrow bracts and bracteoles, often reflexed. Rays rather numerous, slender, spreading-ascending or spreading. Umbellets many flowered, the flowers pedicellate. Flowers white or yellowish-white; petals obovate to obcordate, emarginate, with a narrower inflexed apex; calyx teeth obsolete; stylopodium depressed conical, shorter than the recurved or reflexed styles. Pedicels slender, spreading-ascending. Fruit ovoid, rounded at both ends, compressed laterally and slightly constricted at commissure, the mericarps subterete in transection, glabrous, the ribs prominent, obtuse, unwinged. Vittae small and numerous, usually obliterated in ripe fruit. Commissural face sulcate. Carpophore entire.

Type species: *Conium maculatum* Linn.

Distribution: A small Mediterranean and South African genus of perhaps 5 species; the type species, a widely adventive, poisonous weed, has been reported from India.

1. **Conium maculatum** Linn. Sp. Pl. 243. 1753.

Plants slender to stout, much branched, 0.3–3 m tall, the stem fistulose, conspicuously purple maculate. Leaves broadly ovate, 15–30 cm long, 5–30 cm broad, the ultimate leaflets oblong to ovate, incised or pinnatifid into lanceolate, acuminate lobes; petiole stout, 10–30 cm long, dilated and scarious sheathing at base; upper cauline leaves sessile. Peduncles 3–8 cm long, rarely abortive. Involucre of 4–6 lanceolate to ovate, acuminate bracts 2–5 mm long. Rays 10–20, slender, spreading-ascending, subequal, 1.5–5 cm long, slightly scaberulous. Umbellets 10–20-flowered, the mature pedicels slender, unequal, spreading-ascending, 2–7 mm long. Involucel of 5 or 6 bracteoles, like the bracts, 1.5–3 mm long, connate at base and becoming ± scarious. Flowers white; stylopodium depressed conical. Fruit broadly ovoid, 2–3 mm long, 2–2.5 mm broad, the ribs prominent, obtuse, undulate-crenate, unwinged. A ring of secretory tissue surrounding seed. June.

Type locality: "In Europae cultis, agria, ruderatis," no collector indicated.

Distribution: Mediterranean region; adventive throughout tropical and temperate area of both Old and New Worlds.

Representative specimens: Jammu & Kashmir: Srinagar. *T. A. Rao 9607* (BSD), May 1979, *Mukherjee* (CUH, UC).

28. **AMMI** Linn.
Sp. Pl. 243. 1753.

Apium sect. *Ammi* (Linn.) Calest., Webbia 1: 175. 1905. *Visnaga* Gaertner, Fruct. 92. 1788.

Slender or stout, erect, branching, essentially glabrous, annual or biennial herbs from a slender taproot. Leaves petiolate, membranaceous, pinnate or ternate-pinnate (or rarely simple) with narrow leaflets, to pinnately decompound with linear or filiform ultimate divisions; petiole slender, narrowly sheathing; cauline leaves similar to basal, reduced upward. Inflorescence of lax terminal or axillary, compound umbels, the peduncles elongate. Involucre of numerous entire or divided bracts. Rays numerous, spreading-ascending or spreading, sometimes contracted and incurved at maturity. Involucel of numerous entire bracteoles. Umbellets many flowered, the flowers pedicellate. Flowers white; petals obcordate, with a narrower inflexed apex, the marginal sometimes slightly radiant; calyx teeth minute or calyx obsolete; stylopodium low conical, shorter than the slender, reflexed styles. Pedicels slender, spreading. Fruit oblong to ovoid, rounded at both ends, compressed laterally, constricted at commissure, the mericarps subterete in transection, glabrous, the ribs filiform, acute, subequal, unwinged, much narrower than intervals. Vittae solitary in intervals, 2 on commissure. Commissural face plane. Carpophore entire or bipartite.

Type species: *Ammi majus* Linn.

Distribution: A Mediterranean genus of perhaps half a dozen species, two of which have become widely cultivated and adventive.

1. **Ammi majus** Linn. Sp. P1. 243. 1753.

Carum majus (Linn.) Kozo-Polj., Bull. Soc. Nat. Mosc. II. 29: 198. 1916.

Plants slender, erect, branched, 20–150 cm tall, the inflorescence slightly scaberulous. Leaves oblong to triangular-ovate, 4–20 cm long, 2–15 cm broad, 1–2-pinnate or ternate-pinnate, the leaflets lanceolate to oblanceolate, spinulose-serrate with appressed teeth, occasionally lobed toward base, 1–7 cm long, 5–20 mm broad; petiole slender, 3–13 cm long, narrowly sheathing; cauline leaves similar to basal or pinnately or ternate-pinnately decompound with linear-ultimate divisions. Peduncles slender, 3–15 cm long. Involucre of 10–25 ovate, pinnate or ternate bracts with linear divisions 1–3 cm long, usually reflexed. Rays 20–60, subfiliform, spreading-ascending, subequal, slightly scaberulous, 1.5–8 cm long. Umbellets 30–50-flowered, the mature pedicels slender, unequal, spreading-ascending to somewhat contracted, 3–10 mm long. Involucel of 6–15 linear-lanceolate, acuminate, broadly scarious-margined entire bracteoles 3–6 mm long. Calyx obsolete; stylopodium low conical, shorter than the reflexed styles. Fruit oblong-ovoid, 1.5–2 mm long, ca. 1 mm broad, the ribs prominent, acute. Carpophore bipartite, the halves weak, filiform. Jan.–July.

Type locality: "In Europa australi," no collector indicated.

Distribution: Eurasia; widely cultivated and adventive.

Representative specimens: Uttar Pradesh: FRI Dehradun, *Raizada* (DD). West Bengal: Science College garden, Calcutta, *Mukherjee* (CUH). Meghalaya: Mawphlang, *Kingdon-Ward 18,780* (NY).

Note. *Ammi visnaga* (Linn.) Lam. reportedly has been recently introduced, presumably for experimental purposes. It differs from *A. majus* by its linear to filiform leaf divisions, discoid inflorescence base, contracted mature rays, and entire carpophore. Pimenov & Posorelova (1974) prefer to assign *A. visnaga* to a distinct genus, as *Visnaga daucoides* Gaertner.

29. **APIUM** Linn.

Sp. P1. 264. 1753; C. B. Clarke in Hook. f., Fl. Brit. Ind. 2: 678. 1879.

Apium sect. *Euapium* DC., Prodr. 4: 101. 1830.
Celeri Adans., Fam. P1. 2: 498. 1763.
Helosciadium Koch, Nova Acta Acad. Leop.-Carol. 12: 125. 1825; *Sium* sect. *Helosciadium* (Koch) Gaudich., Fl. Helv. 2: 430. 1826; *Apium* sect. *Helosciadium* (Koch) Bab. Man. Brit. Bot. ed. 8. 156. 1881; *Apium* subgen. *Helosciadium* (Koch) Drude in Engl. & Prantl, Pflanzenfam. 38: 185. 1898. *Mauchartia* Necker, Elem. 1: 172. 1790, *nomen illegit.*; *Helosciadium* sect. *Mauchartia* (Necker) DC., Prodr. 4: 104. 1830; *Apium* sect. *Mauchartia* (Necker) Benth. in Benth. & Hook. f., Gen. P1. 1: 888. 1867.

Prostrate, ascending, or erect, branching, glabrous, biennial or perennial, rarely annual, often aquatic or amphibious herbs from taproots or creeping rootstocks rooting at nodes. Leaves petiolate, membranaceous, simply pinnate to ternate-pinnately compound, the leaflets obovate to linear; petiole slender, sheathing at base; cauline leaves usually numerous, reduced upward. Inflorescence of lax terminal or leaf-opposed compound umbels, the peduncles usually short or abortive. Involucre and involucel lacking, or both present (Sect. *Machuartia*). Rays rather few, spreading-ascending or spreading. Umbellets few flowered, the flowers pedicellate. Flowers white or greenish; petals ovate to suborbicular with a narrower inflexed apex, subequal; calyx of minute lobes, or obsolete; stylopodium low conical to depressed, the styles short. Fruit ovoid-oblong to globose, rounded at both ends, compressed laterally and somewhat constricted at commissure, the mericarps subterete in transection, glabrous, the ribs prominent, usually obtuse, broad and corky or even subwinged. Vittae large, solitary in intervals, 2 on commissure. Commissural face plane. Carpophore stout, entire or shortly bifid.

Type species: *Apium graveolens* Linn.

Distribution: As construed here, a genus of about 20 species, of the temperate zones of both hemispheres; the widely cultivated and adventive type species reaches northwestern India.

1. **Apium graveolens** Linn. Sp. P1. 264. 1753; C. B. Clarke in Hook. f., Fl. Brit. Ind. 2: 679. 1879.

Plants in moist terrestrial soil, annual, biennial, or perennial, erect, branched, 50–150 cm tall from a taproot. Leaves oblong to obovate, 7–18 cm long, 3.5–8 cm broad, simply pinnate with a few pairs of usually trilobed, ovate to suborbicular leaflets 2–5 cm in diameter; petiole 3–25 cm long; cauline leaves mostly 3-parted or trilobed, sessile. Umbels opposite leaves, the peduncles short to abortive. Involucre and involucel lacking. Rays 3–15, slender, 1–4 cm long, spreading-ascending. Umbellets 5–10-flowered, the mature pedicels 1–6 mm long. Flowers greenish-white; calyx teeth minute; stylopodium low conical, the styles short, ascending. Fruit ovoid to orbicular, 1–2 mm in diameter, the mericarps subpentagonal in transection, the ribs prominent, obtuse. Carpophore shortly bifid. Dec.–Jan. on plains (cultivated); July–Sept. in hills.

Type locality: "In Europae humectis praesertim maritimis," no collector indicated.

Distribution: Eurasia; cultivated and/or adventive throughout the temperate zones of the world.

Representative specimens: Jammu & Kashmir: Srinagar, *T. A. Rao 9606* (BSD). Punjab: Salt Range, *Ram 446* (DD). Uttar Pradesh: Saharanpur, *Brothers* (DD); Rajpur, Dehra Dun, *Babu 35,217* (BSD).

30. **CICLOSPERMUM** Lagasca
Amen. Nat. 101. 1821.

Heliosciadium sect. *Cyclospermum* (Lagasca) DC., Prodr. 4: 105. 1830; *Apium* sect. *Cyclospermum* (Lagasca) H. Wolff in Engl., Pflanzenreich 90 ($4\frac{228}{}$): 53. 1927.

Low to slender, spreading to erect, usually profusely branching, glabrous annual, terrestrial herbs from slender taproots. Leaves petiolate, membranaceous, pinnately decompound with narrow ultimate divisions; petioles slender, sheathing at base; cauline leaves reduced upward, petiolate to sessile. Inflorescence of lax terminal or leaf opposed, compound or rarely simple umbels, the peduncles short or abortive. Involucre and involucel lacking. Rays few, slender, spreading-ascending or spreading. Umbellets rather few flowered, the pedicels slender. Flowers white; petals ovate, acute, but without a narrower inflexed apex; calyx teeth minute to obsolete; stylopodium low conical, the styles short to abortive, spreading. Fruit lanceolate-ovoid or oblong to orbicular, rounded at both ends or a little narrowed toward apex, compressed laterally and somewhat constricted at commissure, the mericarps subpentagonal to terete in transection, glabrous to rarely hispidulous, the ribs prominent, filiform to obtuse, broad and corky. Vittae solitary in intervals, 2 on commissure. Commissural face plane. Carpophore shortly bifid.

Type species: *Pimpinella leptophylla* Pers. [*Ciclospermum leptophyllum* (Pers.) Sprague ex Britton & Wilson].

Distribution: A temperate and tropical South American genus of apparently three species, the type species widely adventive.

1. **Ciclospermum leptophyllum** (Pers.) Sprague ex Britton & Wilson, Bot. Porto Rico 6: 52. 1925.

Pimpinella leptophylla Pers., Syn. Pl. 1: 324. 1805; *Aethusa leptophylla* (Pers.) Sprengel, Umbell. Prodr. 22. 1813; *Helosciadium leptophyllum* (Pers.) DC., Mém. Soc. Nat. Hist. Genève 4: 493. 1829; *Apium leptophyllum* (Pers.) F. v. Muell. ex Benth., Fl. Austral. 3: 372. 1866; *Apium ammi* var. *leptophyllum* (Pers.) Chodat & Wilczek, Bull. Herb. Boiss. II. 2: 526. 1902; *Cyclospermum leptophyllum* (Pers.) Sprague, Jour. Bot. 61: 131. 1923, *nomen invalid.*

Sison ammi sensu Jacq., Hort. Vindob. 2: 95. 1773. *Ciclospermum ammi* Lagasca, Amen. Nat. 101. 1821; *Apium ammi* Urban in C. Martius, Fl. Bras. 11(1): 341. 1879; *A. ammi* var. *genuinum* H. Wolff in Engl., Pflanzenreich 90 ($4\frac{228}{}$): 54. 1927, non *Sison ammi* Linn. 1753.

Cnidium tenuifolium Moench, Meth. 98. 1794, nomen illegit.; *Apium tenuifolium* (Moench) Thell. in Hegi, Ill. Fl. Mittel-Eur. 52: 1140. 1926.

Plants low or slender, erect or spreading, 5-60 cm tall. Leaves oblong-ovate, 3.5–10 cm long, 3.5–8 cm broad, pinnately decompound with linear to filiform ultimate divisions 1.5–7 mm long, 0.5–1 mm broad; petiole slender, 2.5–11 cm long, the sheath white-scarious-margined; cauline leaves reduced upward, usually sessile. Peduncles short, 2 cm long to abortive. Rays 1–5, slender, 5–25 mm long. Umbellets usually more than 10-flowered, the mature pedicels slender, 2–15 mm long, spreading. Stylopodium low conical, longer than the spreading styles. Fruit ovoid to orbicular, 1.5–3 mm long, 1–2 mm broad, glabrous, the ribs prominent, filiform to obtuse and corky. Carpophore shortly bifid.

Type locality: "Ins. St. Dominica," West Indies, no collector indicated.

Distribution: Tropical America; disseminated as a pantropical and temperate cosmopolitan weed, reaching the Indian Himalaya.

Representative specimens: Uttar Pradesh: Chakrata, *Mukherjee 52* (UC), *105* (CUH); Bindal Ravines, Dehra Dun, *Babu 33,325* (BSD).

31. **VANASUSHAVA** P. K. Mukherjee & Constance
Kew Bull. 29: 595. 1974.

Low and scandent, sparsely branching, hispidulous perennial herbs from a trailing rootstock. Leaves petiolate, membranaceous, trifoliate, the large petiolulate leaflets ovate to lanceolate, the 2 lateral once cleft nearly to base, the lamina thus appearing palmately 5-foliate; petiole slender, sheathing; cauline leaves similar, reduced upward. Inflorescence of compound umbels terminal on short branches. Involucre usually lacking. Rays few, unequal, spreading-ascending. Involucel of few to several short, narrow, reflexed bracteoles, or occasionally involucel lacking. Umbellets several flowered, polygamous, the outer perfect, the others staminate, all flowers pedicellate. Flowers white; petals oblanceolate to obovate with a narrower inflexed apex, the marginal radiant; calyx teeth lanceolate, unequal, evident only in perfect flowers; stylopodium low conical, the styles slender, reflexed. Pedicels slender, unequal, the fertile longer than the sterile. Fruit ovoid-oblong, slightly compressed laterally, narrowed at apex, shallowly cordate at base, the mericarps compressed dorsally in transection, glabrous, the ribs filiform, low and rounded. Vittae 2 or 3 in intervals, 2–6 on commissure. Commissural face sulcate. Carpophore lacking.

Type species: *Heracleum pedatum* Wight [*Vanasushava pedata* (Wight) P. K. Mukherjee & Constance].

Distribution: A monotypic genus of hill forests of peninsular India.

Note. For the burgeoning literature on *Vanasushava*, see Blasco *et al.* 1978.

1. **Vanasushava pedata** (Wight) P. K. Mukherjee & Constance,
Kew Bull. 29: 595. 1974.

Heracleum pedatum Wight, Ic. Pl. Ind. Orient. 2: 3. 1843; C. B. Clarke in Hook. f., Fl. Brit. Ind. 2: 716. 1879.

Plants slender, trailing, 30–60 cm long, rooting at nodes. Leaves orbicular to obovate, 3–10 cm long, 4–15 cm broad, trifoliate, the terminal leaflet occasionally shallowly trilobed, the lateral deeply cleft, 3–10 cm long, 1–6 cm broad, acuminate, cuneate at base, spinulose-serrate, sparsely strigulose-hirsutulous on both surfaces, hirtellous on rachis; petiole slender, 3–15 cm long, hirsutulous, the sheath scarious-margined, truncate above; cauline leaves like the basal but smaller. Peduncles very slender, 2.5–8 cm long, hirsutulous. Rays 2–5(–7), slender, unequal, 1–3.5 cm long. Umbellets 3–16-flowered (1–4 marginal, perfect, radiant; 2–12 central, sterile, regular), the mature pedicels slender, 3–6 mm long. Involucel of 2–5(–7) linearly lanceolate to lanceolate, acute, green bracteoles 1.5–3 mm long. Stylopodium much shorter than styles. Fruit 3–4 mm long, 2–3 mm broad, the mericarps flattened-reniform in transection. Vittae variable in size and number. June–Oct.

Type locality: "Alpine jungles Shevagherry" [Sivagiri Hills], *Wight 1183* (K holotype; CAL isotype).

Distribution: Peninsular India, Pulney Hills area, 1,500–2,300 m.

Representative specimens: Tamil Nadu: Periyur Shola, *Bourne 2640* (K); Kodaikanal Shola, *Bourne 540* (K), *2639* (CAL, K, MH); Upper Shola Road, *Mukherjee 163* (CUH, UC). Kerala: Lakshmi Estate, Munnar, *Deb 30,758* (MH); Umaiy-Malai, Devicolam, *Shetty 28,349* (MH).

32. PTERNOPETALUM Franchet

Nouv. Arch. Mus. Hist. Nat. II. 8: 246. 1886.

Carum sect. *Cryptotaeniopsis* Franchet, Bull. Soc. Philom. Paris VIII. 6: 119. 1894; *Pimpinella* subgen. *Cryptotaeniopsis* (Franchet) Diels, Bot. Jahrb. Syst. 29: 494. 1900; *Cryptotaeniopsis* (Franchet) Dunn in Hooker's Icon. Pl. pl. 2737. 1902.

Low to slender, delicate to robust, erect, sparsely branching or simple, glabrous or pubescent perennial herbs from a slender rootstock. Leaves petiolate, membranaceous, simply ternate to ternately or ternate-pinnately divided, the leaflets broad or dissected into narrow ultimate divisions; petiole slender, sheathing; cauline leaves few, similar to or differing sharply from basal leaves, or sometimes lacking. Inflorescence of lax, mostly terminal, compound umbels, the peduncles usually elongate. Involucre of a few bracts, or usually lacking. Rays few to very numerous, slender, unequal, spreading or spreading-ascending. Involucel of 2–4 narrow, inconspicuous bracteoles. Umbellets 2–4-flowered, some flowers pedicellate, one or more subsessile. Flowers white or purplish; petals oblanceolate, acute, and sometimes incurved, but without a definite inflexed apex; calyx teeth prominent and subequal to stylopodium, to minute or rarely obsolete; stylopodium conical, the styles slender, erect or spreading. Pedicels slender but short, usually one obsolete. Fruit oblong to broadly ovoid, compressed laterally and constricted at commissure, narrowed toward apex, often cordate at base, the mericarps subterete in transection, glabrous (rarely scaberulous), the ribs filiform, evident, subequal, unwinged. Vittae usually solitary (1–3) in intervals, 2 or 4 on commissure. Commissural face plane to slightly concave. Carpophore bipartite or bifid.

Type species: *Pternopetalum davidii* Franchet.

Distribution: A genus of perhaps two dozen species, extending from China and Japan to northeastern India.

KEY TO SPECIES

1. Involucre evident, of 1–3 bracts; calyx teeth conspicuous, subulate, persistent. .... 1. *P. vulgare*.
1. Involucre lacking; calyx teeth minute or obsolete.
    2. Leaflets oblong, crenate-ciliate; rays 20–50, 3.5–6 cm long. .... 2. *P. senii*.
    2. Leaflets ovate-lanceolate to orbicular, serrate to incised; rays 6–30, 1–3 cm long.
        3. Cauline leaves similar to basal, with ovate-lanceolate to ovate divisions; rays 6–12. .... 3. *P. radiatum*.
        3. Cauline leaves usually dissimilar to basal, with linear to filiform divisions; rays 10–30. .... 4. *P. tanakae*.

1. **Pternopetalum vulgare** (Dunn) Hand.-Mazz., Symbol. Sinicae 7: 719. 1933.

Cryptotaeniopsis vulgaris Dunn in Hooker's Icon. Pl. pl. 2737. 1902; *Deringa vulgaris* (Dunn) Kozo-Polj., Vestn. Tiflissk. Bot. Sad. 11: 6. 1915.

Pimpinella clarkeana Watt ex M. L. Banerji, Jour. Bombay Nat. Hist. Soc. 50: 88. 1951. Based on *Watt 6556* (lectotype of the preceding).

Plants slender, erect, few branched above, 15–45 cm tall, the foliage slightly pubescent, from an oblique rootstock bearing fleshy roots. Leaves slightly chartaceous, ovate, 5–30 cm in diameter, ternate or biternate, the leaflets ovate, 1–5 cm long, 1–3.5 cm broad, coarsely crenate-serrate, the larger often few lobed or parted, glabrous above, setulose on margins and veins beneath; petiole slender, 8–25 cm long; cauline leaves similar, short petiolate or sessile. Peduncles slender, 5–10 cm long. Involucre lacking. Rays 12–30, very slender, spreading or spreading-ascending, 2–5 cm long, subequal. Umbellets usually 3-flowered, the mature pedicels unequal, 4 mm long to obsolete. Involucel of 1–3 linear bracteoles 1–2.5 mm long. Flowers white or bluish; calyx teeth prominent, subulate, persistent. Fruit oblong-ovoid, 2–3.5 mm long, ca. 1 mm broad, glabrous or the filiform ribs sometimes minutely scaberulous. Vittae usually 1–3 in lateral intervals, 4 on commissure. Carpophore bipartite. Apr.–May.

Type locality: "India, Manipur, Ching Sow," *Watt 6556* (K, lectotype).

Distribution: Western China to northeastern India, 2,000–3,000 m.

Representative specimens: Nepal: w/o further locality, *Banerji 446* (CAL). Arunachal Pradesh: Bomdila via Rupa, Kameng, *R. S. Rao 6961* (ASSAM). Manipur: Chingsow, *Watt 6556* (CAL).

2. **Pternopetalum senii** Deb & Dutta, Bull. Bot. Soc. Bengal, J. Sen Memorial, vol. 421. 1969.

Plants slender, erect, usually unbranched, 30–40 cm tall, the foliage slightly pubescent, from a rootstock. Leaves ovate, ternately decompound, 20–30 cm long, 15–25 cm broad, the leaflets oblong, 1–4 cm long, 5–15 mm broad, pinnatifid, crenate, ciliate, glabrous above, hispidulous beneath; petiole slender, to 25 cm long; cauline leaves lacking. Peduncles terminal, elongate. Involucre lacking. Rays 20–50, slender, spreading-ascending, unequal, 3.5–6 cm long. Umbellets 2- or 3-flowered, the mature pedicels unequal, 1–3 mm long. Involucel of 2 or 3 linear bracteoles. Flowers purple; petals oblong; calyx teeth minute, subulate; stylopodium conical, the styles erect, slender. Fruit oblong, slightly compressed laterally, a little constricted at commissure, glabrous, the ribs filiform. Vittae solitary in intervals, 2 on commissure. Carpophore bipartite. July.

Type locality: "On way from Rahu to Wakka, Tirap Frontier Division, North East Frontier Agency," Arunachal Pradesh, *D. B. Deb, 26,39A* (CAL-A, holotype; -B, ASSAM, isotype).

Distribution: Known only from the type locality.

3. **Pternopetalum radiatum** (W. Smith) P. K. Mukherjee, Indian Forester 97: 55. 1971.

Pimpinella radiata W. Smith, Rec. Bot. Surv. India 4: 266. 1911; *Acronema radiata* (W. Smith) H. Wolff in Engl., Pflanzenreich 90 ($4\frac{228}{}$): 324. 1927.

Plants very slender, branched, to 30 cm tall, essentially glabrous. Leaves ovate to triangular-ovate, 2.5–3.5 cm long, 2.5–4 cm broad, ternate-pinnate, the leaflets ovate to ovate-lanceolate, 1–2 cm long, 8–12 mm broad, acute, cuneate at base, coarsely serrate to deeply incised; petiole slender, 2–4 cm long; cauline leaves similar, some of the upper pinnate. Peduncles terminal and lateral, the former rather stout, 3–10 cm long, the latter only 3–15 mm long, shorter than rays or subtending petiole. Involucre lacking. Rays 6–12, spreading, subequal, 1–3 cm long. Umbellets 2–4-flowered, the mature pedicels unequal, 3 mm long to abortive. Involucel of 1 or 2 setiform bracteoles 1–1.5 mm long. Flowers not seen. Fruit oblong-ovoid, ca. 2 mm long, 1 mm broad, the ribs filiform, evident to obscure. Vittae solitary in intervals, 2 on commissure. Commissural face plane to slightly concave, the endocarp very thin. Carpophore bipartite. July–Aug.

Type locality: Sikkim, Yeumtong, *Gammie 992* (CAL, lectotype [Mukherjee 1971]).

Distribution: Sikkim Himalaya, at 3,660–4,330 m.

Representative specimen: Sikkim: Sebu Valley, *Gammie* (CAL).

4. **Pternopetalum tanakae** (Franchet & Savat.) Hand.-Mazz., Symbol. Sinicae 7: 719. 1933.

Chamaele ? tanakae Franchet & Savat., Enum. Pl. Jap. 1: 185. 1875, *nomen nudum*; *Carum tanakae* Franchet & Savat., Enum. Pl. Jap. 2: 371. 1879; *Pimpinella tanakae* (Franchet & Savat) Diels, Bot. Jahrb. Syst. 29: 494. 1900; *Cryptotaeniopsis tanakae* (Franchet & Savat) Boiss., Bull. Herb. Boiss. II. 2: 806. 1902.

Plants very slender, simple or with a single lateral branch, 8–30 cm tall, glabrous, from a slender rootstock. Leaves ovate to triangular-ovate, 1.5–6 cm in diameter, 1-2-pinnate or ternate-pinnate, the leaflets ovate to orbicular, 0.5–2 cm long and broad, serrate to pinnatifid; petiole capillary, 1.5–8 cm long; cauline leaves few, at least the upper pinnate with usually elongate linearly lanceolate to filiform divisions. Peduncles slender, 2–15 cm long. Rays usually 10–30, capillary, spreading, 1.5–3 cm long, subequal. Umbellets 1–3-flowered, the mature pedicels 3 mm long to obsolete. Involucel of 1 or 2 setiform bracteoles up to 1.5 mm long. Flowers white; calyx teeth barely evident. Fruit oblong-ovoid, 2–2.5 mm long, 1.25–2 mm broad, glabrous, the ribs filiform, obscure. Vittae solitary in intervals, 2 on commissure. Carpophore bipartite.

Type locality: "HAB. in Japonia, ubi detexit *Tanaka*."

Distribution: Japan, Korea, and central China to Bhutan.

Specimens examined: Bhutan: Between Gyasa Dzong and Laya, Mo Chu, *Ludlow, Sheriff & Hicks 16,554* (CAL). Assam. Delai Valley, *Kingdon-Ward 8317* (BM).

33. AEGOPODIUM Linn.
Sp. Pl. 265. 1753.

Pimpinella sect. *Aegopodium* (Linn.) Kuntze in Post & Kuntze, Lex. Gen. Phan. 439. 1903.
Apium sect. *Aegopodium* (Linn.) Calest., Webbia: 179. 1905.
Podagraria Hill, Brit. Herb. 405. 1756.

Slender or stout, erect, branching or simple, glabrous or slightly pubescent perennial herbs from a creeping rootstock. Leaves petiolate, membranaceous, ternately or ternate-pinnately compound, with broad, serrate, dentate, or incised leaflets; petiole slender, sheathing; cauline leaves reduced upward. Inflorescence of lax terminal and lateral compound umbels, the peduncles elongate. Involucre and involucel usually lacking, or occasionally of a few caducous narrow bracts or bracteoles. Rays rather few to more numerous, subequal or unequal, spreading-ascending or ascending. Umbellets rather numerous flowered, the flowers pedicellate. Flowers white or rarely reddish; petals oblanceolate to obovate, with a narrower inflexed apex; calyx teeth minute, or calyx obsolete; stylopodium low conical to conical, conspicuously bifid and the halves divaricate, the styles slender, reflexed. Pedicels rather slender, unequal, spreading-ascending. Fruit oblong to ovoid, narrowed toward apex and rounded at base, compressed laterally and constricted at commissure, the mericarps subterete in transection, glabrous, the ribs inconspicuous, filiform, much narrower than intervals. Vittae obscure at maturity and replaced by a continuous subepidermal layer. Commissural face plane. Carpophore bipartite.

Type species: *Aegopodium podagraria* Linn.

Distribution: A Eurasian genus of perhaps a half-dozen species, two of which reach northern India.

KEY TO SPECIES

1. Leaves 2-ternate or ternate-pinnate, the sheaths scarcely inflated; calyx obsolete; fruit oblong to oblong-ovoid. .... 1. *A. alpestre*.
1. Leaves ternately or ternate-pinnately decompound, the sheaths prominent, somewhat inflated; calyx teeth evident; fruit ovoid. .... 2. *A. henryi*.

1. **Aegopodium alpestre** Ledeb. Fl. Alt. 1: 354. 1829.

Carum alpestre (Ledeb.) Kozo-Polj., Bull. Soc. Nat. Mosc. II. 29: 199. 1916. *Pimpinella kashmirica* R. Stewart ex Dunn, Kew Bull. 384. 1924. Based on *Stewart 6267*, "Kashmir; Sonamarg."

Plants slender, erect, simple or branched, essentially glabrous, 20–80 cm tall. Leaves triangular-ovate, 3–15 cm in diameter, biternate or ternate-pinnate, the leaflets oval to lanceolate, 1–4 cm long, 0.3–2 cm broad, acute or acuminate, cuneate at base, mucronate-serrate to deeply incised; petiole slender, 3–30 cm long, with a narrow, scarious-margined sheath; cauline leaves few, reduced upward, the sheaths scarcely dilated. Peduncles slender, mostly terminal, 4–12 cm long. Rays 6–16, slender, unequal, 1–4.5 cm long, essentially glabrous. Umbellets 15–25-flowered, about half staminate, the mature pedicels slender, unequal, spreading-ascending, 2–10 mm long. Calyx obsolete; stylopodium conical, the halves divaricate, shorter than the slender reflexed styles. Fruit oblong to oblong-ovoid, 2.5–4 mm long, 1.5–2 mm broad, glabrous, the ribs inconspicuous, filiform. July–Aug.

Type locality: "Prope Riddersk, circa fontes fl. belaja Uba et Tschaysch," Siberia, *Ledebour*.

Distribution: Japan and eastern Siberia to the Western Himalaya of India and Pakistan, 3,000–4,000 m.

Representative specimens: Jammu & Kashmir: Tjawas nullah, near Sonemarg, *Duthie 13,625* (CAL, DD); Kargail Valley, *Duthie 13,924* (DD); Sonemarg, *R. R. Stewart 6267* (K).

2. **Aegopodium henryi** Diels, Bot. Jahrb. Syst. 29: 497. 1900.

Plants slender, erect, subflaccid, sparsely branched, 50–100 cm tall, from a fleshy rootstock. Leaves subflaccid, paler beneath, suborbicular, ternately or ternate-pinnately decompound, the leaflets ovate-lanceolate, 1–2 cm long, 5–12 mm broad, longly acuminate, truncate or rounded at base, coarsely serrate or dentate; petiole 3–15 cm long, broadly sheathing; cauline leaves numerous, the sheaths conspicuous, ± inflated. Peduncles elongate, terminal and axillary, to 18 cm long. Rays 9–12, ascending or spreading-ascending, unequal, 1.5–4 cm long. Umbellets 15–20-flowered, some staminate, the mature pedicels unequal, 4–10 mm long. Calyx teeth minute; stylopodium conical. Fruit ovoid, 2.5–4 mm long, the ribs filiform. Sept.

Type locality: Patung, Hupeh, China, *Henry 4946*.

Distribution: Western China to northeastern India, ca. 2,500 m.

Representative specimens: Arunachal Pradesh: 40 miles from Kimin, Subansiri, *Panigrahi 19,714* (ASSAM). Nagaland: Dzulake Valley, *Bor 99,562* (DD).

34. **HAPLOSPHAERA** Hand.-Mazz.

Akad. Wiss. Wien, Math.-Naturwiss. Kl., Anz. 57: 143. 1921.

Slender or stout, erect, simple or branching, essentially glabrous, perennial herbs from a fibrous caudex surmounting a sparsely branched rhizome. Leaves petiolate, membranaceous, ternately or pinnately compound, with broad crenate-dentate to pinnatisect leaflets; petioles slender, sheathing; cauline leaves reduced upward. Inflorescence of compact, terminal and lateral compound (or simple ?) umbels, the peduncles elongate. Involucre of narrow bracts, or lacking. Rays short to abortive (?). Umbellets rather few flowered, the flowers pedicellate. Involucel of several narrow bracteoles (in our species). Flowers purplish-brown; petals broadly ovate, with a narrower inflexed apex; calyx minute; stylopodium apparently depressed, the immature styles very short. Pedicels rather short, thick, erect or ascending. Immature fruit obovoid, rounded at apex and narrowed toward base, compressed laterally, sharply constricted at commissure, the mericarps pentagonal and slightly compressed laterally in transection, glabrous, the ribs prominent, subwinged, the dorsal most prominent, all much narrower than intervals. Vittae usually 2 or 3 in intervals, 3–6 on commissure. Commissural face plane (?). Carpophore not seen.

Type species: *Haplosphaera phaea* Hand.-Mazz.

Distribution: A genus of two known species, in Tibet-Bhutan and Yunnan, China respectively.

1. **Haplosphaera himalayensis** Ludlow, Bull. Bot. Brit. Mus. [Nat. Hist.], Bot. 5: 276. 1976.

Plants stout, erect, simple or few branched, glabrous, 1–1.5 m tall. Leaves triangular-ovate, 12–15 cm in diameter, tripinnate, the leaflets broadly ovate to lanceolate, 3–6 cm long, 1.5–5 cm broad, acute or acuminate, petiolulate to sessile, pinnatifid; petiole slender, 10–15 cm long, with an oblong sheath; cauline leaves few, reduced upward, the sheaths slightly dilated. Peduncles slender, terminal and lateral, 5–10 cm long. Involucre lacking. Umbellets 6–18-flowered, the immature pedicels thick, 2–3 mm long. Involucel of 4–8 narrow bracteoles ca. 6 mm long. Calyx teeth inconspicuous, triangular; stylopodium apparently depressed, the immature styles very short. Immature fruit obovoid, 3 mm long, 1.5 mm broad, glabrous, the ribs prominent, subwinged. Aug.

Type locality: "S. E. Tibet: Mira La, Nyang Chu," *Ludlow, Sherriff & Taylor 6087* (BM, holotype).

Distribution: Southeastern Tibet and northeastern Himalaya, openings and forest clearings at 3,900 m.

Representative specimens: Bhutan: Shingbe (Me La), *Ludlow, Sherriff & Hicks 21,102* (BM). Arunachal Pradesh: Kameng, near Shungetshuan, *Hajra 68,310* (ASSAM).

Note. Ludlow (1976) placed his new species in the genus *Haplosphaera* "with some hesitation," because it was described as possessing simple umbels. He remarked that although immature umbels of *H. himalayensis* appear to be simple, their compound nature becomes apparent with maturity. We believe that this may be true of the type species as well, of which we have seen only flowering specimens.

35. **KEDARNATHA** P. K. Mukherjee & Constance
Brittonia 38(2): 147. 1986.

Low, subcaulescent, simple or sparsely branched from base, essentially glabrous perennial herbs from a thick fibrous base. Leaves petiolate, membranaceous, mostly rosulate, narrowly oblong, pinnatisect, the leaflets obovate, dissected into linear-acuminate divisions; petiole broadly scarious-sheathing; cauline leaves consisting of a few opposite pairs. Inflorescence of a lax terminal compound umbel, the peduncle stout, hirtellous at apex. Involucre of several foliaceous, pinnatisect bracts. Rays rather few, stout, unequal, spreading-ascending. Involucel of several linear to lanceolate pinnatisect bracteoles. Umbellets rather few flowered, the flowers pedicellate. Flowers not seen; calyx teeth lanceolate; stylopodium conical, about equaling the short-reflexed styles. Pedicels stout, angled, spreading-ascending. Fruit broadly ovoid, subterete, narrowed at apex, rounded at base, the mericarps lightly compressed dorsally in transection, glabrous, the ribs very prominent, corky thickened, subequal, broader than intervals. Vittae large, solitary in intervals, 3–5 on commissure. Commissural face plane. Carpophore bipartite, the halves remaining appressed.

Type species: *Kedarnatha sanctuarii* P. K. Mukherjee & Constance.

Distribution: Known only from the type collection, obtained in the Himalaya of Uttar Pradesh.

1. **Kedarnatha sanctuarii** P. K. Mukherjee & Constance,
Brittonia 38(2): 147. 1986.

Plants subcaulescent, erect, essentially glabrous, simple or usually with a few weak lateral branches arising from the stout fibrous base, to 25 cm tall. Leaves narrowly oblong, 6–7.5 cm long, 1–1.5 cm broad, pinnatisect, the leaflets obovate in 3–5 opposite, remote sessile pairs, dissected into linearly acuminate ultimate divisions 1–2 mm long, less than 0.5 mm broad; petiole slender, 1–1.5 cm long, broadly scarious-sheathing; cauline leaves (of lateral branches) paired, like the basal. Terminal peduncles stout, 15–20 cm tall, ribbed, hirtellous at apex. Involucre of 5–8 pinnatisect bracts 1–3.5 cm long. Rays of terminal umbel 10–12, stout, ribbed, spreading-ascending, unequal, 2–7 cm long. Involucel of 8–10 linear to lanceolate, pinnatisect or a few entire bracteoles 5–10 mm long. Umbellets 8–12-flowered, the mature pedicels stout, angled, spreading-ascending, 2–5 mm long. Flowers not seen. Fruit broadly ovoid, 4–5 mm long, 2–3 mm broad, subterete, narrowed toward apex, rounded at base, the mericarps lightly compressed dorsally in transection, the ribs prominently corky thickened, each with a small airspace, much broader than intervals. Vittae large, solitary in intervals, 3–5 on commissure. Endosperm grooved beneath intervals. Carpophore halves diverge to form a small "Y" or "O" near apex. Oct.

Type locality: "Uttar Pradesh: on swampy meadow behind. . . . Kedarnath temple," 25 Oct. 1982, *Mukherjee* (CUH, holotype; UC, isotype).

Distribution: Known only from the type collection.

36. SCHULZIA Sprengel

Umbell. Prodr. 30. 1813, *nomen conserv*.; C. B. Clarke in Hook. f., Fl. Brit. Ind. 2: 697. 1879, non *Schultzia* Raf. 1808.

Trachydium subgen. *Pentazygon* C. B. Clarke in Hook. f., Fl. Brit. Ind. 2: 672. 1879, *pars.*

Slender and erect to low and spreading, caulescent to acaulescent, simple or sparsely branching, glabrous perennial herbs from a fibrous caudex surmounting an elongated taproot, or fibers lacking. Leaves petiolate, membranaceous, 2–3-pinnate, the ultimate divisions small and narrow; petiole slender, scarious-sheathing; cauline leaves few and reduced upward, or lacking. Inflorescence of a lax fertile terminal compound umbel, subsessile, umbellets appear as simple umbels, or pedunculate, often with one or more smaller lateral umbels, the peduncles elongate to inconspicuous. Involucre and involucel of several to many, mostly pinnatisect bracts and bracteoles, or involucre 0. Rays few to numerous, subequal to unequal, spreading-ascending or ascending. Umbellets few- to numerous-flowered, the flowers pedicellate. Flowers white; petals oval to narrowly obovate, with a narrower inflexed apex; calyx teeth prominent to obsolete; stylopodium conical, a little shorter than the ascending to recurved styles. Pedicels rather rigid, spreading-ascending. Fruit oblong-ovoid to ovoid, compressed laterally, rounded and slightly narrowed at apex and base, broadest at commissure, the mericarps lightly compressed dorsally in transection, glabrous, the pericarp shiny, the ribs prominent and acute or obtuse, or obscure, narrower than intervals. Vittae small, 2–4 in intervals, 4–8 on commissure. Commissural face plane to concave. Carpophore bipartite or merely lobed at apex, the halves shortly divergent.

Type species: *Sison crinitum* Pallas [*Schulzia crinata* (Pallas) Sprengel].

Distribution: A small genus of central Asia and Siberia, extending to the Himalaya, 3,000–5,300 m.

KEY TO SPECIES

1. Rays subequal; fruit ribs filiform to obscure. .... 1. *S. dissecta.*
1. Rays usually very unequal; fruit ribs prominent, low and broad.
    2. Involucre of foliaceous pinnatisect bracts; calyx teeth persistent; fruiting pedicels 2–5 mm long. .... 2. *S. garhwalica.*
    2. Involucre lacking; calyx lacking; fruiting pedicels 5–10 mm long. .... 3. *S. hameliana.*

1. Schulzia dissecta (C. B. Clarke) Norman,

Jour. Bot. 76. 231. 1938.

Trachydium dissectum C. B. Clarke in Hook. f., Fl. Brit. Ind. 2: 672. 1879.
Ligusticum baltistanicum H. Wolff, Feddes Repert. 27: 310. 1929. Based on *Duthie 11,956*, "Baltistan: Karpuchi-Tal, 13,000–14,000 ft." (BM, lectotype; K, L).

? *Trachydium hirsutulum* C. B. Clarke in Hook. f., Fl. Brit. Ind. 2: 672. 1879; *Physospermopsis hirsutula* (C. B. Clarke) Farille, in Farille, Cauwet-Marc & Malla, Candollea 40(2): 513. 1985. Based on *Hooker*, "SIKKIM; Tungu, alt. 13,000–14,000 ft."

Plants low but slender, simple or branched from a nonfibrous base, 5–30 cm tall. Leaves narrowly oblong, 2–8 cm long, 0.4–1.5 cm broad, bipinnate, the ultimate divisions narrowly linear, 1–2 mm long, usually less than 1 mm broad, acute; petiole 1–4 cm long. Peduncles 5–12 cm long, rarely abortive. Involucre of 5–7 oblong to oval pinnatifid bracts 1–2 cm long. Rays usually 10–20, slender, quadrangular, spreading-ascending or ascending, subequal, 1–3 cm long. Umbellets 8–20-flowered, the mature pedicels quadrangular, unequal, spreading-ascending, 3–10 mm long. Involucel of 5–7 oblanceolate to obovate, pinnatifid bracteoles 4–8 mm long, equal to or exceeding flowering umbellets. Calyx obsolete. Fruit ovoid, 2–3 mm long, ca. 1.5 mm broad, the mericarps dorsally subcompressed in transection, the ribs filiform, obscure. Vittae very small, several in intervals and on commissure. Commissural face plane. Carpophore flattened, merely bifid at apex. July–Aug.

Type locality: "SIKKIM: Tungu, 13,000–14,000 ft.", *Hooker* (*Hooker & Thomson*, Herb. Ind. Orient., *Cortia 3*).

Distribution: Himalaya: Nepal and Sikkim to Kashmir and Baltistan, 4,000–5,000 m.

Representative specimens: Nepal: Gossain Than, *Wallich 589A* (CAL, *pars*). Sikkim: Tungu, *Hooker* (BM, CAL, K, P).

Note. Two flowering specimens from the Upper Mu Chu District of Bhutan obtained by *Sinclair and Long* (*5376, 5469*, both E), marked by slender unbranched habit and flaring cauline leaf sheaths, appear to belong to *Schulzia*. If they represent *S. dissecta* (C. B. Clarke) Norman, then Nasir (1972, pages 111–112) would seem to be mistaken in referring Kashmir material and the taxon *Ligusticum baltistanicum* H. Wolff to that species.

Clarke described *Trachydium hirsutulum* from a Hooker collection made at Tungu, Sikkim, mentioning among other features the "hirsute ultimate [leaf] segments" and "leaves with fine hairs on the midrib." Norman (1938b, p. 231) believed this to be "only a burnt-up and stunted specimen of *T. obtusiusculum*," and noted that: "The type-sheet, in addition to specimens of *T. obtusiusculum*, comprises three specimens of *Potentilla coriandrifolia* Don." Farille (1985, p. 514), referring to Norman's disposition, remarked: "Un seul échantillon appartient effectivement à l'espèce de Clarke (lectotype de fait), les autres échantillons étant étrangers à la famille des Apiaceae. Les quelques débris de feuilles et les deux fruits qui subsistent aujourd'hui scientifiquement insuffisants et il y aurait nécessité à néotypifier cette espèce." He did not, however, proceed to neotypify the taxon. The specimens cited by Farille under *P. hirsutula*, examined by us, are uniformly glabrous and do, in our opinion, constitute a distinct species, which we take pleasure in designating *P. farillei*. One of us (PKM) saw the Kew holotype of *T. hirsutulum* and believes it to represent *Schulzia*, in all likelihood *S. dissecta* (C. B. Clarke) Norman, but because of the dubious character of the type material, we recommend that the epithet "*hirsutula*" be abandoned in this connection.

2. **Schulzia garhwalica** (H. Wolff) P. K. Mukherjee & Constance, Edinburgh Jour. Bot. 48(1): 43. 1991.

Trachydium garhwalicum H. Wolff, Feddes Repert. 27: 124. 1929; *Chamaesciadium garhwalicum* (H. Wolff) Norman, Jour. Bot. 76: 232. 1938.

Trachyspermum wolffianum H. Wolff, Feddes Repert. 27: 180. 1929. Based on same type material as the preceding.

Plants low, subcaulescent or caulescent, simple or few branched from a fiber-bearing base, the stems erect or ascending, 5–40 cm tall. Leaves narrowly oblong, 1.5–5 cm long, 0.5–1 cm broad, bipinnate, the ultimate divisions linear, 2–5 mm long, up to 1 mm broad, acute; petiole slender, 1–3 cm long; cauline leaves few, similar, or usually lacking. Peduncles rather stout, 2–10 cm long, or inconspicuous. Involucre of 3–6 foliaceous, bipinnatisect bracts 1.5–4 cm long, nearly indistinguishable from leaves. Fertile rays mostly 8–12, stout, quadrangular, spreading-ascending, unequal, 1–10 cm long. Umbellets 15–25-flowered, the mature pedicels slender, quadrangular, spreading-ascending, 2–5 mm long. Involucel of ca. 10 linear to obovate, acute, entire to pinnatifid bracteoles 5–10 mm long. Calyx teeth prominent, linear. Fruit ovoid-oblong, 3–4 mm long, 1.5–2 mm broad, the ribs prominent, subacute. Vittae 2 or 3 in intervals, usually 4 on commissure. Commissural surface plane. Carpophore stout, bipartite, the halves diverging at apex. Aug.–Oct.

Type locality: "Tehri Garhwal, Taulea under Srikanta," *Duthie 1136a* (DD, lectotype; CAL).

Distribution: Western Himalaya, from Himachal Pradesh to Uttar Pradesh, 3,000–5,300 m.

Representative specimens: Himachal Pradesh: Kulu Valley, Bhirgu Valley, Bhirgu Lake, *Vaid 24,669* (DD). Uttar Pradesh: Mandraola, *Duthie 1136* (DD); Hemkund, *Bhattacharyya 29,500* (BSD); Kedarkanta, *Native collector 530* (CAL); Chitona, Tehri, *Koelz 22,024* (G).

3. **Schulzia hameliana** (Farille & Malla) P. K. Mukherjee & Constance, Edinburgh Jour. Bot. 48(1): 43. 1991.

Trachydium hamelianum Farille & Malla, Bull. Soc. Bot. France 131, Lettres Bot. 1984(1): 64. 1984; *Chamaesciadium hamelianum* (Farille & Malla) Farille, in Farille, Cauwet-Marc & Malla, Candollea 40(2): 557. 1985.

Plants caulescent, slender, erect, glabrous, 3–20 cm tall from a fibrous caudex surmounting a branched taproot. Leaves mostly basal, narrowly oblong, pinnatisect with 3–5 pairs of leaflets, the lower petiolulate, trisect, the others sessile, irregularly palmately parted to entire, the ultimate divisions subulate, mucronate; petiole about equal to or longer than the blade; cauline leaves 0. Peduncles terminal, short to abortive, usually with some smaller lateral compound umbels. Involucre 0. Rays usually 10–30, stout, unequal, 5–10 cm long, strongly divergent. Umbellets 8–10-flowered, the pedicels unequal, 5–10 mm long. Involucel of 6–8 very narrow, 1–2-pinnatisect bracteoles about equal to the pedicels. Flowers white; petals lanceolate to ovate, with a narrower inflexed apex, abruptly clawed; calyx teeth 0; stylopodium conical, about equal to the reflexed styles. Fruit ovoid, rounded at apex and base, the mericarps slightly compressed dorsally in transection, glabrous, the ribs low and broad, the lateral a little more prominent than the dorsal, the pericarp firm. Vittae 2 or 3 in intervals, 4 on commissure. Commissural surface plane. Carpophore bipartite, the halves strongly diverging at apex. Oct.

Type locality: "Népal central: versant sud de Lamjung Himal," 3,900 m, 26 Oct. 1978, *A. Bourrit & M. A. Farille 7852* (G, JAY, isotypes).

Distribution: Himalaya of Nepal, 3,900 m.

Representative specimens: Nepal: Lamjung Himal, *Bourrit & Farille 7879* (BM, G); Rambrong, Lamjung Himal, *Stainton, Sykes & Williams 6031* (BM); Pang-Sing, *Lall Dhwoj 50* (BM); Ganesh Himal, Shia Khola, *Gardner 1299* (BM).

37. **KERAYMONIA** Farille
in Farille, Cauwet-Marc & Malla, Candollea 40(2): 528. 1985.

Low, acaulescent to caulescent, the stem short to abortive, simple or branching from the sheath-covered base, glabrous or subglabrous perennial herbs from a strong taproot. Leaves petiolate, membranaceous, pinnately or ternate-pinnately decompound, with linear to oblong, deeply lobed to pinnatifid ultimate divisions; petiole slender, very broadly scarious-sheathing, persistent; cauline leaves, if any, like the basal. Inflorescence of a lax sessile to short-pedunculate terminal compound umbel and usually 1–4 slender pedunculate lateral umbels. Involucre of 2 to several pinnatisect bracts, those of terminal umbels like and often indistinguishable from the leaves, those of lateral umbels greatly reduced. Rays of the terminal umbel few to numerous, elongate, those of lateral umbels few and much shorter, spreading to ascending, those of terminal umbels appearing like peduncles bearing simple umbels. Umbellets rather many flowered, the flowers pedicellate. Involucel of several narrow, entire or trisect or pinnatisect bracteoles. Flowers white or purplish-red, slightly irregular; petals oval to obovate, plane or slightly incurved but without an inflexed apex, rounded at apex, clawed at base, the peripheral flowers slightly radiant; calyx teeth evident, ovate, petaloid, unequal, persistent; stylopodium pulvinate, massive and swollen at anthesis, the styles rather short, flattened, reflexed. Pedicels strongly accrescent, rather slender when mature, unequal, spreading-ascending. Fruit oblong to oblong-ovoid, subterete or slightly compressed laterally, the pericarp smooth and shiny, the mericarps subterete in transection, glabrous, the ribs prominently and subequally winged, the wings much shorter than the body and narrower than intervals. Vittae 2–5 in intervals, 4–8 on commissure. Commissural face broadly concave to sulcate. Carpophore bipartite, with slender halves.

Type species: *Keraymonia nipaulensis* Cauwet-Marc & Farille.

Distribution: A small genus known thus far only from the high Himalaya of Nepal, 4,500–5,000 m.

KEY TO SPECIES

1. Terminal umbel distinctly pedunculate, with an evident involucre; peduncles and rays scaberulous at summit. .... 1. *K. nipaulensis*.
1. Terminal umbel sessile or nearly so and the involucre 0, or indistinguishable from foliage leaves; peduncles and rays glabrous.
    2. Leaf divisions somewhat acute; flowering umbellets lax. .... 2. *K. cortiformis*.
    2. Leaf divisions acute or apiculate; flowering umbellets capitate. 3. *K. triradiata*.

1. **Keraymonia nipaulensis** Cauwet-Marc & Farille,
Candollea 40(2): 528. 1985.

Plants rather slender, spreading to erect, 5–40 cm tall, shortly caulescent, with 1 to several flowering stalks arising from a taproot. Leaves mostly rosulate or clustered toward base, oval to triangular-ovate, 3–15 cm long, 1.5–10 cm broad, pinnately or ternate-pinnately decompound, the ultimate divisions linear to linearly lanceolate or oblong, 5–10 mm long, to 1 mm broad, acute, apiculate, often deeply lobed or cleft; petioles slender, 3–15 cm long, broadly scarious-sheathing. Inflorescence of a lax terminal compound umbel, sessile to shortly pedunculate, the umbellets appear as simple umbels, and 1 to several longly pedunculate lateral umbels; the peduncles scaberulous at summit. Involucre of the terminal umbel of several foliaceous bracts resembling the leaves, the bracts of lateral umbels 2 or 3 and much reduced. Rays of terminal umbels 8–15, unequal, spreading or spreading-ascending, 10–20 cm long, scaberulous at apex, the rays of lateral umbels 4–6, 2–5 cm long in anthesis. Umbellets 10–30-flowered, the mature pedicels angled, unequal, spreading-ascending, 5–10 mm long. Involucel of 3–10 linear to oblanceolate, entire to pinnatisect bracteoles 5–10 mm long. Flowers white or dark red; petals oval to obovate; calyx teeth oblong-oval, to 0.5 mm long; stylopodium hemispherical, dark, the styles to 1 mm long. Fruit oblong to oblong-ovoid, 4–6 mm long, 3–4 mm broad, a little narrowed at apex, rounded at base. July–Sept.

Type locality: "C. Nepal: massif de l'Annapurna, Muktinath range," 4500 m, 18 Sept. 1981, *Farille 81395*.

Distribution: Known only from the high Himalaya of Nepal at 4,500–4,800 m.

Representative specimens: Nepal: Maharigaon, *Polunin, Sykes & Williams 126* (AAH, BM); Rambrong, Lamjung Himalaya, *Farille 81409D* (JAY), *Stainton, Sykes & Williams 6163* (AAH, BM); Annapurna, Lamjung Himal, *Farille 81551* (JAY); Seng Kola, *Stainton, Sykes & Williams 3805* (AAH, BM).

2. **Keraymonia cortiformis** Cauwet-Marc & Malla, Candollea 40(2): 531. 1985.

Plants low, spreading, 5–15 cm tall, acaulescent or subcaulescent, with 1 or more flowering stalks arising from a taproot. Leaves rosulate, oval to triangular-ovate, 1–2 cm long and broad, bipinnate or ternate-pinnate, the ultimate divisions linearly oblong, somewhat acute 1–3 mm long, 1 mm broad; petioles slender, 1–2.5 cm long, broadly scarious-sheathing. Inflorescence of lax terminal compound umbel, sessile or subsessile, umbellets appearing as simple umbels, and (usually ?) 1 or more smaller lateral umbels. Involucre of the terminal umbel of several foliaceous bracts indistinguishable from the leaves. Rays of terminal umbel 5–10, unequal, spreading or spreading-ascending, unequal, 2–4 cm long in anthesis. Umbellets 15–25-flowered, the immature pedicels unequal, 2–5 mm long. Involucel of 6–10 linear to oblanceolate, entire or trifid, narrowly scarious-margined bracteoles 2–8 mm long. Flowers white; petals oval to obovate; calyx teeth ovate, unequal, ca. 0.5 mm long; stylopodium hemispherical, dark, the immature styles to 1 mm long. June–Sept.

Type locality: "C. Népal: haute Kali Gandaki, entre Muktinath et Thorung-la," 19 Sept. 1981, *Farille 81456B*.

Distribution: Known only from the high Himalaya of Nepal, at 4,900–5,000 m.

Representative specimen: Nepal: Soldanggaon, *Stainton, Sykes & Williams 32* (AAH, BM, CAL).

Note. The describers note that this is a "Taxon voisin de *Keraymonia nipaulensis*." The single specimen examined by us is too immature to indicate whether or not it is specifically distinct, or merely a more compact form from a still higher altitude.

3. Keraymonia triradiata Cauwet-Marc & Farille, Candollea 40(2): 531. 1985.

Plants low, acaulescent or subcaulescent, spreading or erect, 10–15 cm tall, the main stem to 6 cm long or abortive, and with 1 or 2 weak lateral flowering stalks arising from near the base, from a taproot. Leaves rosulate or crowded, oval, 3–8 cm long, 2–4 cm broad, bipinnate or ternate-pinnate, the ultimate divisions linear or linearly lanceolate, acute, apiculate, 2–4 mm long, ca. 0.5 mm broad; petioles slender, 2–4 cm long, flaring at base into broad scarious sheath 1-2 cm long and 1 cm broad; cauline leaves like the basal or 0. Inflorescence of lax terminal compound umbel, sessile or subsessile, umbellets appear as compact simple umbels. Involucral bracts indistinguishable from the leaves. Rays of the terminal umbel 2–10, spreading, 4–12 cm long, webbed, the rays of lateral umbels 3–7, 1–2.5 cm long. Umbellets 15–20-flowered, the mature pedicels strongly accrescent, unequal, angled, spreading-ascending, 3–10 mm long, webbed. Involucel of 3–10 linear, entire or apically cleft bracteoles 1–15 mm long. Flowers white to purplish-red, a little irregular; petals obovate, somewhat acute, clawed; calyx teeth ovate; stylopodium hemispherical, about equaling the short styles. Fruit oblong to oblong-ovoid, 4–8 mm long, 3–4 mm broad, a little narrowed at apex, rounded at base, the pericarp smooth and shining, the ribs prominently winged. Vittae 2 or 3 in intervals, 4 on commissure. Seed face deeply concave. Aug.–Oct.

Type locality: "C. Népal: Massif de l'Annapurna, Lamjung Himal, versant nord du col de Namun," 4500 m, 24 Sept. 1981, *Farille 81551*.

Distribution: Known only from the high Himalaya of Nepal, at 4,200–4,500 m.

Representative specimens: Eastern Himalaya: Chumolari, *Rohmoo 529* (AAH, E). Nepal: Between Pass and Dhampus Pass below Marpha peak, Dhaulagiri Himal, American Women's Climbing Expedition, *W. Jones V-50* (COLO, UC).

38. TRACHYDIUM Lindley

In Royle, Ill. Bot. Himal. Mount. 232. 1835; C. B. Clarke in Hook. f., Fl. Brit. Ind. 2: 671. 1879.

Trachydium subgen. *Pentazygon* C. B. Clarke in Hook. f., Fl. Brit. Ind. 2: 692. 1879, *pro parte*.

Low, acaulescent to short caulescent, glabrous to hispidulous perennial herbs, simple or with 1 or 2 smaller branches arising from base, the stem, when present, stout, fistulose, and erect, more commonly lacking, with a fibrous caudex or lacking fibers, from a stout elongated taproot. Basal leaves rosulate, petiolate, membranaceous, 1-3-pinnate, with narrow to broad, lobed to pinnatifid ultimate divisions; petiole scarious-sheathing; cauline leaves like the basal, or lacking. Inflorescence of lax terminal compound umbel, sessile, umbellets appear as simple umbels, or shortly pedunculate, often with 1 or more smaller, sterile lateral umbels. Involucre and involucel with foliaceous, entire to pinnatisect, often scarious-margined bracts and bracteoles. Rays few to numerous, markedly unequal, spreading or spreading-ascending. Umbellets rather many flowered, the flowers pedicellate. Flowers white or pinkish; petals oblanceolate to obovate, with a narrower inflexed apex; calyx teeth evident, obtuse, or 0; stylopodium low conical, the styles recurved or spreading. Fruit ovoid to suborbicular, slightly narrowed at apex, rounded at base, the mericarps subterete in transection, the pericarp loose, spongy, and tuberculate, the ribs triangular, inflated, rugulose, lustrous. Vittae usually 1 or 2 in intervals, 2 on commissure. Commissural face sulcate. Carpophore entire.

Type species: *Trachydium roylei* Lindley.

Distribution: A small genus occurring throughout the Himalaya from Tibet and West Bengal to Pakistan, 3,000–4,600 m.

Note. The accepted circumscription and content of this genus differs drastically from author to author. Norman (1938b) concluded that "most—probably all—of the plants that have been assigned to *Trachydium* since Lindley's day really belong to other genera". He then proceeded to transfer most of them to *Chamaesium* or *Physospermopsis*, so that "*Trachydium* remains a monotypic genus". Kljuykov, Pimenov, and Tikhomirov (1976a, 1976b) further divested *Trachydium* by referring other taxa to *Aulacospermum*. Yet Rechinger f. and Riedl (1963, pp. 130–132) continued to employ *Trachydium* for Afghan material; Nasir (1972, p. 141) referred to *Trachydium* as "a genus of 40 species extending from Iran and Central Asia to China," and Pu (1979, in Shan & Sheh, pp. 194–206) keys, describes, and illustrates six species from China. In *Flora Iranica* (Hedge, Lamond & Rechinger f. 1987, 162: 180–187) Rechinger f. remarked that, "Die Beziehungen zwischen *Trachydium* und *Aulacospermum* sind unklar," emphasizing the isolation of *T. roylei* Lindley, and effectively transferred *Aulacospermum* sect. *Pseudotrachydium* back to *Trachydium*.

The treatment adopted here reflects that of Norman, except that we have tentatively assigned one taxon in addition, whose fruit is insufficiently known to justify its reference elsewhere.

KEY TO SPECIES

1. Stem base conspicuously fibrous; herbage glabrous; bracteoles pinnatisect, exceeding umbellet; petals subequal. .... 1. *T. roylei.*
1. Stem base not fibrous; herbage sparsely hispidulous; bracteoles entire, much shorter than umbellet; marginal petals radiant. .... 2. *T. subnudum.*

1. **Trachydium roylei** Lindley in Royle, Ill. Bot. Himal. Mount. 232. 1835; C. B. Clarke in Hook. f., Fl. Brit. Ind. 2: 672. 1879.

Plants acaulescent or shortly caulescent, 5–25 cm tall. Leaves oblong or oblong-lanceolate, 2–6 cm long, 0.5–2 cm broad, 2–3-pinnate, the ultimate divisions linearly lanceolate, apiculate, 1–3 mm long, up to 1 mm broad; petiole 2–4 cm long. Peduncle stout, fistulose, up to 6 cm long, or usually abortive. Involucre of 5–10 pinnatisect, foliaceous, oblong to oval bracts 1–5 cm long, scarcely distinguishable from foliage leaves. Rays 12–25, spreading or spreading-ascending, 2–20 cm long. Umbellets 20–40-flowered, the mature pedicels unequal, spreading-ascending, 3–8 mm long. Involucel of usually 8–10 oblanceolate to obovate, pinnatisect, conspicuously scarious-margined bracteoles 5–15 mm long, 2–10 mm broad, exceeding flowering umbellets. Stylopodium conspicuously purplish-black, the styles ca. 1 mm long. Fruit ovoid, 3–5 mm long, 2–2.5 mm broad, the pericarp lax, tuberculate, the ribs hollow and inflated, rugulose. July–Sept.

Type locality: Pir Panjal Pass, Kashmir, *Royle*.

Distribution: Western Himalaya, from Uttar Pradesh to Kashmir and Pakistan, 3,000–4,500 m.

Representative specimens: Jammu & Kashmir: Tragbol, *Clarke 29,267* (CAL). Himachal Pradesh: Rupam Pass, *Nair 24,244* (BSD). Uttar Pradesh: Rishiganga Valley, *T. A. Rao 2904* (BSD); Byans, *Duthie 5605* (CAL).

2. **Trachydium subnudum** H. Wolff, Feddes Repert. 27: 125. 1929.

Trachydium subnudum C. B. Clarke in Herb. Kew, June 1905, *nomen nudum*; *Carum clarkei* Lipsky in Herb. Kew, July 1905, *nomen nudum*; *Chamaesciadium subnudum* (H. Wolff) Norman, Jour. Bot. 76: 233. 1938.
Trachydium verrucosum Shan & Pu, Fl. Reipubl. Popul. Sinicae 55(1): 299. 1979. Based on *Chang & Lan 1885*, "Tibet, La-sa, Nai-go-ma-Shan alt. 5000 m," 8 Aug. 1865.
Trachydium lamondianum Farille & Malla in Farille, Cauwet-Marc & Malla, Candollea 40(2): 533. 1985. Based on *Rohmoo Lepcha 52 [252]*, "East Himalaya: Phasi [Phari]" (AAH, isotype).

Plants short caulescent, 5–10(20) cm tall, the foliage sparsely hispidulous, from a non-fibrous caudex surmounting a massive vertical taproot. Leaves triangular, 3–10 cm long, 3–7 cm broad, ternate-pinnate, the ultimate divisions oval to orbicular, 3–8 (–10) mm long, 2–6 mm broad, dentate and pinnately incised; petiole 1–2 cm long, wholly sheathing, the sheath scarious, moderately dilated; cauline leaves, if any, like the basal. Umbels terminal, the peduncles 5 cm long to abortive, scaberulous at summit. Involucre of 1 to several foliaceous bracts like the foliage leaves. Rays 3–10, 1–5 cm long, spreading-ascending, scaberulous at apex. Umbellets 20–30-flowered, the immature pedicels 1–5 mm long, flattened, spreading-ascending. Involucel of 1 to many linear entire bracteoles 1–3 mm long, much shorter than umbellet. Petals oblanceolate to obovate, clawed, the marginal evidently radiant; calyx 0; stylopodium low conical, the styles very short. Immature fruit ovoid, tuberculate. Vittae solitary in intervals, 2 on commissure. Seed face concave. Carpophore bipartite. Aug.–Sept.

Type locality: TIBET: Khambajong, Sept. 1903, *Prain* (K, lectotype; CAL).

Distribution: Eastern Himalaya and Tibet, ca. 5,000 m.

Representative specimen: (Tibet: Khambajong, *Younghusband 310* (CAL)).

Note. The involucre and involucel, and apparently also the immature fruit, suggest that this taxon may not be generically compatible with *T. roylei*.

39. **PHYSOSPERMOPSIS** H. Wolff
Notizbl. Bot. Gart. Berlin-Dahlem 9: 276. 1925.

Trachydium sensu C. B. Clarke in Hook. f., Fl. Brit. Ind. 2: 673. 1879, *pro parte*, non Lindley 1835.
Haploseseli H. Wolff & Hand.-Mazz., Symbol. Sinicae 7: 721. 1933.

Usually low but slender, erect, sparsely branching or simple, glabrous or scaberulous, sometimes glaucous, perennial herbs, from a vertical or oblique caudex often bearing fibrous leaf bases. Leaves petiolate, membranaceous to subcoriaceous, pinnate or ternate-pinnate, with broad, dentate to pinnatifid leaflets, or entire; petiole stout, narrowly but conspicuously sheathing; cauline leaves few, reduced upward, with conspicuous sheaths. Inflorescence of lax terminal and usually some axillary compound umbels, the peduncles usually elongate. Involucre of 1 to several narrow or broader entire or lobed to pinnatifid bracts, or rarely lacking. Rays few, unequal or subequal, spreading-ascending. Umbellets several to many flowered, some staminate, the flowers pedicellate. Involucel of 1 to several narrow, entire or trifid bracteoles. Flowers white, cream-colored, green, or purplish; petals obovate or oblanceolate, with a narrower inflexed apex; calyx teeth evident, linear to lanceolate, or calyx obsolete; stylopodium low conical to depressed, forming a massive disk, the styles slender to very short, longer to shorter than stylopodium, spreading or reflexed. Pedicels slender, spreading-ascending or ascending. Fruit ovoid to globose, shortly constricted at apex, rounded or cordate at base, compressed laterally, constricted at commissure and subdidymous, the pericarp loose and often somewhat inflated, the mericarps slightly compressed dorsally in transection, glabrous, the ribs narrow, smooth, undulate, or roughened, subequal, narrower than intervals, unwinged. Vittae usually 2 or 3 in intervals, 2 or 4 on commissure. Commissural face plane to slightly concave (sulcate ?). Carpophore bipartite.

Type species: *Arracacha delavayi* Franchet [*Physospermopsis delavayi* (Franchet) H. Wolff].

Distribution: An Asian genus of 10 or 12 species, extending from China and Tibet to the Eastern Himalaya.

KEY TO SPECIES

1. Leaves oblong to oval; bracteoles lacking conspicuous white margins.
    2. Flowers green; caulescent and some cauline leaves usually present. ....1. *P. obtusiuscula*.
    2. Flowers reddish-purple or white or pinkish; acaulescent or nearly so and cauline leaves lacking.
        3. Flowers reddish-purple; umbellets 4–10-flowered; leaflets crenate-dentate. .. ..2. *P. bhutanensis*.
        3. Flowers white or pinkish; umbellets 10–20-flowered; leaflets pinnately lobed or dissected.
            4. Leaves 0.5–1.5 cm broad; rays slender; pedicels flattened, winged. ....3. *P. Kingdon-wardii*.

4. Leaves 1.5–2.5 cm broad; rays stout; pedicels terete or angled. .... 4. *P. farillei.*

1. Leaves triangular-ovate; bracteoles narrowly but conspicuously white-margined. ... . 5. *P. rubrinervis.*

1. **Physospermopsis obtusiuscula** (DC.) Norman, Jour. Bot. 76: 231. 1938.

Ligusticum obtusiusculum Wallich, Cat. 543. 1839, *nomen nudum*; *Hymenolaena obtusiuscula* DC., Prodr. 4: 246. 1830; *Trachydium obtusiusculum* (DC.) C. B. Clarke in Hook. f., Fl. Brit. Ind. 2: 673. 1879; *Pleurospermum obtusiusculum* (DC.) Hiroe, Umbell. World, 741. 1979.

Trachydium obtusiusculum var. *stricta* C. B. Clarke in Hook. f., Fl. Brit. Ind. 2: 673. 1879. Based on *Hooker*, "Sikkim, Chola Pass, alt. 10,000–12,000 ft." (*Hooker & Thomson*, Herb. Ind. Orient., *Trachydium 6*) (GH, isotype).

Plants low, erect, sparsely branched above, glabrous, 10–35 cm tall, from a vertical caudex bearing fibrous leaf bases. Leaves oblong, 6–12 cm long, 1.5–5 cm broad, simply pinnate, the leaflets ovate, pinnately incised to pinnatifid; petiole slender, 4–8 cm long, the basal sheath moderately inflated; cauline leaves few, reduced upward, with conspicuously inflated sheaths. Peduncles slender, 10–25 cm. Involucre of 4–6 oblong to ovate, prominently sheathing, incised to pinnatifid bracts 2–6 cm long. Rays (3–)6–15, very slender, unequal, 2–10 cm long. Umbellets 10–15-flowered, the mature pedicels slender, unequal, 3–4 mm long. Involucel of about 5 oblong to ovate; pinnately lobed bracteoles 3–6 mm long. Flowers green; calyx obsolete; stylopodium massive, low conical, the styles very short, recurved. Fruit ovoid, 2.5–4 mm long, 2–2.5 mm broad, slightly narrowed toward apex, rounded at base, the ribs prominent, smooth but slightly undulate. Vittae mostly 2 in intervals, 2 on commissure. July–Oct.

Type locality: "ad Gossain-Than Nepalensium," *Wallich 543*.

Distribution: Western China to Himalaya of Arunachal Pradesh, Sikkim, and Nepal, 3,500–5,000 m.

Representative specimens: Nepal: Gossain Than, *Wallich 542* (K-*pars*), *543* (CAL); Annapurna Himal, Seti Khola, *Stainton, Sykes & Williams 6566* (AAH, BM). Sikkim: Zemu Valley, *Smith & Cave 1227* (CAL); Jongri, *Anderson 610* (CAL); Dzongri, *Mukherjee 319* (CUH, UC); Chola Pass, *Hooker* (GH, K). Bhutan: Chi La, on way to Paro, *Sen Gupta 712* (CAL); Tharizam Chu, Upper Mo Chu Distr., *Sinclair & Long 5312* (E), *5327* (E). Arunachal Pradesh: Tawang-Natula road, Kameng, *Nair 64,848* (ASSAM).

2. **Physospermopsis bhutanensis** Farille & Malla in Farille, Cauwet-Marc & Malla, Candollea 40(2): 516. 1985.

Plants low, acaulescent to shortly caulescent, glabrous, 3–12 cm tall from a fibrous base surmounting a branched taproot. Leaves oblong, 1.5–3 cm long, 0.5–2 cm broad, pinnate with 2–4 pairs of oval to orbicular, crenate-dentate or incised leaflets 3–10 mm in diameter, the lobes obtuse; petiole slender, 0.5–3 cm long, conspicuously dilated into oval, scarious sheath. Peduncle terminal, slender, to 7 cm long, or obsolete and umbellets appear as simple pedunculate umbels. Involucre of 5–8 foliaceous, petiolate, spatulate, pinnatifid bracts to 2 cm long. Rays 5–10, very slender, unequal, 1–8 cm long, spreading or spreading-ascending. Umbellets 4–12(–15)-flowered, the unequal pedicels 1–5 mm long, angled. Involucel of 2–7 bracteoles like the bracts, equal to or exceeding the umbellet. Flowers reddish-purple; petals oval to obovate with a slightly incurved apex, clawed; calyx teeth minute; stylopodium low conical, the styles short. Immature fruit ovoid, 1–1.5 mm in diameter, truncate at apex, rounded at base, the ribs obtuse and rugose, the pericarp apparently lax. Vittae 2 or 3 in intervals, 2–4 on commissure. Commissural face slightly concave. May–Aug.

Type locality: Sikkim: "Chamnago, 12,000–13,000 feet," 19 Jul. 1913, *Cooper 299* (E, holotype).

Distribution: Eastern Himalaya, Bhutan to Nepal, 3,300–4,500 m.

Representative specimens: Nepal. Khunde tree-line, Khumbu Himal, *Byers & Byers 323* (COLO); Lamjung Himal, *Stainton, Sykes & Williams 6339* (BM); Chilime Kharka, *Polunin 1181* (BM); Hongu Khola, *McCosh 351* (BM); Kyangsu Kharka, *Malla 9187* (BM).

3. **Physospermopsis kingdon-wardii** (H. Wolff) Norman, Jour. Bot. 76: 231. 1938.

Trachydium kingdon-wardii H. Wolff, Feddes Repert. 27: 124. 1929; *Pleurospermum kingdon-wardii* (H. Wolff) Hiroe, Umbell. World, 747. 1979.

Plants low, acaulescent or nearly so, glabrous, 3–10 cm tall, from a fibrous base surmounting a simple taproot. Leaves oblong, 1–3 cm long, 0.5–1.5 cm broad, pinnate with 2–5 pairs of orbicular to oblong, pinnately lobed or pinnatifid leaflets 3–8 mm in diameter, the lobes abruptly acute; petiole 0.5–3 cm long, conspicuously dilated into oval scarious sheath; cauline leaves apparently 0. Peduncles terminal, to 2 cm long to obsolete and umbellets appear as simple pedunculate umbels, usually with one or more axillary compound umbels on peduncles 4–6 cm long. Involucre of ca. 5 foliaceous, petiolate, spatulate, pinnate bracts 2–3 cm long. Rays usually 10–15, slender, unequal, 2–7 cm long, spreading or spreading-ascending. Umbellets 10–20-flowered, the immature pedicels flattened, winged, 2–5 mm long. Involucel of 2–5 scarious entire or apically lobed, lanceolate to spatulate bracteoles 2–8 mm long, about equal to umbellet. Flowers white or pink; petals obovate with slightly incurved apex, clawed; calyx teeth obscure, ovate; stylopodium low conical, the styles slightly longer than stylopodium, recurved, flattened. Immature fruit ovoid, 1.5–2 mm in diameter, rounded at apex and base, the ribs obscure, rugose, the pericarp lax. Vittae, seed face, and carpophore not seen. July–Aug.

Type locality: "Tibet: A-tun-tai, 14500 feet," 1913, *F. Kingdon-Ward 992* (E, neotype).

Distribution: Eastern Himalaya in Tibet and Bhutan, 4,400–4,500 m.

Representative specimens: Bhutan: Gafoola, Upper Pho Chu, *Ludlow, Sherriff & Hicks 16,751* (BM, E); Kengla, *Cooper 4254* (E). Eastern Himalaya: Chaerlung, *Rohmoo Lepcha 387* (AAH, BM); Chumolari, *Rohmoo Lepcha 490* (AAH, BM).

4. **Physospermopsis farillei** P. K. Mukherjee & Constance, Edinburgh Jour. Bot. 48(1): 41. 1991.

Plants low, acaulescent or nearly so, glabrous, 5–13 cm tall from a fibrous base surmounting a stout taproot. Leaves oblong or oblong-ovate, 2–4 cm long, 1.5–2.5 cm broad, pinnate with 2–4 pairs of oblong to orbicular, highly pinnately dissected leaflets 3–7 mm in diameter, the ultimate divisions lanceolate to linear, acute or somewhat acute; petiole 1–1.5 cm long, conspicuously dilated into oval scarious sheath; cauline leaves apparently 0. Peduncles terminal, short, to 2 or 3 cm long, or obsolete and umbellets appear as simple pedunculate umbels, occasionally with 1 to several much smaller lateral compound umbels on peduncles 5–6 cm long. Involucre of ca. 5 foliaceous, petiolate, oblong, pinnatifid bracts 2–5 cm long. Rays usually 8–10, stout, very unequal, 2–12 cm long, spreading or spreading-ascending, those of lateral umbels 1–3 cm long. Umbellets 10–20-flowered, the pedicels stout, 3–7 mm long, angled. Involucel of 1–5 oblong to ovate bracteoles, the larger incised or pinnatifid, up to 1 cm long and usually exceeding umbellet. Flowers white; petals oval to obovate with a slightly incurved apex, shortly clawed; calyx teeth minute; stylopodium low conical, the styles short. Fruit oblong-oval, 2.5–3 mm long, 2–2.5 mm broad, a little narrowed toward the truncate apex, rounded at base, the mericarps a little compressed dorsally in transection, glabrous, the ribs triangular (a little rugose when immature), the pericarp lax. Vittae 2 or 3 in intervals, 4 on commissure. Seed face concave. Carpophore bipartite. Aug.–Sept.

Type locality: "C. Népal: Lamjung Himal, au pied sud de col de Namun, vers 4000–4150 m," 26 Sept. 1981, *M. A. Farille 81,604* (E, holotype).

Distribution: Himalaya of Tibet and Nepal, 4,100–4,600 m.

Representative specimens: (Tibet: Tibet side of Yak La, *Cooper 644* (E)). Nepal: Lamjung Himal, *Farille 81,600* (BM), *81,612* (E), *81,613 bis* (E); Khumbu Himal, *Byers & Byers 561* (COLO).

5. **Physospermopsis rubrinervis** (Franchet) Norman, Jour. Bot. 76: 231. 1938.

Trachydium rubrinerve Franchet, Bull. Soc. Philom. Paris VIII. 6: 112. 1894: *Pleurospermum rubrinerve* (Franchet) Hiroe, Umbell. World, 747. 1979.

Physospermopsis muktinathensis Farille & Malla in Farille, Cauwet-Marc & Malla, Candollea 40(2): 512. 1985. Based on *Stainton, Sykes & Williams 7222*, "C. Népal: Maharung (sud of Mustang), 14,500 feet!" (BM, holotype).

Plants low, erect, flexuous, glabrous and glaucous, the herbage reddish-green, 5–25 cm tall, from a vertical caudex bearing fibrous leaf bases surmounting a taproot. Leaves triangular-ovate, 3–8 cm in diameter, pinnate, the 3–5 leaflets coriaceous, ovate-cuneate, coarsely incisely lobed, unequal; petiole 5–8 cm long, slender, elongate, longer than limb, rather broadly sheathing at base; cauline leaves like the basal, reduced upward. Peduncles terminal, and a few smaller lateral. Involucre of 3–6 broad, foliaceous, petiolulate, unequal, deeply incised bracts. Rays 6–14, unequal, 5–10 cm long, spreading-ascending. Umbellets 10–15-flowered, the pedicels 2–10 mm long, filiform, unequal. Involucel of several oblong to linear, scarious white-margined, entire or incised bracteoles, exceeding the umbellets. Flowers yellowish-green; calyx obsolete; stylopodium massive, low conical, the styles rather short, diverging. Fruit ovoid, 3.5–5 mm long, 2.5–3 mm broad, rounded at apex and base, the epicarp loose, colliculate. Vittae 1 or 2 (rarely 3) in intervals, 2 on commissure. Seed face deeply concave. Aug.

Type locality: CHINA. "Yun-nan in pascuis ad collum montis Lapinchan, alt. 3200 m", 31 Aug. 1888, *Delavay 3235*.

Distribution: Western China to Himalaya of Nepal, 3,200–3,800 m.

Representative specimen: Nepal: Kali Gandaki River, Muktinath Range, *Farille 81397* (BM, JAY).

40. VICATIA DC.

Prodr. 4: 243. 1830; C. B. Clarke in Hook. f., Fl. Brit. Ind. 2: 670. 1879.

Sphallerocarpus subgen. *Vicatia* (DC.) Kozo-Polj., Bull. Soc. Nat. Mosc. II. 29: 202. 1916.

Slender, erect, branching, glabrous to scaberulous perennial herbs from taproots. Leaves petiolate, membranaceous, oblong to orbicular, bipinnate to pinnately or ternate-pinnately decompound, the ultimate divisions linear to oblong or ovate; petiole slender, dilated and sheathing at base; cauline leaves like the basal, reduced upward. Inflorescence of lax terminal and often a few lateral compound umbels, the peduncles elongate. Involucre usually lacking. Rays few to subnumerous, unequal, spreading-ascending. Involucel of several inconspicuous linear bracteoles. Umbellets several flowered, the flowers pedicellate. Flowers white or greenish-white to red; petals oval or obovate, with a narrower inflexed apex; calyx obsolete; stylopodium low conical, the styles rather short, recurved or reflexed. Pedicels short and stout. Fruit oblong-ovoid to ovoid, compressed laterally, often slightly constricted at commissure, narrowed toward apex or not, rounded at base, the mericarps pentagonal, subterete to slightly compressed dorsally in transection, glabrous, the ribs filiform to prominent, obtuse or acute but unwinged, narrower than intervals, the lateral ribs ± commissural. Vittae 2 or 3(–5?) in intervals, 2–4 on commissure. Commissural face concave to sulcate. Carpophore bifid to bipartite.

Type species: *Vicatia coniifolia* DC.

Distribution: A small genus of the Himalaya from Bhutan and Tibet to Pakistan and central Asia.

KEY TO SPECIES

1. Leaves orbicular, ternate-pinnately compound; rays and pedicels scaberulous. ....1. *V. thibetica*.
1. Leaves oblong to ovate, 1–2-pinnate to pinnately decompound; rays and pedicels mostly glabrous.
    2. Leaves 1–2-pinnate; flowers greenish-white to pinkish. ....2. *V. wolffiana*.
    2. Leaves pinnately decompound; flowers usually red. ....3. *V. coniifolia*.

1. Vicatia thibetica Boiss., Bull. Soc. Bot. France 53: 423. 1906.

Plants slender, erect or ascending, corymbosely branched above, scaberulous, 30–45 cm tall, the root not seen. Basal leaves orbicular, 15–25 cm in diameter, 1–2-ternate-pinnately compound, the primary divisions petiolulate, the leaflets oblong to ovate, 2–5 cm long, 1–3 cm broad, pinnatifid and pinnately incised, the ultimate divisions oblong-lanceolate or -ovate, 1–5 mm long, 1–3 mm broad, apiculate; petiole 5–10 cm long, narrowly sheathing; cauline leaves bipinnatisect, the sheaths scarcely dilated. Peduncles slender, terminal and lateral, 5–25 cm long, scaberulous at summit. Involucre 0, or of a single linear bract. Rays 8–15, slender, unequal, quadrangular, scaberulous at least below, 2–5.5 cm long.

Umbellets 8-15-flowered, the mature pedicels, like the rays, 2-4 mm long. Involucel of 1-5 inconspicuous linear bracteoles 1-3 mm long, or 0. Flowers white or pinkish; petals obovate to obcordate with a shorter inflexed apex, clawed, slightly unequal; calyx teeth 0; stylopodium low conical, exceeded by the reflexed or spreading styles. Fruit ovoid, ca. 5 mm long, 3 mm broad, rounded at apex and base, compressed laterally and constricted at commissure, the ribs conspicuous. Vittae apparently 2 or 3 in intervals and 4 on commissure. Carpophore bipartite. Commissural face deeply sulcate. Aug.-Sept.

Type locality: "Yargong [Tibet], haies, lieux ombragés et forets, juin-août," *Soulié*.

Distribution: Himalaya of Nepal and Tibet, 3,500-4,200 m.

Representative specimens: Nepal: Kali Gandaki, *Farille 81-375* (E); Tegar, N of Mustang, *Stainton, Sykes & Williams 2198* (BM, E); Mugu, Mugu Khola, *Polunin, Sykes & Williams 3014* (BM, E).

2. **Vicatia wolffiana** (H. Wolff) Norman, Jour. Bot. 73: 229. 1940.

Tongoloa wolffiana H. Wolff in Feddes Repert. 27: 181. 1929.

Pimpinella saxifraga var. *dissectifolia* C. B. Clarke in Hook. f., Fl. Brit. Ind. 2: 685. 1879. Based on *Clarke*, "KASHMIR; Tilail, alt. 13,000 ft., and WESTERN TIBET; KARAKORUM, alt. 14,000 ft."

Plants low but slender, often branched from base, the stems nearly naked above, 5-30 cm tall. Leaves oblong, 5-20 cm long, 1-4 cm broad, 1-2-pinnate, the leaflets oval, 0.5-2 cm in diameter, the ultimate divisions linear to oblanceolate, 1-10 mm long, 0.5-5 mm broad, mostly acute, apiculate, sometimes minutely scaberulous on margins; petiole slender, 2-8 cm long; cauline leaves 1 or 2, with slightly dilated sheaths. Peduncles elongate, rigid, 5-30 cm long, often a little scaberulous at summit. Involucre of a single inconspicuous bract, or lacking. Fertile rays 5-10, rather stout, quadrangular, spreading-ascending, markedly unequal, 1-5 cm long. Umbellets 5-10-flowered, the mature pedicels stout, quadrangular, unequal, 1-4 mm long. Involucel of 3-6 inconspicuous, linear-lanceolate, entire bracteoles 1-4 mm long. Flowers greenish-white to pinkish; petals obovate, with a short inflexed apex; calyx lobes barely evident, ovate; stylopodium low conical, the styles of about equal length, recurved. Fruit oblong-ovoid to ovoid, 2-3 mm long, 1.5-2 mm broad, narrowed toward apex, the mericarps pentagonal, slightly compressed dorsally in transection, glabrous, the ribs low, filiform. Vittae 2 or 3 in intervals, 4 on commissure. Commissural face sulcate. Carpophore shortly bifid.

Type locality: "N.W. Himalaya; Tihri-Garhwal; Ourie Gadh, in Nila Valley," *Duthie 1134* (fide Norman, 1940, p. 230).

Distribution: Himalaya, Uttar Pradesh and Nepal, to Kashmir and Pakistan, 3,000-5,000 m.

Representative specimens: Jammu & Kashmir: Tilail, *Clarke 30,257* (CAL); Astor Valley, Alampi-la, *Duthie 12,176* (CAL, DD). Uttar Pradesh: Tihri Garhwal, *Duthie 1134* (CAL, DD); Garhwal, near Bhuna, *Bhattacharyya 37,259* (BSD).

3. **Vicatia coniifolia** DC., Prodr. 4: 243. 1830; C. B. Clarke in Hook. f., Fl. Brit. Ind. 2: 671. 1879.

Sison ? coniifolium Wallich, Cat. 519. 1828, *nomen nudum*; *Sphallerocarpus coniifolius* (DC.) Kozo-Polj., Bull. Soc. Nat. Mosc. II. 29: 202. 1916.

Chaerophyllum gracillimum Klotzsch in Klotzsch & Garcke, Bot. Ergebn. Reise Prinz Waldemar 149. 1862. Based on *Hoffmeister*, "in Himalaya."

Chaerophyllum millefolium Klotzsch in Klotzsch & Garcke, Bot. Ergebn. Reise Prinz Waldemar 149. 1862; *Vicatia millefolia* (Klotzsch) C. B. Clarke in Hook. f., Fl. Brit. Ind. 2: 671. 1879; *Sphallerocarpus millefolius* (Klotzsch) Kozo-Polj., Bull. Soc. Nat. Mosc. II. 29: 202. 1916. Based on *Hoffmeister*, "Himalaya."

Vicatia stewartii C. B. Clarke in Hook. f., Fl. Brit. Ind. 2: 671. 1879. Based on *J. L. Stewart 888*, "GURWHAL, alt. 6500 ft."

Plants slender, erect or ascending, often branching from base, the stems sparsely leafy above, 10–40 cm tall. Leaves oblong to ovate, 2–10 cm long, 1–5 cm broad, pinnately decompound, the ultimate divisions linear-lanceolate, 1–5 mm long, mostly less than 1 mm broad, acute, apiculate, venose beneath; petiole slender, 2–10 cm long, dilated and broadly scarious-sheathing at base; cauline leaves 2 or 3, with prominently dilated oblong to oval sheaths. Peduncles elongate, rigid, 3–15 cm long, mostly glabrous at summit. Involucre of an occasional linear bract, or lacking. Fertile rays mostly 3–8, slender, quadrangular, spreading-ascending, unequal, mostly glabrous, 1–4.5 cm long. Umbellets 3–12-flowered, the fertile pedicels stout, quadrangular, spreading-ascending, 1–5 mm long. Involucel of 2–6 linear, entire, acute bracteoles 2–5 mm long. Flowers usually red; petals narrowly obovate, with a narrower inflexed apex; calyx obsolete; stylopodium low conical, the styles recurved, shorter than stylopodium. Fruit oblong-ovoid to ovoid, 3–5 mm long, 1.5–2.5 mm broad, narrowed toward apex, the mericarps pentagonal, subterete in transection, glabrous, the ribs filiform, acute. Vittae 2 or 3 in intervals, 2–4 on commissure. Commissural face sulcate. Carpophore shortly bifid. May–Oct.

Type locality: "Nepalea, ad Kamaon," *Wallich 591*.

Distribution: Himalaya from Arunachal Pradesh to Kashmir, 2,500–4,500 m; China, Pakistan, and Afghanistan.

Representative specimens: Jammu & Kashmir: Tjwas nullah near Sonamarg, *Duthie 13,637* (DD). Himachal Pradesh: Zozi Pass, Ladak, *J. L. Stewart* (DD). Uttar Pradesh: Deoban, *Gamble 24,400* (DD); Naini Tal, *Duthie 4046* (DD). Nepal: Padmara, NE of Jumla, *Polunin, Sykes & Williams 4052* (AAH, BM); w/o further locality, *Wallich 591* (CAL). Bhutan: Paro, *Mukherjee 5878* (CAL). Arunachal Pradesh: Shyamrong, Kameng Distr., *Hajra 64,881* (ASSAM).

41. MEEBOLDIA H. Wolff
Feddes Repert. 19: 313. 1924.

Sinodielsia H. Wolff, Notizbl. Bot. Gart. Berlin 9: 278. 1925; Liou, Fl. Reipub. Popul. Sinica 55(1): 108. 1979; Kljuykov, Feddes Repert. 97: 753–757. 1986.

Slender, caulescent, erect, branching or simple, ± scaberulous perennial herbs from thickened, digitately branched roots. Leaves petiolate, membranaceous, ovate, pinnately or ternately decompound, with petiolulate, pinnatifid leaflets, the ultimate divisions very narrow; petiole slender, dilated and sheathing at base; cauline leaves few, similar to the basal, reduced upward, with narrow sheaths. Inflorescence of lax terminal and usually some lateral compound umbels, the peduncles elongate, scaberulous at apex. Involucre of 1 to several small, narrow bracts, or lacking. Rays few, quadrangular, subequal, spreading-ascending to divaricate. Umbellets rather few flowered, the flowers pedicellate. Involucel of several linear or linear-lanceolate bracteoles shorter than the pedicels. Flowers white (yellowish ?); petals obovate, subentire, with a narrower inflexed apex, shortly clawed at base; calyx teeth evident, subulate, persistent; stylopodium conical, the styles slender, recurved. Pedicels slender, unequal, spreading-ascending. Fruit oblong-ovoid to ovoid, narrowed toward apex, rounded at base, lightly compressed laterally, ± constricted at commissure, the mericarps subterete to slightly compressed dorsally in transection, glabrous, the ribs filiform, acute, much narrower than intervals. Vittae 2–4 in intervals, 2–5 on commissure. Commissural face concave to sulcate. Carpophore divided to middle or below.

Type species: *Ptychotis achilleifolia* DC. [*Meeboldia achilleifolia* (DC.) P. K. Mukherjee & Constance].

Distribution: A genus of two or more species of the Eastern Himalaya, from Yunnan and Tibet to Bhutan, Nepal, and Uttar Pradesh, 2,000–4,500 m.

1. Meeboldia achilleifolia (DC.) P. K. Mukherjee & Constance, Edinburgh Jour. Bot. 48(1): 44. 1991.

Ptychotis achilleifolia DC., Prodr. 4: 109. 1830; *Pimpinella achilleifolia* (DC.) C. B. Clarke in Hook. f., Fl. Brit. Ind. 2: 684. 1879; *Vicatia achilleifolia* (DC.) P. K. Mukherjee, Bull. Bot. Surv. India 24(1–4): 43. 1983.

Meeboldia selinoides H. Wolff, Feddes Repert. 19: 313. 1924. Based on *Meebold 3402*. "Himalaya: in valle Naini, 6500′ s. m.," Oct.

Sinodielsia digitata Kljuykov, Feddes Repert. 97: 756, fig. 2 (mislabeled). 1986. Based on *Ludlow, Sherriff & Taylor 7263*, "Flora of Eastern Bhutan, Chungar, N 27°03′, E 91°27′, 6500 ft.", 13 Nov. 1938 (BM).

Plants slender, erect, the stems few leaved above, 20–120 cm tall. Leaves ovate, 4–8 cm in diameter, pinnately decompound, the ultimate divisions linear-lanceolate, acute, apiculate, 1–3 mm long, 1 mm broad, conspicuously venose beneath; petiole slender, 3–12 cm long; cauline leaves moderately sheathing. Peduncles rigid, 5–12 cm long, scaberulous at summit. Involucre of 1 to several inconspicuous linear bracts or lacking. Fertile rays 6–15, slender, spreading-ascending, scaberulous, 1.5–5 cm long. Umbellets 10–20-flowered, the mature pedicels slender, quadrangular, unequal, 2–12 mm long. Involucel of 5–8 linear or linear-lanceolate bracteoles 2–5 mm long. Flowers greenish-white to pinkish; petals slightly unequal; calyx prominent, the teeth 0.5–0.8 mm long; stylopodium sharply conical, the styles longer than stylopodium. Fruit oblong-ovoid, 2–5(?) mm long, 1–1.5 mm broad, strongly narrowed at apex, the mericarps pentagonal, glabrous, the ribs prominent, acute. Commissural face sulcate. Carpophore apparently bipartite. Aug.–Oct.

Type locality: "Nepalia ad Kamaon," *Wallich 568.*

Distribution: Himalaya, from Bhutan and Nepal to Uttar Pradesh.

Representative specimens: Uttar Pradesh, Kumaon, *Wallich 568B* (CAL); Naini Tal, *Gillett, Pangtey & Gupta 2614* (CUH, UC). Nepal: Nepalea, *Wallich 568A* (CAL); Dhunche, *Mukherjee 237* (CUH, UC); Gurjackani, *Stainton, Sykes & Williams 3618* (AAH, BM), *4462* (AAH, BM). Bhutan: Chungar, *Ludlow, Sherriff & Taylor 7263* (BM).

42. **CARUM** Linn.

Sp. Pl. 263. 1753; C. B. Clarke in Hook. f., Fl. Brit. Ind. 2: 680. 1879.

Slender and erect, caulescent, branching, glabrous perennial or biennial herbs from a taproot. Leaves petiolate, membranaceous, pinnately divided to decompound, the ultimate divisions narrow to filiform; petiole slender, sheathing at base; cauline leaves reduced upward, often sessile, rarely lacking. Inflorescence of lax terminal and lateral compound umbels, the peduncles elongate. Involucre and involucel of a few narrow bracts or bracteoles, or lacking. Rays rather few, spreading-ascending. Umbellets rather few flowered, the flowers pedicellate. Flowers white to reddish; petals obovate with a narrowed inflexed apex; calyx teeth minute, or calyx obsolete; stylopodium conical, the styles slender. Fruit oblong to oblong-ovoid, rounded at both ends, compressed laterally but scarcely narrowed at commissure, the mericarps pentagonal or subterete in transection, glabrous, the ribs prominent, filiform, obtuse, subequal, narrower than intervals, unwinged. Vittae usually solitary (1–3?) in intervals, 2 (4?) on commissure. Commissural face plane. Carpophore bipartite or bilobed.

Type species: *Carum carvi* Linn.

Distribution: A genus of very uncertain limits, comprising some two dozen species of Eurasia, especially the Mediterranean area.

KEY TO SPECIES

1. Basal leaves 8–15 cm long; upper cauline leaves pinnately dissected. ... 1. *C. carvi*.
1. Basal leaves 4–7 cm long; upper cauline leaves 3-parted. ....2. *C. diversifolium*.

1. **Carum carvi** Linn. Sp. Pl. 263. 1753; C. B. Clarke in Hook. f., Fl. Brit. Ind. 2: 680. 1879.

Plants slender, biennial, 30–100 cm tall. Leaves oblong to oval, 8–15 cm long, 3–8 cm broad, pinnately decompound, the ultimate divisions linear, entire or divided; petiole slender, 5–13 cm long, scarious-sheathing; cauline leaves pinnately dissected, reduced upward, the upper sessile with broadly dilated, scarious-margined sheaths. Peduncles slender, 4–12 cm long. Involucre and involucel of a few filiform bracts or bracteoles, or commonly lacking. Rays 3–13, spreading-ascending, unequal, 1–5 cm long. Umbellets 2–15-flowered, the mature pedicels unequal, 2–13 mm long. Flowers white or pinkish; stylopodium conical. Fruit oblong, 3–5 mm long, 2–2.5 mm broad, the ribs prominent, filiform. Apr.–Sept.

Type locality: "In Europae borealis," no collector indicated.

Distribution: Eurasia; widely cultivated and adventive.

Representative specimens: Jammu & Kashmir: Dras Valley, *Duthie 13,700* (CAL). Himachal Pradesh: Lahul, Darcha, *Bor 13,168* (DD); Kardang, Lahul, *Mukherjee 333* (CUH, UC). Punjab: Muree, *Hunter 10,218* (DD). Uttar Pradesh: Ganges Valley, *Duthie 1140* (CAL, DD); Nipchang Valley, Kumaon, *Duthie 2948* (CAL).

2. Carum diversifolium (DC.) C. B. Clarke in Hook. f., Fl. Brit. Ind. 2: 681. 1879.

Sison diversifolium Wallich, Cat. 592. 1829, *nomen nudum*; *Falcaria diversifolia* DC., Prodr. 4: 110. 1830.

Plants slender, much branched, sparingly leafy above, ca. 90 cm tall. Leaves 4–7 cm long, the ultimate divisions oblong to linear; upper cauline leaves 3-parted. Involucre and involucel of several linear bracts or bracteoles, or lacking. Rays 8–14, unequal, 2–5 cm long. Umbellets many flowered. Flowers white; calyx obsolete; stylopodium conical, the styles slender. Immature fruit oblong, ca. 2 mm long, 1 mm broad, the ribs slender.

Type locality: "Nepalia," *Wallich 592*.

Distribution: Known only from the type collection.

Representative specimen: Nepal, *Wallich 592* (CAL).

Note. In the Calcutta specimen examined, the basal leaves are badly preserved, but the upper leaves match the published description. The proper generic placement of this material remains doubtful to me, as it was to de Candolle, C. B. Clarke, and H. Wolff. (P. K. M.).

Doubtful Species:

Carum villosum Haines, Jour. Asiat. Soc. Bengal II. 15: 309, pl. 9. 1920, *nomen nudum*; Bot. Bihar & Orissa 408. 1922.

Haines states: "A third Umbellifer, tentatively called by me *Carum villosum*, may be a variety of Wallich's *Carum anethifolium* and is therefore not described pending further comparison with material at Kew. This plant was obtained from the Sameshwar Hills in North Champaran. . ." (1920, p. 309). The published photograph is difficult to interpret, but the other three taxa treated under "*Carum*" in the later work are all referable to *Trachyspermum.* I was able to examine the young fruits and determined them to be hairy, thus confirming this generic attribution. Two collections were examined by me: Bihar: Ramnagar Hills, *Haines 4744* (K); sandstone hills, Ramnagar, *Haines 4744* (K). (P. K. M.). The formal generic transfer is refrained from till examination of specimens with mature fruits from the type locality.

43. **ACRONEMA** Edgew.

Proc. Linn. Soc. 1: 252. 1846; Liou & Shan, Acta Phytotax. Sinica 18(2): 194–204. 1980; Liou, Fl. Reipub. Popul. Sinica 55(2): 113–134. 1985.

Carum sect. *Acronema* (Edgw.) Franchet, Bull. Soc. Philom. Paris VIII. 6: 122. 1894; *Pimpinella* sect. *Acronema* (Edgw.) Hand.-Mazz., Akad. Wiss. Wien, Math.-Naturwiss. Kl., Anz. 62: 26. 1925.
Pimpinella sensu C. B. Clarke in Hook. f., Fl. Brit. Ind. 2: 684–690. 1879, *pro parte*, non Linn. 1753.

Low or slender, erect, caulescent, usually branching, essentially glabrous to slightly scaberulous perennial herbs from a globose or fusiform tuberous root. Leaves petiolate, membranaceous, simply ternate or pinnate, to pinnately or ternate-pinnately decompound, the ultimate divisions linear to orbicular; petiole slender, sheathing; cauline leaves reduced upward, the uppermost often sessile, with elongate, entire divisions, or reduced to sheaths. Inflorescence of lax terminal and usually also lateral, compound umbels, the peduncles slender. Involucre lacking, or of an occasional narrow, setaceous bract. Rays usually few, rarely more numerous, slender, subequal to unequal, spreading or spreading-ascending. Involucel of few short, inconspicuous narrow bracteoles, or lacking. Umbellets few or more flowered, the flowers pedicellate. Flowers white, pinkish, or purplish-red, all regular; petals linear to ovate, plane and without an inflexed apex, acuminate to longly caudate, subequal; calyx teeth obsolete, rarely evident; stylopodium conical to depressed and forming a thick flattened disk, or obsolete, the styles short to slender, erect to recurved or reflexed. Pedicels short to slender, usually unequal. Fruit ovoid to oblong-elliptic, compressed laterally, often narrowed at apex, rounded at base, constricted at commissure and sometimes didymous, the mericarps subterete, pentagonal, or a little compressed dorsally in transection, often a little divergent at apex, glabrous, the ribs filiform to obscure or obsolete. Vittae usually 2 or 3 (rarely 1) in intervals, 2 on commissure. Commissural face plane. Carpophore entire, bifid, or bipartite.

Type species: *Helosciadium ? tenerum* DC. [*Acronema tenerum* (DC.) Edgew.].

Distribution: A genus of two dozen or so species chiefly of the Eastern Himalaya of China and India, one extending to the Western Himalaya.

Note. The genera *Acronema* Edgew. and *Sinocarum* H. Wolff ex Shan & Pu comprise taxa transferred from the Linnean genera *Carum* and *Pimpinella* by Wolff and Handel-Mazetti and augmented by a number of additional taxa described by Wolff, Mukherjee, Farille & Malla, Liou & Shan, and others. Although Wolff transferred 5 species from *Carum* to *Sinocarum* in *Das Pflanzenreich* (1927) and described four more in the latter genus in *Feddes Repertorium* (1929), he never described the genus. Actual description and validation of the genus *Sinocarum* was finally accomplished by Shan & Pu in 1980. Hence, none of the transfers or new species attributed to Wolff prior to that date are valid.

Farille *et al.* (1985) point out that Asiatic *Carum* in the sense of Franchet is to be divided “entièrment sur la morphologie et biologie florales le conceptual générique. C’est en effect à ces niveaux que les processes de différencion de ‘surrévolution’ se sont accomplis, le fruit restant par contre typiquement caröide (non évolué)” (p. 551). Thus,

Acronema is to be distinguished by its subequal petals terminating in a long caudate tip from *Sinocarum*, with ± blunt or rounded petals, of which the outer is slightly radiant, and *Dactylaea*, which has lobed or fimbriate petals (not reported from our area). It is not clear to us how sound a taxonomic judgment this approach represents. The extreme petal conditions cited are certainly striking, but there appear to be intermediate conditions between *Acronema* and *Sinocarum*, as thus defined, and perhaps between the latter and *Pimpinella*. Reassessment of the generic lines may be called for when better material is available.

Despite the reference to floral biology above, we know almost nothing about the biology of these plants. The presence of tuberous roots is supposed to distinguish *Acronema* and *Sinocarum* from both *Carum* and *Pimpinella*, but the underground parts of several of the species are not known, as are some key stages in their life cycles.

Seeds of *Acronema bellum* (*Mukherjee 308*), *Sinocarum sikkimense* (*Mukherjee 309, 317*), and *S. wolffianum* (*Bartholomew 195*, *Mukherjee 320*) germinated to produce seedlings with a single cotyledon, suggesting that monocotyly may be a significant character.

KEY TO SPECIES

1. Leaves simply pinnate or ternate.
    2. Leaflets short, broad, trilobed.
        3. Stylopodium conical; carpophore entire. .... 1. *A. muscicolum.*
        3. Stylopodium depressed, pulvinate; carpophore bipartite. .... 2. *A. pseudotenerum.*
    2. Leaflets elongate, linearly-subulate, entire. .... 3. *A. johrianum.*
1. Leaves bipinnate, ternate-pinnate, or pinnately or ternately decompound.
    4. Fruit oblong or lanceolate-ovoid; calyx teeth evident.
        5. Flowers white; leaves bipinnate; umbellets 8–16-flowered .. 4. *A. bellum.*
        5. Flowers dark red; leaves ternate-pinnate; umbellets 3–8-flowered. .... 5. *A. dyssimetriradiatum.*
    4. Fruit broadly ovoid to globose; calyx teeth 0.
        6. Rays and pedicels capillary, flexuous.
            7. Leaf divisions scaberulous on margins and veins. .... 6. *A. nervosum.*
            7. Leaf divisions glabrous on margins and veins.
                8. Inflorescence with well-developed lateral umbels.
                    9. Rays and pedicels divaricate, subequal .... 7. *A. hookeri.*
                    9. Rays and pedicels spreading-ascending, unequal. .... 8. *A. paniculatum.*
                8. Inflorescence almost exclusively terminal, lateral umbels very few or 0.
                    10. Flowers red; stylopodium conical .... 9. *A. sichuanense.*
                    10. Flowers white; stylopodium depressed. .... 10. *A. handelii.*

6. Rays and pedicels slender but rigid; carpophore entire or shortly bifid. .... 11. *A. tenerum.*

1. **Acronema muscicolum** (Hand.-Mazz.) Hand.-Mazz.,
Symbol. Sinicae 7: 715. 1933.

Pimpinella muscicola Hand.-Mazz., Wien, Math.-Naturwiss. Kl., Anz. 63: 26. 1925.

Plants slender, ascending or erect, 3–15 cm tall, from an ovoid or globose tuber. Basal leaves 2–5, soon withering, orbicular, 0.5–3 cm in diameter, deeply trilobed or ternate, the lobes or leaflets cuneate-orbicular, sessile or short petiolulate, shallowly trilobed and crenately incised; petiole slender, slightly dilated at base, 1–6 cm long, much longer than lamina; cauline leaves with reduced blades and dilated, oblong sheaths. Peduncles slender, 1–4 cm long. Involucre lacking. Rays 2–7, capillary, ascending, unequal, 3–30 mm long. Umbellets 4–7-flowered, the mature pedicels capillary, 2–4 mm long. Involucel of 1–3 linear bracteoles 1–2 mm long, or lacking. Petals white, ovate, with a filiform projection up to 2 mm long; stylopodium conical. Fruit (mature fruit not seen) apparently subglobose, didymous, ca. 1.2 mm long, 1 mm broad. Vittae minute. Commissural face plane. Carpophore entire. May.

Type locality: "NW-Y[unnan]: Auf moosbedeckten Diabasfelsen der ktp. St. in Tannenwa"ldern des Nguka-la sw von Dschungdien ("Chungtiend"), 3750–4800 m," China, 25 Aug. 1915, *Handel-Mazzetti 7807*.

Distribution: Western China and Indian Himalaya from Arunachal Pradesh to Kumaon, 3,500–5,000 m.

Representative specimens: Uttar Pradesh: Above Kamiri, *Duthie 4058* (DD). Nepal: Rambrong Lamjung Himal, *Stainton, Sykes & Williams 6121* (AAH, BM). Sikkim: Zemu Valley, *Smith & Cave 1667* (CAL); Samdong, *Hooker* (K); Lachen, *Hooker* (K); w/o further locality, *Hooker* (CAL). Arunachal Pradesh: Tawang-Mela road, Kameng Distr., *Hajra 64,837* (ASSAM).

2. **Acronema pseudotenerum** P. K. Mukherjee, Actes II sympos.
Internat. Ombell. Perpignan, 52. 1978.

Plants slender, erect, few branched, up to 10 cm tall from a tuberous root. Basal leaves few, soon withering, the lower cauline leaves simply pinnate, 1–1.5 cm long, the 5 leaflets longly petiolulate, trilobed, paler beneath; petiole slender, 1–1.5 cm long; cauline leaves strongly reduced upward, the uppermost with dilated oblong sheaths. Peduncles stout, quadrangular, ca. 4.5 cm long. Involucre lacking or caducous. Rays 4–6, quadrangular, up to 1.5 cm long. Umbellets ca. 6-flowered, the mature pedicels quadrangular, subequal, ca. 3.5 mm long. Involucel of 2 or 3 linear bracteoles 1–2 mm long. Flowers not seen; stylopodium depressed, forming a flattened disk, the styles ascending or reflexed. Fruit ovoid, ca. 1 mm in diameter, the ribs slender. Vittae solitary in intervals. Commissural face plane. Carpophore bipartite. Aug.

Type locality: Sikkim: Momay Sandong (Lachung), 16,500 ft., 19 Aug. 1892, *Gammie 916* (CAL, holotype).

Distribution: Known solely from the type collection.

3. **Acronema johrianum** Babu, Brittonia 25: 159. 1973.

Plants slender, erect, simple, 15–20 cm tall, from a globose tuber. Leaves few, ovate, 10–15 cm in diameter, pinnate, the leaflets linearly subulate; petiole slender, 5–10 cm long; cauline leaves reduced upward. Peduncles slender, 2.5–5 cm long. Involucre and involucel lacking. Rays 6 or 7, slender, striate, 1.5–2.5 cm long. Umbellets 6–10-flowered, the mature pedicels unequal, 3–5 mm long. Petals white, elliptic, caudate, up to 2.5 mm long; calyx teeth small, triangular, persistent; stylopodium low conical, the styles short, spreading. Fruit elliptic, ca. 2 mm long, 1.5 mm broad, the ribs acute. Vittae 2 or 3 in intervals, 2 on commissure. Commissural face plane. Carpophore bipartite. Sept.–Nov.

Type locality: Nepal: Ghopte-Gossainthan, 4290 m, 9 Sept. 1965, *Shrestha & Sakya 3740* (CAL, holotype).

Distribution: Himalaya of Nepal, ca. 4,000 m.

Representative specimens: Nepal: Kyangchin, Langtang, *Mukherjee 274* (CUH); Kyangchin, *Mukherjee 280* (CUH, UC).

4. **Acronema bellum** (C. B. Clarke) P. K. Mukherjee,
Bull. Bot. Surv. India 24(1–4): 42. 1983.

Pimpinella bella C. B. Clarke in Hook. f., Fl. Brit. Ind. 2: 686. 1879.

Plants slender, erect, branched above, 30–100 cm tall, from an elongated, cylindrical, branched tuberous root. Leaves ovate-elliptic, 15–30 cm in diameter, bipinnate, the leaflets pinnatifid with oblong to oval ultimate divisions 5–15 mm long, 2–10 mm broad, whitish beneath; petiole slender, 8–20 cm long, narrowly sheathing at base; cauline leaves reduced upward, all petiolate. Umbels terminal and axillary, the peduncles slender but rigid, 4–12 cm long. Involucre usually lacking. Rays 8–16, slender, angular, spreading-ascending, unequal, 1.5–5 cm long. Umbellets 8–16-flowered, the mature pedicels capillary, subequal, spreading, 5–10 mm long. Involucel of 2–5 linear bracteoles 3–5 mm long. Flowers white; petals lanceolate, acuminate, 1.5–3 mm long; stylopodium depressed, forming a flattened disk, the styles spreading. Fruit narrowly elliptic-oblong, 2–3 mm long, 1–1.5 mm broad, narrower toward apex and constricted just below stylopodium, the mericarps terete, subpentagonal in transection, glabrous, the ribs slender, obscure. Vittae solitary in dorsal intervals, 2 or 3 in lateral intervals, 2 on commissure. Commissural face usually plane. Carpophore bipartite. July–Nov.

Type locality: "SIKKIM; alt. 10,000–12,000 ft., "*Thomson, Clarke, Gamble*; "on the watershed between Sikkim and Nepal."

Distribution: Eastern Himalaya, in West Bengal and Sikkim, 3,500–5,000 m.

Representative specimens: West Bengal: Darjeeling, Syngaleleh, *Clarke 12,584A* (CAL, E, K); Sandakphu, *Gamble 78A, B, C* (K), *s. n.* (CAL); Phalut, *Thomson* (CAL, K), *Clarke 25,719* (K), *Mukherjee 308* (CUH); Tharay, *Ribu 308* (CAL); Sandakphu-Phalut, *Mukherjee 293* (CUH, UC); Lachen, *Hooker* (K); Tongu, Lama Kangra, *Hooker* (K). Sikkim: Jaktoph, *King's collector* (CAL, K); Zemu Valley, *Smith & Cave 2805* (CAL); Changu, *W. Smith 3361* (CAL).

5. **Acronema dyssimetriradiatum** Farille & Malla in Farille, Cauwet-Marc & Malla, Candollea 40(2): 557. 1985.

Plants low, erect, glabrous, simple, 6–25 cm tall, from a swollen taproot. Leaves mostly basal, triangular-ovate, 3–6 cm in diameter, ternate-pinnate, the leaflets 9–15, oval to orbicular, obtuse at apex, truncate to cuneate at base, the lowest pairs shortly petiolulate or all subsessile, 2–10 mm in diameter, pinnately incised to pinnatifid, the lobes mucronulate; petiole slender, 2–7 cm long, with a broadly dilated scarious sheath; cauline leaves 1 or 2, or 0, like the basal, the sheath broadly dilated. Peduncles slender, terminal, 5–18 cm long, glabrous. Involucre usually 0. Rays 2–4, slender, 1–6 cm long, angled, very unequal, ascending. Umbellets 3–8-flowered, the pedicels very unequal, 3–12 mm long, spreading-ascending. Involucel of 2–5 linear, purplish bracteoles, 2–5 mm long. Flowers dark red; petals ovate, apparently plane, shortly acuminate, subequal, not clawed; calyx lobes evident, ovate; stylopodium low conical, about equal to the reflexed styles. Fruit lanceolate-ovoid, 3–3.5 mm long, 2 mm broad, compressed laterally and constricted at commissure, narrowed toward apex, rounded at base, the mericarps slightly compressed dorsally in transection, the ribs filiform, inconspicuous. Vittae 3–4 in intervals, usually 4 on commissure. Seed face concave. Carpophore bipartite. July–Sept.

Type locality: "C. Népal: Annapurna, Lamjung himal, versant nord du col de Namun, 4000–4200 m," 24 Sept. 1981, *M. A. Farille 81566* (BM, E, isotypes).

Distribution: Himalaya of Nepal, 4,000–4,300 m.

Representative specimen: Nepal: Rambrong, Lamjung Himal, *Stainton, Sykes & Williams 6155* (AAH, BM).

6. **Acronema nervosum** H. Wolff, Feddes Repert. 27: 315. 1930.

Plants very slender, erect, simple, 10–30 cm tall, from a tuberous root. Leaves ovate, 1.5–2 cm in diameter, ternate-pinnately divided, the leaflets oblong to ovate, trilobed to pinnatifid, 0.3–1 cm long, the ultimate divisions ovate, apiculate, scaberulous on margins and veins; petiole slender, 1–8 cm long; cauline leaves 1 or 2, sessile, usually with linear divisions. Umbels terminal and lateral, the peduncles filiform, 1–5 cm long. Involucre and involucel lacking. Rays 1–5, filiform, spreading-ascending, unequal, 5–15 mm long. Umbellets 3–7-flowered, the immature pedicels capillary, 3–5 mm long. Flowers purplish-red or cream-colored; petals ovate to orbicular, longly caudate; calyx 0; stylopodium low conical, the styles short, erect. Immature fruit elliptic, 1–1.25 mm long, 1 mm broad, the ribs obscure. Vittae, commissure, and carpophore not seen. July.

Type locality: "Sikkim, Changu, 2000′," July 1910, *W. W. Smith 3361*.

Distribution: Central Himalaya, Sikkim to Nepal, 2,000–4,000 m.

Representative specimen: Nepal: Rambrong, Lamjung Himal, *Stainton, Sykes & Williams 6188* (AAH, BM).

7. **Acronema hookeri** (C. B. Clarke) H. Wolff in Engl., Pflanzenreich 90 (4^{228}): 323. 1927.

Pimpinella hookeri C. B. Clarke in Hook. f., Fl. Brit. Ind. 2: 686. 1879; *Carum hookeri* (C. B. Clarke) Franchet, Bull. Soc. Philom. Paris VIII. 6: 122. 1894.

Plants very slender, erect, branched above, 20–100 cm tall, from a globose tuber. Basal and lower cauline leaves triangular-ovate, 5–10 cm long, 3–5 cm broad, ternate-pinnately compound, the leaflets various; petiole slender, up to 15 cm long, narrowly sheathing at base; upper cauline leaves often ternate with elongated linear divisions. Peduncles very slender, capillary, 1–9 cm long. Involucre and involucel usually lacking. Rays 3–6, capillary, 0.3–7 cm long, unequal. Umbellets 2–6-flowered, the mature pedicels capillary, unequal, spreading, 2–20 mm long. Flowers white or pinkish; petals ovate, with a filiform projection up to 3 mm long; stylopodium obsolete, the styles short, erect or spreading. Fruit ovoid, narrowed toward apex, ca. 1.5 mm in diameter, the ribs obscure. Vittae 2 or 3 in intervals. Commissural face plane. Carpophore bipartite.

KEY TO VARIETIES

1. Basal leaf segments oval to ovate-lanceolate, 5–35 mm long, serrate to deeply pinnately incised. .... a. var. *hookeri*.

1. Basal leaf segments narrowly linear, elongated, 50–80 mm long, only 1.5–2 mm broad. .... b. var. *graminifolium*.

7a. var. **hookeri**

July–Oct.

Type locality: "Sikkim, Northern Valleys, Samdong and Lachen," 8–11,000 ft., *Hooker & Thomson*, Herb. Ind. Orient., *Acronema 2*).

Distribution: Eastern Himalaya; Tibet.

Representative specimens: West Bengal: Phalut, *Thomson* (CAL). Sikkim: Lachen, *Hooker* (K); Samdong, *Hooker* (K); Yakcha, above Lachoong, *Smith & Cave 2597* (CAL).

7b. var. **graminifolium** (W. Smith) H. Wolff in Engl., Pflanzenreich 90 ($4\frac{228}{}$): 323. 1927.

Pimpinella hookeri var. *graminifolia* W. Smith, Rec. Bot. Surv. India 4: 200. 1911; *Acronema graminifolium* (W. Smith) Liou & Shan, Acta Phytotax. Sinica 18(2): 197. 1980.

July–Oct.

Type locality: Sikkim, Pan-Ka-bee-see-Moo, 24 July 1884, *Dungboo 247*.

Distribution: Eastern Himalaya; Tibet.

Representative specimens: Sikkim: Yakcha, above Lachoong, *Smith & Cave 2714* (CAL); Zemu Valley, *Smith & Cave* (CAL); Sikkim, *Dungboo 247* (K).

8. **Acronema paniculatum** (Franchet) H. Wolff in Engl., Pflanzenreich 90 ($4\frac{228}{}$): 323. 1927.

Carum paniculatum Franchet, Bull. Soc. Philom. Paris VIII. 6: 122. 1894.

Stature, foliage, etc. intermediate between that of *A. hookeri* and *A. tenerum*; differs, however, in spreading lateral umbels, divaricate rays and pedicels, semiglobose mericarps, inconspicuous ribs, 3-vittate intervals, and usually bipartite carpophore.

Type locality: "Yun-nan, ad cacumina montis Che-tcho-tze, supra Tapin-Tze," 10 Oct. 1892, *Delavay 487* (P, holotype).

Distribution: Cannon (1979), who reports its occurrence in Bhutan and Nepal, suggests that it is closely related to *A. tenerum* and may possibly be conspecific with it. We have seen no specimens cited by him but the type specimen at (P) shows depressed stylopodium and bipartite carpophore, differing at least in these characters from *A. tenerum.*

9. **Acronema sichuanense** Liou & Shan, Acta Phytotax. Sinica 18(2): 199. 1980.

Plants slender, erect, 0.5–30 cm tall, from an ovoid root. Leaves broadly triangular to orbicular, 1–2 cm in diameter, 1–2-ternate, the leaflets cuneate to obovate, 0.5–1.5 cm long, 0.3–1 mm broad, entire or 2–3-lobed at apex, apiculate; petiole slender, 1–2.5 cm long, oblong-sheathing; cauline leaves few, reduced. Umbels terminal, the peduncles slender, 1–2 cm long. Involucre lacking. Rays 3–6, filiform, 1–4.5 cm long, spreading-ascending, unequal. Umbellets 3–10-flowered, the immature pedicels capillary, unequal, 2–5 mm long. Involucel of 1 or 2 minute bracteoles, or lacking. Flowers red; petals lanceolate, longly caudate; calyx 0; stylopodium conical, the styles slender, spreading. Immature fruit ovoid, 1–1.25 mm long, 1 mm broad, the ribs filiform, evident. Vittae 2 or 3 in intervals, 2–4 on commissure. Commissural face plane. Carpophore not seen. Sept.

Type locality: CHINA: "Sichuan: Zhuosijia," *P. H. Li 10551*.

Distribution: Eastern Himalaya; Sichuan, 4,800 m.

Representative specimen: Sikkim: Temu La, *Rohmoo Lopcha 371* (AAH, E).

10. **Acronema handelii** H. Wolff in Engl., Pflanzenreich 90 ($4\frac{228}{}$): 322. 1927.

Pimpinella handelii (H. Wolff) Hiroe, Umbell. World, 831. 1979.

Plants slender, flaccid, unbranched, glabrous, 15–20 cm tall, from an ovoid-globose tuber. Leaves few, suborbicular, biternate, the primary divisions semiorbicular, 4–6 cm in diameter, petiolulate, the secondary obovate, trilobed to the middle, 7 mm long, 5 mm broad; petiole filiform, 5–8 cm long, 2–3 times longer than blade; cauline leaves 1 or 2, trifoliate or bipinnate. Umbels shortly pedunculate. Involucre and involucel lacking. Rays ca. 5, capillary, divaricate, equal to unequal, ca. 3 cm long. Umbellets 5–7-flowered, the immature pedicels capillary, ca. 5 mm long. Petals yellowish-white (?), lanceolate, with filiform projection up to 2 mm long; stylopodium depressed, patelliform. Fruit (immature) subglobose, ca. 1 mm in diameter, the ribs evident. Vittae, seed, and carpophore not seen. July–Aug.

Type locality: "China: Nordöstliches Yünnan, zwischen den Ortschaften Bödö und Alo, sudöstlich von dem Dorfe Dschungcien (Chungtien), in der gemässigten kalken Region, in Wäldern auf der Westseite, auf Kalk und Mergel, 3800–4000 m ü. M.", *Handel-Mazzetti 4588*.

Distribution: Western China to Assam, Bhutan, and Nepal, 3,800–4,000 m. No specimens have been examined.

11. Acronema tenerum (DC.) Edgew., Trans. Linn. Soc. 20: 51. 1846.

Sison tenerum Wallich, Cat. 593. 1829, *nomen nudum*: *Helosciadium ? tenerum* DC., Prodr. 41: 105. 1830; *Pimpinella tenera* (DC.) C. B. Clarke in Hook. f., Fl. Brit. Ind. 2: 686. 1879; *Carum tenerum* (DC.) Franchet, Bull. Soc. Philom. Paris VIII. 6: 122. 1894. *Pimpinella tenera* var. *dissecta* C. B. Clarke in Hook. f., Fl. Brit. Ind. 2: 686. 1879; *Acronema tenerum* var. *dissectum* (C. B. Clarke) H. Wolff in Engl., Pflanzenreich 90 ($4\frac{228}{}$): 321. 1927. Based on the same collection as the typical variety ["With the type."]

Plants low and rigid to slender, erect, few branched above, 10–40 cm tall, from a small globose tuber. Basal and lower cauline leaves oval or ovate, 2–8 cm long, 2–6 cm broad, bipinnate or ternate-pinnate, the ultimate divisions linearly oblong to oblanceolate or oval, 1-5 mm long, 1-3 mm broad, obtuse, apiculate; petiole very slender, 2–10 cm long, narrowly sheathing at base; cauline leaves gradually reduced upward. Peduncles slender but rigid, 0.5–5 cm long. Involucre usually lacking. Rays 4–9, very slender, unequal, 3–20 mm long, spreading. Umbellets 3-7-flowered, the mature pedicels slender but rigid, 3–5 mm long. Involucel of 2–5 linearly lanceolate, caducous bracteoles 1–2 mm long. Flowers white, cream-colored (?), or red; petals ovate, with a filiform projection 1-1.5 mm long; stylopodium conical, the styles slender, divaricately recurved. Fruit ovoid, 1-1.5 mm in diameter, narrowed toward apex, rounded at base, the mericarps slightly compressed dorsally, slightly divergent at apex, the ribs obscure. Vittae minute, 2 in intervals. Commissural face plane. Carpophore entire or shortly bifid. July–Oct.

Type locality: "In Nepaliae monte alto Sheopore," *Wallich 593*.

Distribution: Himalaya, from Sikkim to Himachal Pradesh, 2,500–4,000 m; also China.

Representative specimens: Himachal Pradesh: above Dharamsala, *Edgeworth* (CAL). Uttar Pradesh: Kali Valley, *Duthie 2950* (CAL); Ghangaria, *Bhattacharyya 29,618* (BSD). Nepal: Searpagaon, *Mukherjee 260* (CUH); Syabrubensi to Searpagaon, *Mukherjee 254* (CUH, UC). West Bengal: Darjeeling, Singaleleh, *Clarke 554* (CAL). Sikkim: Lachoong, *Gammie* (CAL).

44. **SINOCARUM** H. Wolff ex Shan & Pu

H. Wolff in Engl. Pflanzenreich 90 ($4^{\underline{228}}$): 164. 1927, *nomen nudum*;
Shan & Pu, Acta Phytotax. Sinica 18(3): 374–375. 1980; Pu, Fl.
Reipub. Popul. Sinicae 55(2): 30–38. 1985.

Carum sect. *Eu-Carum* Franchet, Bull. Soc. Philom. Paris VIII. 6: 124. 1894; Diels, Notes Roy. Bot. Gard. Edinb. 25: 287–288. 1912.
Pimpinella sensu C. B. Clarke in Hook. f., Fl. Brit. Ind. 2: 686. 1879, *pro parte*, non Linn. 1753.
Similisinocarum Cauwet-Marc & Farille, Bull. Soc. Bot. France 131, Lettres Bot. 1984 (1): 68. 1984.

Low or subcespitose to slender, caulescent, usually simple, usually glabrous perennial herbs from a globose or tuberous root. Leaves petiolate, membranaceous, 1–3-pinnatisect, the ultimate divisions linear to obovate; petiole slender, sheathing; cauline leaves reduced upward, often with fewer, narrower divisions, and narrow to moderately dilated sheaths. Inflorescence of lax terminal and occasionally also lateral, compound umbel, the peduncles short to slender. Involucre usually lacking, or occasionally of one or more narrow bracts. Rays rather few, slender, subequal to slightly unequal, spreading-ascending. Involucel of few, inconspicuous, narrow bracteoles, or lacking. Umbellets few flowered, the flowers pedicellate. Flowers cream-white or purplish, irregular; petals orbicular to obovate, plane and without an inflexed apex, obtuse or rounded at base, often undulate or slightly lobed, clawed at base, the peripheral flowers radiant; calyx teeth 0, or minutely triangular; stylopodium low conical or pulvinate to conical, the styles very short to exceeding the stylopodium, erect to reflexed. Pedicels short to rather slender, subequal to unequal. Fruit oblong-ovoid to obovoid, compressed laterally, usually truncate at apex, rounded at base, the mericarps subterete or pentagonal, glabrous, the ribs filiform, subequal. Vittae 1–3 in intervals, 2–6 on commissure. Commissural face usually plane. Carpophore shortly bifid to bipartite.

Type species: *Carum coloratum* Diels [*Sinocarum coloratum* (Diels) H. Wolff ex Shan & Pu].

Distribution: A small genus of perhaps 12–15 species extending from SW China (Yunnan, Szechaun, Tibet) and northern Burma to Assam, and through the Eastern Himalaya to Nepal.

Note. Two new species are added tentatively without formalities of naming etc., because perfect fruits for them could not be seen.

KEY TO SPECIES

1. Ultimate leaf divisions linear to linearly lanceolate or linearly oblong; calyx obscure or lacking.
    2. Leaves ternately decompound; rays 15–25; stylopodium sharply conical; vittae 1 or 2 in intervals. .... 1. *S. clarkeanum*.

2. Leaves bi- or tripinnate; rays 5–12; stylopodium low conical or pulvinate; vittae 3 in intervals (not seen in *S. coloratum*).

3. Basal leaves oblong, 1–2 cm long; bracteoles 3–6. .... 2. *S. Wolffianum.*

3. Basal leaves oblong-ovate, 2–14 cm long; bracteoles usually 0.

4. Plants from a globose tuber; rays unequal, filiform; petals lanceolate-elliptic, acute. .... 3. *S. sikkimense.*

4. Plants from a taproot; rays subequal, slender but scarcely filiform; petals oval to orbicular, somewhat acute to obtuse. .... 6. *S. coloratum.*

1. Ultimate leaf divisions ovate to obovate; calyx evident, or if lacking, the fruit ribs thick and obtuse.

5. Plants essentially glabrous.

6. Rays 3–7, subequal; bracteoles 1.5–4 mm long. .... 4. *S. normanianum.*

6. Rays 7–10, unequal; bracteoles up to 2 mm long, or lacking.

7. Umbellets ca. 16-flowered; bracteoles ca. 2 mm long; stylopodium sharply conical; fruit ribs slender. .... 5. *S. acronemifolia.*

7. Umbellets 8–12-flowered; bracteoles less than 1 mm long or lacking; stylopodium low conical; fruit ribs thick. .... 5. *S.* sp. 1.

5. Plants thinly soft-hirsute throughout. .... *S.* sp. 2.

1. **Sinocarum clarkeanum** P. K. Mukherjee & Constance, Edinburgh Jour. Bot. 48(1): 42. 1991.

Pimpinella acronemifolia C. B. Clarke in Hook. f., Fl. Brit. Ind. 2: 686. 1879, *pro parte*; *Acronema acronemifolium* (C. B. Clarke) H. Wolff in Engl., Pflanzenreich 90 ($4\frac{228}{}$): 322. 1927, *pro parte*.

Plants low and bushy, 2–5 branches ascending from base, 10–30 cm tall, from an elongated cylindrical or fusiform, sometimes branched tuberous root 3–10 cm long. Basal and lower cauline leaves ovate-lanceolate to triangular-ovate, 5–25 cm in diameter, ternately decompound, with linearly oblong entire or apically lobed ultimate divisions 1.5–8 mm long, up to 1 mm broad; petiole 1.5–15 cm long, dilated and sheathing at base; cauline leaves reduced upward. Peduncles stout, fistulose, sometimes puberulent, 2.5–20 cm long. Involucre and involucel of 2–5 linear bracts or bracteoles 2–5 mm long. Rays 15–25, somewhat rigid, unequal, ascending but diverging above, 2–10 cm long. Umbellets ca. 20-flowered, the mature pedicels somewhat rigid, subequal, 2–8 mm long. Flowers white; petals obovate, acute or obtuse, a little unequal; stylopodium sharply conical, the styles diverging. Fruit elliptic-oblong, ca.1.5 mm long, 0.6–0.8 mm broad, subdidymous, glabrous, the mericarps diverging slightly at apex, the ribs filiform. Vittae 1 or 2 in intervals, 2 on commissure. Commissural face slightly concave. Carpophore bipartite. July–Oct.

Type locality: "SIKKIM, ... Singaleleh, alt. 10,000 ft." [Darjeeling, 3650 m, 7 Oct. 1870], *Clarke 12,545* (K, holotype).

Distribution: Sikkim Himalaya, 3,300–4,000 m.

Representative specimens: Sikkim: Phedap, *Ribu 294* (CAL); Megu, *Thomson* (CAL), *Prain's collector 199* (CAL); Sikkim Himalaya, *Gammie* (CAL).

2. **Sinocarum wolffianum** (H. Wolff) P. K. Mukherjee & Constance, Edinburgh. Jour. Bot. 48(1): 42. 1991.

Acronema wolffianum H. Wolff, Feddes Repert. 27: 328. 1930; *Pimpinella feddeana* Hiroe, Umbell. World, 832. 1979.

Plants low, slender, erect, simple or with 1 or 2 lateral branches from base, 5-15 cm tall, from a globose or ovoid tuber. Basal (?) and lower cauline leaves oblong, 1-2 cm long, up to 1 cm broad, bipinnate, the leaflets palmately or ternately divided into linearly lanceolate, acute ultimate divisions 1-2 mm long; petiole dilated, broadly oblong, 1-2 cm long; cauline leaves confined to base. Peduncles slender, 3-6 cm long, scaberulous beneath umbel. Involucre lacking, or of an occasional setaceous bract. Rays 5-12, quadrangular, 10-15 mm long, spreading-ascending, subequal. Umbellets (?5-)10-20-flowered, the mature pedicels quadrangular, spreading-ascending, subequal, 2-5 mm long. Involucel of 3-6 linear bracteoles 1-2 mm long. Flowers white; petals oval to obovate, obtuse or acute; stylopodium low conical, purple, much shorter than the flattened, reflexed styles. Fruit oblong-ovoid, 1-1.5 mm long, 1 mm broad, slightly narrowed toward apex, the maricarps pentagonal in transection, the ribs acute, filiform. Vittae 3 in intervals, 2 on commissure. Commissural face plane. Carpophore shortly bifid. Aug.-Oct.

Type locality: "Himalaya, Sikkim: Jongri, 12,000′", *Clarke 25,796*.

Distribution: Eastern Himalaya in Bhutan and Sikkim, 2500-4000 m.

Representative specimens: Sikkim: Bijan, Jaktop, *King's collector* (CAL); Zongri, *Mukherjee 320* (CUH). Bhutan: Pajoding Monastery, Thimbu, *Bartholomew 195* (BM, UC).

3. **Sinocarum sikkimense** (P. K. Mukherjee) P. K. Mukherjee & Constance, Edinburgh Jour. Bot. 48(1): 43. 1991.

Acronema sikkimense P. K. Mukherjee, Actes II Sympos. Internat. Ombell. Perpignan, 52. 1978.

Pimpinella acronemifolia C. B. Clarke in Hook. f., Fl. Brit. Ind. 2: 689. 1879, *pro parte*; *Acronema acronemifolium* (C. B. Clarke) H. Wolff in Engl., Pflanzenreich 90($4\frac{228}{}$): 322. 1927, *pro parte*.

Plants slender, erect, simple or with 1 or 2 branches above, 7-30 cm tall, from a globose tuber. Leaves oblong-ovate, 2-10 cm long, bipinnate, the ovate or oblong leaflets trilobed, with linear ultimate divisions 1-6 mm long, ca. 0.5 mm broad; petiole slender, 2-7 cm long, narrowly sheathing at base; cauline leaves several, reduced upward, bipinnate, conspicuously sheathing. Peduncles slender, 1-5 cm long. Involucre lacking. Rays 6-10, filiform, 1-3 cm long, spreading-ascending, unequal. Umbellets 10-15-flowered, the mature pedicels filiform, unequal, 2-5 mm long. Involucel of 1-3 linear bracteoles 1-2 mm long, or lacking. Flowers white, pinkish, or purplish-red; petals lanceolate-elliptic, acute; stylopodium low conical, forming a flattened disk, the styles slender, recurved. Fruit ovoid, 1-2 mm long, ca. 1 mm broad, didymous, glabrous, the ribs evident, raised. Vittae 3 in intervals. Commissural face plane. Carpophore bipartite. Aug.-Nov.

Type locality: "Sikkim, Lachoong, 3650-4000 m," *Hooker* (CAL, holotype).

Distribution: Eastern Himalaya; Tibet, 3,000-4,000 m.

Representative specimens: Nepal: above Saurwala Khola, *Stainton, Sykes & Williams 3578* (AAH, BM). West Bengal: Phalut, *Mukherjee 309* (CUH, UC). Sikkim: Le-lap-la,

Dungboo (CAL), *Mukherjee 317* (CUH), Tankra Mountain, *Gammie 571* (CAL).

4. **Sinocarum normanianum** (Cauwet-Marc & Farille) Farille in Farille, Cauwet-Marc & Malla, Candollea 40(2): 561. 1985.

Similisinocarum normanianum Cauwet-Marc & Farille in Farille, Cauwet-Marc & Malla, Bull. Soc. Bot. France 131, Lettres Bot. 1984 (1): 68. 1984.

Plants low, subcaulescent, erect, simple or few branched from base, essentially glabrous, 8–25 cm tall from a tuberous root. Leaves mostly basal, oblong to ovate, 2.5–8(–10) cm long, 1.5–5(–8) cm broad, pinnate, the leaflets 3–7, ovate to orbicular, obtuse at apex, rounded to truncate or cuneate at base, the lowest pair petiolulate, the others sessile, 0.5–3 cm in diameter, crenate-dentate to ternately or pinnately incised or lobed with rounded mucronulate teeth or lobes, the margins minutely scaberulous-ciliolate, paler or purplish and venose beneath; petiole slender, 1.5–5 cm long, with a conspicuous oblong, usually purplish, scarious sheath; cauline leaves 1–4 and like the basal but with more broadly dilated sheaths, or cauline leaves 0. Peduncles very slender, mostly terminal, 3–15 cm long, scaberulous at summit. Involucre of a single linear scarious bract 1–8 mm long, or usually 0. Rays 3–7, filiform, 1–2 cm long, angled, a little scaberulous at summit. Umbellets 5–12-flowered, the immature pedicels unequal, spreading. Involucel of 2–5 linear to linearly lanceolate, acuminate, scarious or broadly scarious-margined bracteoles 1.5–4 mm long. Flowers white; petals oval to obovate, apparently plane, obtuse to somewhat acute, slightly radiant, clawed; calyx teeth minute but evident, ovate; stylopodium rounded conical (hemispherical ?), about equal to the short styles; ovary purple. Immature fruit oblong-ovoid, compressed laterally and constricted at commissure, the mericarps subterete in transection, the ribs thick and obtuse. Vittae, seed face, and carpophore not seen.

Type locality: "Nepal, au nord de Kathmandu, Pong-Sing [Pangsing?], 15000 feet," 1929, *Lall Dhwoj 50* (BM, holotype & isotype; E, isotype).

Distribution: Himalaya of Nepal, 3500–4500 m.

Representative specimens: Nepal: Central Nepal, *Lall Dhwoj 32* (E); Mancy Dara, *Lall Dhwoj 617* (BM, E-pars); Langtang, *Polunin 1500* (BM); Thade, *Shrestha & Shakya 3773* (BM).

5. **Sinocarum acronemifolium** (C. B. Clarke) P. K. Mukherjee & Constance, Edinburgh Jour. Bot. 48(1): 43. 1991.

Pimpinella acronemifolia C. B. Clarke in Hook. f., Fl. Brit. Ind. 2: 686. 1879, *pro parte*; *Acronema acronemifolium* (C. B. Clarke) H. Wolff in Engl., Pflanzenreich 90(4$\frac{228}{}$): 322. 1927, *pro parte*.

Plants low, slender, mostly unbranched, 8–20 cm tall, from a globose tuber. Basal and lower cauline leaves only 1–3, ovate, 2–3-pinnate or ternate-pinnate, the ultimate divisions ovate to obovate, trilobed, 1.5–2.5 mm long, 0.5–1.5 mm broad; petiole oblong, inflated, up to 1.5 cm long; cauline leaves few and with broadly inflated petioles, or lacking, the uppermost reduced to bladeless sheaths. Peduncles rather slender, 5–15 cm long. Involucre of one or more linear, foliaceous bracts up to 1 cm long. Rays 7–10, slender, unequal, ascending but diverging above. Umbellets 10–15-flowered, the mature pedicels very unequal, 2–6 mm long. Involucel of 3–5 linear, unequal bracteoles ca.

2 mm long. Flowers white; petals obovate, obtuse, slightly unequal; calyx teeth evident, triangular; stylopodium sharply conical. Fruit elliptic-oblong, 1.5 mm long, 1 mm broad, subdidymous, the mericarps diverging slightly at apex, the ribs slender. Vittae minute, 2 or 3 in intervals. Commissural face plane. Carpophore slightly bifid. Nov.

Type locality: "SIKKIM, alt. 11,000 ft., Chola," *Hooker*.

Distribution: Himalaya, 3,650–4,000 m.

Representative specimens: Sikkim: Chola, Hooker (CAL); *Vicatia?* w/o precise locality, *Hooker* (CAL).

6. **Sinocarum coloratum** (Diels) H. Wolff ex Shan & Pu, Fl. Reipub. Popul. Sinica 55(2): 33. 1985.

Carum coloratum Diels, Notes Roy. Bot. Gard. Edinb. 25: 287. 1912.

Plants low, slender, few branched, 8–25 cm tall, from an apparently unthickened taproot. Leaves oblong-ovate, 4–15 cm long, 1–3 cm broad, 2–3-pinnate, the ultimate divisions linearly oblanceolate, entire or trilobed, acutish, to 5 mm long, 1–1.5 mm broad; petiole oblong, slender, 5–10 cm long, slightly inflated; cauline leaves several, with oblong, ± inflated petioles. Peduncles slender, 4–8 cm long. Involucre and involucel usually lacking or inconspicuous. Rays 5–12, slender, subequal, spreading-ascending, 1–2.5 cm long. Umbellets 10–15-flowered, the immature pedicels unequal, to 5 mm long. Flowers white; petals oval to orbicular, obtuse to somewhat acute, often lobed; calyx teeth minute or obsolete; stylopodium low conical, shorter than styles. Fruit oblong-ovoid, glabrous, the ribs evident. Vittae, seed face, and carpophore not seen. Aug.–Sept.

Type locality: CHINA. Yunnan: "cliffs on the eastern flank of the Lichiang Range. Lat. 27°25′N," *G. Forrest 2874, 3060* (E, BM).

Distribution: Eastern Himalaya; Yunnan, 3,200–4,000 m.

Representative specimens: CHINA. Yunnan: Lichiang Range, Aug. 1923. *J. F. Rock 9851* (UC, topotype). INDIA. Nagaland: Jaffu, Kohima, *Angami 1107* (CAL).

Sinocarum sp. 1.

Plants low, erect, simple, essentially glabrous, 5–20 cm tall, from a swollen, sometimes cormlike taproot. Leaves mostly basal, ovate, 1.5–4 cm in diameter, ternate-pinnate or pinnate, the leaflets 9–21, orbicular, obtuse at apex, rounded to cuneate at base, 1–3 mm in diameter, mucronate-dentate distally; petiole slender, 0.5–2 cm long, with a dilated orbicular scarious sheath; cauline leaves 1 or 0. Peduncles very slender, all terminal (?), 3.5–15 cm long, glabrous to slightly scaberulous at summit. Involucre of 1 or 2 minute bracts or lacking. Rays 7–9, filiform, 1–2 cm long, unequal, angled, glabrous or a little scaberulous at apex. Umbellets 8–12-flowered, the immature pedicels unequal, 1–3 mm long, spreading-ascending. Involucel of 1–5 minute linear, scarious bracteoles, to 1 mm long, or lacking. Flowers white; petals orbicular to obovate, apparently plane, rounded at apex, slightly radiant, shortly clawed; calyx lobes 0; stylopodium low conical, exceeded by the spreading styles. Immature fruit ovoid, compressed laterally and constricted at commissure, rounded at apex and base, the mericarps subterete in transection, the ribs thick and obtuse. Vittae, seed face, and carpophore not seen. Himalaya of Nepal and Bhutan, 3,800–4,200 m: Mancy Dara, 13000–14,000 feet, 1930, *Lall Dhwoj 617* (BM, E-pars); w/o further data, *Wigram 95* (E).

Sinocarum sp. 2

Plants low, subcaulescent, thinly hirtellous throughout, the single flowering scape 3–10 cm tall from a vertical taproot. Leaves clustered at base, oblong, 2–3 mm long, 5–10 mm broad, pinnate, the 3–5 pairs of opposite leaflets oval or orbicular, 3–5 mm in diameter, obtuse at apex, rounded or cuneate at base, all sessile, few dentate or shallowly lobed, the terminal 3-lobed, the teeth or lobes rounded; petiole slender, 0.5–4 cm long, abruptly scarious-sheathing at base; cauline leaf solitary, like the basal or reduced, the sheath prominent, usually inserted at or near base of umbel and serving as an involucre (?). Flowering scape or peduncle solitary, terminal, 0.5–4 cm long. Involucre 0, or represented by the single cauline leaf. Rays 2 or 3(–5), filiform, 1.5–3 cm long, subequal to unequal. Umbellets 8–12-flowered, the immature pedicels unequal, spreading, sometimes glabrate, flattened. Involucel usually lacking, or occasionally of 1–3 scarious bracteoles. Flowers "whitish," the petals lanceolate to ovate, apparently plane, acute to acuminate (?), slightly radiant, clawed; calyx lobes evident, ovate, scarious, 0.2–0.3 mm long; stylopodium pulvinate, purplish; anthers purple. Fruit and carpophore not seen. Himalaya of central Nepal at ca. 4,000 m: above Sauwala Khola, 13,500 feet, 1954, *Stainton, Sykes & Williams 3591* (AAH, BM).

45. TONGOLOA H. Wolff

Notizbl. Bot. Gart. Berlin-Dahlem 9: 279. 1925.

Slender, erect, few branched or simple, essentially glabrous perennial herbs from sometimes swollen taproots. Leaves petiolate, membranaceous, ternately or pinnately decompound with usually narrow ultimate divisions; petiole slender, sheathing at base; cauline leaves reduced upward. Inflorescence of lax terminal and usually also lateral compound umbels, the peduncles ± elongate. Involucre and involucel lacking, or of few small, persistent or caducous bracts or bracteoles. Rays few to rather numerous, subequal or unequal, the flowers pedicellate. Flowers white or reddish-purple; petals oblong to obovate, plane or often slightly incurved at tip, narrowed into short claw at base; calyx teeth small, obtusely ovate, or lacking; stylopodium low conical to depressed, the styles usually short, spreading or reflexed. Pedicels slender, spreading-ascending. Fruit oblong-ovoid to globose, rounded or truncate at apex, cordate at base, slightly compressed laterally and constricted at commissure, the mericarps subterete to slightly compressed dorsally in transection, glabrous, the ribs conspicuous or inconspicuous, filiform, narrower than intervals. Vittae 2 or 3 in intervals, usually 2 on commissure. Commissural face plane to sulcate. Carpophore bifid or bipartite.

Type species: *Tongoloa gracilis* H. Wolff.

Distribution: A genus of perhaps a dozen high montane plants, mostly of western China and Tibet, but extending to the Himalaya of Bhutan.

Note. A poorly defined genus apparently related, on the one hand, to *Sinocarum* and *Pimpinella*, and on the other to *Vicatia*. In contrasting the new genus with *Pimpinella*, Wolff (1925) stressed differences in habit, leaf division, and shape and color of petals. *Tongoloa* shares with *Sinocarum* slightly zygomorphic flowers, but the plants are larger and stouter, the petal base is definitely clawed, the fruits tend to be broader, and the vittae more numerous. In *Vicatia* there is usually a stronger taproot development, an inconspicuous involucel is commonly present, the flowers are regular or nearly so, the petals have a narrower inflexed apex and little or no basal claw, and the fruits tend to be appreciably longer than broad. Clearly, none of the three genera *Sinocarum*, *Tongoloa*, and *Vicatia* are well differentiated, crucial information is lacking about some aspects of most taxa currently placed in each assemblage, and material is scant. Detailed examination is likely to continue or even exacerbate the interchange of taxa before a fully satisfactory arrangement is achieved.

Two new species are added but in the absence of their mature stages formal naming and validation are refrained from.

KEY TO SPECIES

1. Peduncles 4–10 cm long; ultimate leaf segments linear, 0.5–2 mm broad.
   2. Petals purplish-red; stems with 2 or 3 leaves; rays 5–8. .... 1. *T. gracilis.*
   2. Petals white; stems with several cauline leaves; rays 8–15.
      3. Leaves 3-pinnate, the ultimate divisions linearly filiform, subequal, apparently interlocking. .... 2. *T. loloensis.*
      3. Leaves 2-pinnate or ternate pinnate, the ultimate divisions linearly acuminate, markedly unequal, the terminal elongate. .... *T.* sp. 1.
1. Peduncles ca. 30 cm long; ultimate leaf segments lanceolate-oblong, 5–10 mm broad .... *T.* sp. 2.

1. **Tongoloa gracilis** H. Wolff, Notizbl. Gart. Berlin-Dahlem 9: 279. 1925.

Pimpinella tilia Hiroe, Umbell. World 834. 1979.

Plants slender, erect, simple or few branched, glabrous, 5–75 cm tall, from a slender napiform perennial root. Leaves ovate-lanceolate to ovate-triangular, 5–10 cm in diameter, 2-3-pinnate, the ultimate divisions linearly oblong, somewhat acute, 3–15 mm long, 0.5–2 mm broad; petiole slender, 10–30 cm long, narrowly sheathing; cauline leaves 2 or 3, similar to the basal, with somewhat inflated oblong sheaths. Peduncles mostly terminal, slender, 4–10 cm long. Involucre and involucel lacking, or early deciduous. Rays 5–18, slender, quadrangular, unequal, 2–4 cm long. Umbellets ca. 15-flowered, the immature pedicels slender, 5–8 mm long. Petals "red" ["dark purplish tipped white"], oblong-oval or oblong, obtuse at apex, long clawed at base; calyx teeth reduced but evident, ovate, obtuse; stylopodium depressed conical, the styles short, reflexed. Immature fruit oblong-ovoid, 2 mm long, truncate, constricted at commissure, the mericarps subterete in transection, glabrous, the ribs conspicuous. Vittae 2 or 3 in intervals, 2 on commissure. Seed face sulcate.

Type locality: "Eastern Tibet: Tongolo," *Soulié 2606*.

Distribution: High Himalaya of Bhutan, 4,080 m; E. Tibet.

Specimen examined: Bhutan: Upper Mo Chu Distr., above Laya, 20 Sept. 1984, *Sinclair & Long 5170* (E, photo).

2. **Tongoloa loloensis** (Franchet) H. Wolff in Engl., Pflanzeur. 90 ($4\frac{228}{}$): 318. 1927.

Carum loloense Franchet, Bull. Soc. Philom. Paris 6(8): 125. 1894; *Pimpinella loloensis* (Franchet) Boiss., Bull. Herb. Boiss. 2(2): 809. 1902; *Trachydium loloensis* (Franchet) Hiroe, Umbell. World 710. 1979.

Plants rather slender, erect, sparingly branched, essentially glabrous, 30–90 cm tall, from a short, thickened, woody rhizome. Leaves ovate-triangular, 6–12 cm in diameter, 3-pinnate, the ultimate divisions linearly filiform, rigid, mucronate, 2–8 mm long, ca. 0.5 mm broad; petiole longer than the blade in basal leaves, moderately dilated-sheathing; cauline leaves rather numerous, with strongly inflated sheaths. Peduncles terminal and lateral, the lateral much smaller, 5–10 cm long, scaberulous or villous at summit. Involucre of

1-4 subulate (deciduous ?) bracts, or 0. Rays 8–15, slender, angled, slightly scaberulous at summit, 3–5 cm long. Umbellets 12–15(30)-flowered, the pedicels slender, 5–10 mm long. Involucel of several subulate bracteoles. Petals oval to obovate, rounded at apex, clawed at base; calyx teeth minute or lacking; stylopodium shortly conical, broadened at base, much shorter than the slender styles. Fruit ovoid, ca. 2 mm in diameter, ± narrowed at apex, obtuse, cordate at base, constricted at commissure, the mericarps subterete in transection, glabrous, the ribs moderately prominent, unwinged. Vittae 2 or 3 in intervals. Seed face concave. Carpophore bipartite. Oct.

Type locality: CHINA. "Yun-nan in palude supra Pee-tsao-lo, vicus e gente Lolo dictus," Oct. 1885, *Delavay 2016*; "ad Oua-pin-keou, in monte Ma-eul chan", 23 Oct. 1889, *s. n.*; "in monte Hee-chan-men", 28 Oct. 1885, *s. n.*

Distribution: E. Himalaya to Bhutan and Nepal; W. China, Tibet, 2,500–3,200 m.

Representative specimen: Bhutan: Paro Valley, Taktsang Monastery, 17 Oct. 1974, *Bartholomew 114* (UC).

Tongoloa sp. 1

Plants very slender, erect, profusely branched throughout, 20–35 cm tall. Leaves ovate-triangular, 5–10 cm in diameter, ternate-pinnate or bipinnate with markedly unequal linearly acuminate divisions, the terminal elongate, 3–30 mm long, ca. 1 mm broad; petiole very slender, about equal to the blade, narrowly sheathing; cauline leaves numerous, subtending short flowering lateral branches, the uppermost with dilated oval sheaths. Peduncles slender but rather short, 5–9 cm long. Involucre and involucel apparently 0. Rays 8–12, subfiliform, 2–3 cm long. Umbellets 8–15-flowered, the pedicels slender. Petals white, the anthers brown. Specimen too young to observe other characters. Himalaya of Nepal, between Gorgianhot and Kumrigaon, SE of Jumla, 4,500 ft., *Polunin, Stainton & Williams 4872* (BM). Cannon (1979) referred this single Nepalese collection to *T. elata* H. Wolff and suggested that this taxon "... may be conspecific with *Carum pityophilum* Diels". It does not agree with (Chinese) specimens we have seen of the former, and descriptions of the latter — without fruit — are too incomplete to establish its status.

Tongoloa sp. 2

Plants rather stiffly erect, to 55 cm tall. Leaves broadly triangular, 10 cm in diameter, bipinnate with lanceolate-oblong leaflets 1.5–5 cm long, 5–10 mm broad, shallowly pinnately incised; petiole of basal leaf ca. 25 cm long, abruptly dilated and broadly sheathing at base; cauline leaves 1 or more, similar to the basal, but with narrower leaf divisions and short, more strongly inflated sheaths. Peduncles stout, ca. 30 cm long. Rays 7, 4–5 cm long, spreading-ascending. Involucre of a single foliaceous bract, or 0. Umbellets 15–25-flowered, the immature pedicels capillary, ca. 5 mm long. Involucel of several filiform bracteoles to 5 mm long. Petals purple (?). Mature stages of this taxon are unknown. Bhutan: Saka-La to Ha, 9000–12,000 ft., *B.J. Gould 1052* (K, photo). It is just conceivable that this may bear some relationship to the foregoing; alternatively, it may very well not belong to this genus.

46. **Pimpinella** Linn.

Sp. Pl. 1: 263. 1753; C. B. Clarke in Hook. f., Fl. Brit. Ind. 2: 684. 1879.

Reutera Boiss., Elench. 46. 1838; *Pimpinella* sect. *Reutera* (Boiss.) Kuntze in Post & Kuntze, Lex. Gen. Phan. 439. 1903; *Apium* sect. *Reutera* (Boiss.) Calest., Webbia 1: 176. 1905.
Tragoselinum Miller, Dict. ed. 4. 1754; *Pimpinella* sect. *Tragoselinum* (Miller) DC., Prodr. 4: 119. 1830; *Apium* sect. *Tragoselinum* (Miller) Calest., Webbia 1: 176. 1905.
Tragium Sprengel, Pl. Umbell. Prodr. 26. 1813; *Pimpinella* sect. *Tragium* (Sprengel) DC., Prodr. 4: 120. 1830; *Apium* sect. *Tragium* (Sprengel) Calest., Webbia 1: 176. 1905.
Ledebouria Link, Enum. Hort. Berol. 1: 286. 1821; *Pimpinella* sect. *Ledebouria* (Link) Kuntze in Post & Kuntze, Lex. Gen. Phan. 439. 1903.

Low or slender, usually erect, caulescent, branching, glabrous to hirsute perennial, biennial, or annual herbs, or suffrutescent, from a slender taproot. Leaves petiolate, membranaceous, simple or ternate or pinnate to pinnately or ternately decompound, the leaflets or ultimate divisions various; petiole sheathing at base; cauline leaves often heteromorphic and dissected. Inflorescence of lax terminal and lateral compound umbels. Involucre and involucel of a few narrow bracts or bracteoles, or usually lacking. Umbellets few to many flowered, the flowers pedicellate. Flowers white, yellow, red, or purple; petals broad or narrow, with a narrower inflexed apex (rarely without), glabrous or pubescent dorsally, the outer sometimes radiant; calyx teeth evident and usually minute, or calyx obsolete; stylopodium conical or low conical, the styles short or slender. Pedicels usually slender, spreading. Fruit oblong to globose, usually narrowed at apex, rounded or cordate at base, compressed laterally, often constricted at commissure and didymous, the mericarps subterete, pentagonal, or slightly compressed dorsally in transection, glabrous to variously pubescent or squamose, the ribs filiform to prominent and subwinged, much narrower than intervals. Vittae usually 2 or 3 in intervals, 2 or 4 on commissure, or forming a circle surrounding seed. Commissural face plane or concave. Carpophore bifid or bipartite or entire.

Type species: *Pimpinella saxifraga* Linn.

Distribution: A large genus of Eurasia and Africa, comprising some 150 species. With about two dozen species each, this and *Bupleurum* are the largest Indian genera of Umbelliferae. Eight species are largely confined to peninsular India, eleven occur in the Eastern Himalaya and/or Assam, and two are endemic to the Western Himalaya. There appear to be additional taxa, for which existing material is too immature or otherwise inadequate to permit proper description. Altitudinal range in India extends from 300 to 4,500 m.

Note. Two new species are described here as they appear distinct. Formal naming and validation is refrained from for them since both the two are not seen with mature fruits.

KEY TO SPECIES

1. Mature fruit glabrous and smooth, or if roughened, the ribs prominent and carpophore entire.
   2. At least the basal leaves simple, the cauline simple or ternate.
      3. Basal leaves triangular-lanceolate, acuminate, often hastate; flowers greenish-yellow. .... 1. *P. hastata.*
      3. Basal leaves ovate-cordate to orbicular, obtuse or at most somewhat acute at apex; flowers white or purple.
         4. Basal leaves merely crenate, callous margined; flowers purple .... 2. *P. atropurpurea.*
         4. Leaves usually crenate-serrate or -dentate and without a callous margin; flowers white.
            5. Petiole stout; involucel lacking .... 3. *P. leschenaultii.*
            5. Petiole very slender; involucel present.
               6. Carpophore stout, entire; cauline leaves few and much reduced .... 4. *P. pulneyensis.*
               6. Carpophore slender, bipartite; cauline leaves large, trifoliate .... 5. *P. tongloensis.*
   2. Basal and usually also lower cauline leaves ternately or pinnately compound.
      7. Basal leaves simply ternate or pinnate with 3–7 leaflets.
         8. Umbellets 8–15-flowered, the pedicels hirtellous. ....6. *P. inundata.*
         8. Umbellets 3–8-flowered, the pedicels glabrous.
            9. Plants simple; flowers purple; peduncles 10–30 cm long, scaberulous at summit; rays webbed. .... 7. *P. pimpinellisimulacrum.*
            9. Plants branching; flowers white; peduncles filiform, 2–5 cm long, glabrous; rays distinct to base.
               10. Cauline leaves with oblong to linear divisions and inconspicuous sheaths; rays subequal; involucel lacking. ....8. *P. flaccida.*
               10. Cauline leaves with ovate divisions and dilated sheaths; rays very unequal; involucel of 1–3 bracteoles. .... *P.* sp. 1
      7. Basal leaves bipinnate, biternate or ternate-pinnate, or if simply ternate or pinnate, the rays and pedicels filiform and unequal.
         11. Plants stout; rays and pedicels scaberulous. ....9. *P. nervosa.*
         11. Plants slender; rays and pedicels essentially glabrous.
            12. Rays and pedicels filiform and unequal.
               13. Annual; involucre lacking; stylopodium obsolete; peninsular India. .... 10. *P. heyneana.*
               13. Perennial; involucre present; stylopodium low conical; Western Himalaya. .... 13. *P. acuminata.*
            12. Rays and pedicels stout to slender, or if filiform, then subequal; north-eastern India or Eastern Himalaya.

14. Calyx teeth lanceolate, persistent; pedicels 10–15 mm long; carpophore entire. .... 11. *P. sikkimensis.*

14. Calyx obsolete; pedicels 2–6 mm long; carpophore bipartite.

15. Rays rigid, very unequal, to 1.5 cm long; pedicels only 2–2.5 mm long; stylopodium inconspicuous; fruit ellipsoid, 1.5–2 mm in diameter, the ribs prominent. .... 12. *P. wallichii.*

15. Rays slender, subequal, 1.5–4 cm long; pedicels 4–6 mm long; stylopodium conical; fruit ovoid, 3–3.5 mm long, 1.5–2 mm broad, the ribs obscure. .... 14. *P. evoluta.*

1. Mature fruit distinctly roughened or hairy, or if glabrate, the petals pubescent dorsally.

16. Basal leaves 2–3-ternate or 2–3-pinnate; fruit 3.5–4.5 mm long, 1.5–2 mm broad. .... 18. *P. tomentosa.*

16. Basal leaves simple, or simply ternate or pinnate, or ternate-pinnate or pinnate-pinnatifid; fruit 1–3 mm long, 1–2.5 mm broad.

17. Basal leaves simple, cordate.

18. Involucre and involucel lacking or essentially so (or an involucel occasionally present in *P. diversifolia*).

19. Plants glabrous; lower cauline leaves simple. .... 15. *P. tirupatiensis.*

19. Plants variously pubescent; lower cauline leaves, or some of them, ternate or pinnate.

20. Rays and pedicels filiform.

21. Stylopodium low conical; leaflets cartilaginous-dentate; peninsular India. .... 19. *P. wallichiana.*

21. Stylopodium obsolete; leaflets mucronate-dentate; Himalaya. .... 21. *P. urceolata.*

20. Rays and pedicels stouter, not filiform.

22. Plants low and cespitose, 15–40 cm tall; stylopodium conical; fruit oblong-elliptic, 2–3 mm long, 1–1.5 mm broad, pubescent. .... 17. *P. stracheyi.*

22. Plants slender and erect, 50–150 cm tall; stylopodium low conical; fruit ovoid-globose, 1.5–2.5 mm in diameter, scaberulous-papillose to glabrate. .... 23. *P. diversifolia.*

18. Involucre and involucel consistently present.

23. Leaf margins crenate-serrate with subcartilaginous teeth; fruit scaberulous-tuberculate; peninsular India. .... 16. *P. candolleana.*

23. Leaf margins crenate-serrate to coarsely serrate or incised, but teeth not subcartilaginous; fruit scaberulous-papillose to glabrate; Himalayan. .... 23. *P. diversifolia.*

17. Basal leaves ternate, pinnate, ternate-pinnate, or pinnate-pinnatifid.

24. Lower cauline leaves pinnate; involucre and involucel present; peninsular India.

25. Plants low and cespitose, perennial, 15–60 cm tall; carpophore entire. .... 20. *P. adscendens.*

25. Plants stout and erect, biennial, 90–120 cm tall; carpophore bipartite. . . . . 22. *P. bracteata.*

24. Lower cauline leaves simple, ternate, ternate-pinnate, or pinnate-pinnatifid; involucre and involucel present or lacking; Himalayan.

26. Plants 50–150 cm tall, branched; basal leaves simple or ternate; rays 5–20; umbellets 8–16-flowered; flowers white, greenish, or yellowish. . . . . 23. *P. diversifolia.*

26. Plants 20–30 cm tall, unbranched; basal leaves ternate-pinnate or pinnate pinnatifid; rays 4 or 5; umbellets 4- or 5-flowered; flowers deep-red. . . . . *P.* sp. 2.

1. **Pimpinella hastata** C. B. Clarke in Hook. f., Fl. Brit. Ind. 2: 687. 1879.

Plants slender, erect, laxly corymbosely branched and sparsely leafy above, pubescent or glabrate, 90–120 cm tall, from a fusiform root. Leaves triangular-lanceolate, 4–8 (–15?) cm long, 2–3 (–5?) cm broad, simple, acuminate at apex, hastate or cordate at base, crenate-serrate or serrate, setulose on veins above, villous beneath; petiole 4–16 (–20?) cm long; upper cauline leaves often trilobed, with ovate-cordate divisions, the uppermost reduced to narrower segments. Peduncles 3–5 cm long. Involucre and involucel usually lacking, occasionally of several linear bracts or bracteoles. Rays 5–8, hirtellous, 2.5–4 cm long. Umbellets ca. 10-flowered, not all flowers maturing, the mature pedicels 2.5–3 mm long. Flowers greenish-yellow; petals obovate, with a narrower inflexed apex; calyx obsolete; stylopodium low conical, the styles very short. Fruit broadly ovoid, 1.5–3 mm in diameter, didymous, the mericarps compressed dorsally in transection, glabrous, the ribs obscure. Vittae 2 in intervals and on commissure. Commissural face planoconcave. Carpophore bifid to bipartite. Apr.–Sept.

Type locality: "KHASIA MTS.," *Hooker* (*Hooker & Thomson*, Herb. Ind. Orient., *Pimpinella 17*).

Distribution: Hills of northeastern India, 1,000–2,500 m; Burma.

Representative specimens: Manipur: Sirohipurar, *Watt 6477* (CAL); Khongani, *Watt 6758* (CAL). Meghalaya: Smyrang, *Clarke 40,266* (CAL); Khasia Mts., *Hooker & Thomson* (CAL).

2. **Pimpinella atropurpurea** C. Y. Wu ex Shan & Pu, in Shan *et al.* Acta Phytotax. Sin. 21(1): 81. 1983.

Pimpinella chateriana Cannon & Farille, Candollea 40: 554. 1985. Based on *Farille 81,552*, "C. Népal: massif de l'Annapurna, Lamjung himal, versant nord du col de Namun," 4500 m, 24 Sept. 1981.

Plants slender, erect, simple or few branched and sparsely leafy above, pubescent in inflorescence, 10–40 cm tall, from fascicled fibrous roots. Leaves reniform or orbicular-cordate, 1–4.5 cm in diameter, simple, crenate, callous-margined, the sinus broad to narrow, glabrous or ± pubescent; petiole slender, 4–15 cm long, narrowly oblong-sheathing; cauline leaves like the basal but with more prominent sheaths, the uppermost often 3-lobed. Umbels terminal and sometimes also lateral, the peduncles slender, 3–8 cm long. Involucre of 1-3 linear or foliaceous bracts 0.5–1 cm long, or lacking. Rays 4–12, unequal, pubescent, 0.5–3 cm long, spreading-ascending. Umbellets 5–15-flowered, not all maturing, the immature pedicels 3–5 mm long, spreading-ascending, pubescent. Involucel of 1–5 minute linear bracteoles, or lacking. Flowers dark purple ("dark brown-red"); petals ovate, acute, apparently without an inflexed apex, clawed at base; calyx obsolete; stylopodium low conical, the styles sharply reflexed, ca. 0.5 mm long. Fruit ovoid-globose, 1.5–2 mm in diameter, didymous, the mericarps terete in transection, glabrous or puberulent, the ribs filiform. Vittae 3 in intervals, 4–6 on commissure. Commissural face plane. Carpophore bifid ca. 1/5. July–Dec.

Type locality: CHINA. "Yunnan: Zhenkang," 2 Aug. 1938, *T.T. Yü17,136*.

Distribution: Eastern Himalaya, from Yunnan to Nepal, at 2,900–4,500 m.

Representative specimen: Nepal: Annapurna Himal. Seti Khola, *Stainton, Sykes & Williams 6538* (AAH, BM).

3. **Pimpinella leschenaultii** DC., Prodr. 4: 122. 1830; C. B. Clarke in Hook. f., Fl. Brit. Ind. 2: 687. 1879.

Plants slender, erect, branched and sparsely leafy above, ± hirsute or tomentose, 20–120 cm tall, perennial from a fusiform root. Basal leaves subrosulate, ovate- to orbicular-cordate, 3–10 cm in diameter, simple, obtuse at apex, deeply cordate at base, crenate to crenate-serrate or -dentate, palmately 3–5-veined, setulose on veins and margins above, sparsely tomentose beneath; petiole stout, 6–12 cm long, hirsute, dilated and broadly sheathing at very base; cauline leaves very few, the intermediate 3- or 5-lobed with broad lobes, the uppermost with linear segments, wholly sheathing. Peduncles rather slender, 2–5 cm long, hirtellous. Involucre and involucel lacking, or of 1 to several inconspicuous, linear, hirsutulous bracts or bracteoles. Rays 6–16, slender, spreading-ascending, subequal, hirtellous, 2–3 cm long. Umbellets 6–12-flowered, the mature pedicels hirsutulous, subequal, 3–5 mm long. Flowers white; petals obovate, glabrous; calyx obsolete; stylopodium conical, the styles slender. Fruit oblong-ovoid, 2–4 mm long, 1.5–2 mm broad, a little narrowed at apex, compressed laterally, the mericarps compressed dorsally in transection, essentially glabrous at maturity, the ribs slender. Vittae 2 or 3 in intervals, 2–4 on commissure. Commissural face planoconcave. Carpophore bipartite. Mar.–Aug.

Type locality: "in Indiae orientalis montibus Nellygerry," *Leschenault*.

Distribution: Hills of peninsular India; Sri Lanka.

Representative specimens: Tamil Nadu: Nilgiri, *Walker-Arnott 1140* (MH), *Wight* (CAL, MH); Kodaikanal, *Bourne 414* (CAL); Coonoor, *Gamble 11,514* (CAL); Peratmalai, Annamalai Hills, *Fischer 3571* (CAL).

4. **Pimpinella pulneyensis** Gamble, Kew Bull. 228. 1919.

Plants slender, erect, branched, nearly leafless above, hirtellous to glabrate (?), 30–60 cm tall, perennial or biennial from a slender, fusiform taproot. Lower leaves ovate- to orbicular-cordate, 2–6 cm in diameter, simple, obtuse at apex, deeply cordate at base, crenate-serrate with cartilaginous teeth, subpalmately 7–10 veined, mostly glabrate above, hirtellous beneath; petiole very slender, 5–30 cm long, hirsute, dilated and scarious-sheathing at base; cauline leaves very few, strongly reduced upward, the upper ternate with linear divisions and inconspicuous sheaths. Peduncles terminal and lateral, rather slender, 3–10 cm long, hirsutulous. Involucre lacking, or of 1 or 2 diminutive linear bracts. Rays 7–12(–20?), slender, spreading-ascending, subequal, hirsutulous, 2–3.5 cm long. Umbellets 8–15-flowered, the mature pedicels hirsutulous, unequal, ascending, 2–5 mm long. Involucel of 3–8 linear, hirsutulous bracteoles 2–4 mm long. Flowers white; petals obovate, hirsute dorsally; calyx obsolete; stylopodium conical, the styles slender, reflexed. Fruit ovoid, 2–2.5 mm long, 1.5–2 mm broad, sharply narrowed at apex, rounded at base, strongly compressed laterally, constricted at commissure and subdidymous, the mericarps pentagonal and slightly compressed dorsally in transection, strongly rugose to glabrate (?), the ribs prominent, acute. Vittae usually 3 in intervals, 2 on commissure. Commissural face plane. Carpophore broad, stout, entire. May–Nov.

Type locality: Pulney Hills, Bear Shola, Kodaikanal, Tamil Nadu, *Bourne 24*.

Distribution: Apparently endemic to Pulney Hills, Peninsular India.

Representative specimens: Tamil Nadu: Kodaikanal, *Bourne 24* (CAL); Kodaikanal Station, *A. Saulière 50* (UC).

5. **Pimpinella tongloensis** P. K. Mukherjee,
Bull. Bot. Surv. India 12: 7. 1970.

Pimpinella urceolata Watt ex Banerji, Jour. Bombay Nat. Hist. Soc. 50: 89. 1951, *pro parte*.

Plants slender, erect, branched, leafy above, softly pubescent or pilose throughout, 30–50 cm tall, annual (?) from a slender taproot. Basal leaves few, ovate- to orbicular-cordate, ca. 1.5 cm in diameter, simple (those subtending lateral branches similar); petiole slender, 3–8 cm long, pubescent, sheathing at base; cauline leaves ovate-triangular, trifoliate, the leaflets ovate-acuminate, rounded, cordate, or oblique at base, crenate-dentate, pubescent on veins beneath, 2.5–6.5 cm long, 1.5–4 cm broad, on petiolules 1–2.5 cm long, the uppermost leaves reduced. Peduncles slender, pubescent, up to 10 cm long. Involucre lacking. Rays 8–12, slender but rigid, spreading, unequal, pubescent, 1–3 cm long. Umbellets 8–10-flowered, the mature pedicels pubescent, subequal, spreading, 3–6 mm long. Involucel of 3–5 linear bracteoles ca. 3 mm long. Flowers white; petals obovate, glabrous; calyx teeth minute, triangular; stylopodium low conical. Fruit ovoid, ca. 2 mm long, 1.5 mm broad, subterete, strongly narrowed at apex, the mericarps compressed dorsally in transection, glabrous, the ribs obscure. Vittae 2 or 3 in intervals, 2–4 on commissure. Commissural face plane. Carpophore bipartite. July–Oct.

Type locality: Sikkim, *Kurz* (CAL, holotype).

Distribution: Sikkim and Darjeeling Himalaya, 2,000–2,500 m.

Representative specimens: West Bengal: Tonglo, *Gamble 16,378* (CAL), *Thomson* (CAL).

6. **Pimpinella inundata** (Farille & Malla) P. K. Mukherjee &
Constance, Edinburgh Jour. Bot. 48(1): 44. 1991.

Ligusticum inundatum Farille & Malla in Farille, Cauwet-Marc & Malla, Candollea 40(2): 551. 1985.

Plants low to slender, caulescent, weakly erect to ascending, simple or sparsely branched from base, more or less hirtellous on all vegetative parts, 5–40 cm tall, from a slender simple or branched taproot. Leaves orbicular to oblong-oval, 1–2.5 cm long,

1-2 cm broad, (trifid) ternate or simply pinnate with 3 or 5 leaflets, the leaflets orbicular, 0.5-1.5 cm in diameter, crenate or mucronulate-dentate, incised, sessile or subsessile, rounded or truncate at base; petiole slender, 1-5 cm long, narrowly scarious-sheathing; cauline leaves few, the lower like the basal, petiolate, the upper reduced, with much narrower leaflets and sessile or subsessile on dilated oval, scarious-ciliate-margined sheaths. Peduncles slender, terminal, 2-12 cm long. Involucre of 1 or 2 linear or foliaceous bracts, or 0. Rays 4-10, slender, subequal to ± unequal, spreading-ascending, hirsutulous, 1-3 cm long. Umbellets 8-15-flowered, the mature pedicels subequal, spreading-ascending, hirsutulous, 2-5 mm long. Involucel of 1-3 linear, hirsutulous bracteoles up to 1 mm long. Flowers white or purplish-red (sometimes with a central white stripe); petals a little unequal, ovate to obovate, plane or nearly so, obtuse or somewhat acute with a very short narrower apex, shortly clawed at base; calyx teeth 0; stylopodium conical (mammillate), about equal to or a little shorter than the reflexed styles. Fruit oblong-ovoid, 1-1.5 mm in diameter, narrower toward apex, rounded at base, compressed laterally, constricted at commissure, the mericarps a little compressed dorsally, the ribs filiform, acute. Vittae usually 3 in intervals, 4-6 on commissure. Commissural face plane. Carpophore cleft about 1/3. June-Oct.

Type locality: "C. Népal: Dhorpatan, 2840 m," 17 Aug. 1981, *Farille 81,177* (E, isotype).

Distribution: Himalaya of Nepal, 2,600-4,900 m.

Representative specimens: Nepal: Thimgaon, Muktinath Himal, *Stainton, Sykes & Williams 1355* (BM, E); Kali Gandaki, S of Tikucha, *Stainton, Sykes & Williams 1733* (BM, E); above Dorpatan, *Stainton, Sykes & Williams 3478* (BM, E); near Maikot, *Stainton, Sykes & Williams 4770* (BM, E, G); near Tarakot, Bheri River, *Polunin, Sykes & Williams 2386* (BM, E); near Balangra Pass, *Polunin, Sykes & Williams 2573* (BM, E).

7. **Pimpinella pimpinellisimulacrum** (Farille & Malla) Farille in Farille, Cauwet-Marc & Malla, Candollea 40(2): 554. 1985.

Similisinocarum pimpinellisimulacrum Farille & Malla in Farille, Cauwet-Marc & Malla, Bull. Soc. Bot. France 131, Lettres bot. 1984(1): 70. 1984.

Plants slender, caulescent, erect, simple, sparsely pubescent in foliage and scaberulous in inflorescence, 10-30 cm tall, perennial from a slender taproot. Leaves 2-5, mostly basal, orbicular to oval, 2-3 cm long, 3-4 cm broad, ternate, the leaflets 3-(rarely 5-) parted, dentate-lobed, the terminal deeply 3-parted, the venation conspicuous beneath; petiole slender, 2-3 cm long, narrowly sheathing; cauline leaves 1 or 2 and like the basal, or lacking. Peduncles slender, terminal, scaberulous at apex, 10-30 cm long. Involucre of 1 or 2 linear, entire (rarely bi- or trifid) bracts, or lacking. Rays 4-6 (-10), slender, subequal, ascending, scaberulous at apex, webbed, up to 3 cm long. Umbellets 6-8-flowered, the mature pedicels unequal, ascending, to 5 mm long. Involucel of 3-6 very acute, entire, spreading or reflexed bracteoles 4-6 mm long. Flowers purple; petals broadly obovate, obtuse, with a very short narrower apex, shortly clawed at base; calyx teeth acute and up to 2 mm long, or 0; stylopodium convex, the styles short, divaricate or reflexed on the stylopodium. Immature fruit ovoid, ca. 2 mm long, compressed laterally, broadest at commissure, narrowed toward apex, rounded at base, the mericarps a little compressed dorsally, the ribs filiform, acute. Vittae 3 in intervals, 4-6 on commissure. Commissural face plane. Carpophore not seen. Sept.

Type locality: "Népal, haute vallée de la Kali Gandaki, 'spigolo' au SSE de Muktinath (Muktinath-Range, au nord de l'Annapurna) . . . 4500 m," 18 Sept. 1981, *Farille 81,422*.

Distribution: Himalaya of Nepal, 4,500 m.

Specimens examined: Nepal, near Seng Khola, 11,500 feet, 6 Oct. 1954, *Stainton, Sykes & Williams 4704* (AAH, BM, E).

8. **Pimpinella flaccida** C. B. Clarke, Jour. Linn. Soc. Bot. 25: 28. 1889.

Carum flaccidum (C. B. Clarke) Franchet, Bull. Bot. Soc. Philom. Paris VIII. 6: 126. 1894.

Plants slender, flaccid, erect, branched and sparsely leafy above, ± pilose at least on younger parts, annual (?), from a slender taproot. Leaves triangular-ovate, pinnate with 3–7 broad, ovate-triangular to oblong-ovate, petiolulate leaflets, acute at apex, truncate to cuneate at base, crenate-serrate; petiole slender, narrowly sheathing at base; cauline leaves reduced upward, with oblong to linear divisions and inconspicuous sheaths. Peduncles terminal and lateral, very slender. Involucre and involucel lacking. Rays 3 or 4, very slender, spreading or divaricate, subequal, 0.5–2 cm long. Umbellets 3–6(–10?)-flowered, the mature pedicels slender, spreading, 6–8 mm long. Flowers not seen; calyx obsolete; stylopodium conical, the styles short, erect. Fruit ovoid, ca. 2 mm in diameter, subterete, the mericarps glabrous or minutely pilose (?), the ribs filiform. Vittae not seen. Carpophore apparently bipartite.

Type locality: "Kohima, alt. 5300 feet," Nagaland, *Clarke 41,139*.

Distribution: Extreme northeastern India and adjacent China.

Representative specimen: Nagaland: Kohima, *Clarke 41,139* (K).

9. **Pimpinella nervosa** C. B. Clarke in Hook. f., Fl. Brit. Ind. 2: 684. 1879.

Plants rather stout, erect, branched and sparsely leafy above, scaberulous and ± puberulent, 40–120 cm tall, perennial from a fusiform taproot. Basal and lower cauline leaves ovate, biternate, the leaflets ovate-lanceolate, mucronate-dentate or incised, glabrous above, scaberulous on the prominent veins beneath, up to 1.5–5 cm long, ca. 2 cm broad; petiole equal to lamina, narrowly sheathing at base; upper cauline leaves reduced, subsessile. Peduncles terminal and lateral, stout, 10–25 cm long, scaberulous at apex. Involucre lacking. Rays 6–16, slender, ascending, subequal, scaberulous, 2–8 cm long. Umbellets 15–25-flowered, not all maturing, the mature pedicels scaberulous, unequal, spreading, 2–6 mm long. Involucel of 6–8 linear, puberulent bracteoles 4–8 mm long. Flowers greenish-white or purplish (?); petals oval to obovate, glabrous; calyx teeth minute, triangular, or calyx obsolete; stylopodium low conical, the styles moderately slender, reflexed. Fruit oblong-ovoid, 2.5–3 mm long, 2–2.5 mm broad, narrowed at apex, rounded at base, compressed laterally, constricted at commissure and subdidymous, the mericarps pentagonal and compressed dorsally in transection, glabrous, the ribs prominent, filiform. Vittae 2 or 3 in intervals, 2 on commissure. Commissural face planoconcave. Carpophore bipartite. Aug.–Nov.

Type locality: "KHASIA MTS.; Nonkreem, alt. 5500 ft.," *Hooker & Thomson*.

Distribution: Khasia Hills of northeastern India, 1,500–2,000 m.

Representative specimens: Nagaland: Takubama, Naga Hills, *Chand 3390* (UC), *Koelz 26,628* (UC). Meghalaya: Myrang, Khasia, *Clarke 45,465G* (CAL); Shillong, *Clarke 44,623* (CAL); Khasia, Nonkrem, *Hooker & Thomson* (K).

10. **Pimpinella heyneana** (DC.) Kurz, Jour. Asiat. Soc. Bengal 46: 115. 1877; C. B. Clarke in Hook. f., Fl. Brit. Ind. 2: 684. 1879.

Pimpinella heyneana Wallich, Cat. 566B. 1829, *nomen nudum*; *Helosciadium ? heyneanum* DC., Prodr. 4: 106. 1830.
Anethum trifoliatum Roxb., Fl. Ind. ed. 2. 2: 96. 1832. Based on *Roxburgh*, "Circars," fide C. B. Clarke.
Pimpinella multiradiata Santapau, Kew Bull. 485. 1949. Based on *Santapau 8054*, "near Tata's pipe, 600 m, Khandala."

Plants very slender, erect, branched and sparsely leafy above, 50–150 cm tall, sparsely hirtellous to glabrate, the sheaths ciliate, annual from a slender taproot. Basal and lower cauline leaves orbicular, 5–12 cm in diameter, trifoliate or 1-2-tripartite or ternate, the leaflets ovate-lanceolate to lanceolate, acuminate, cuneate at base, sharply serrate to incised, glabrate or slightly hirsutulous on veins beneath, 1–5 cm long, 0.5–2 cm broad; petiole slender, to 6 cm long, the narrow sheath ciliate; cauline leaves reduced upward, the uppermost ternate with linearly lanceolate divisions. Peduncles terminal and lateral, slender, 3–15 cm long, glabrous. Involucre lacking. Rays 5–15(–30), filiform, spreading, unequal, glabrous, 1–4.5 cm long, slightly webbed. Umbellets 3-10-flowered, the mature pedicels filiform, spreading, unequal, glabrous, 2–10 mm long. Involucel of 1–3 filiform bracteoles 2–7 mm long, or involucel lacking (?). Flowers white; petals oval, glabrous; calyx obsolete; stylopodium obsolete, the short styles reflexed. Fruit ellipsoid to globose, 1–2 mm in diameter, rounded at both ends, compressed laterally, constricted at commissure and didymous, the mericarps pentagonal-subterete in transection, glabrous, the ribs evident, filiform, obtuse, light-colored. Vittae usually 2 in intervals, 2 on commissure. Commissural face plane. Carpophore filiform, bifid to bipartite. Feb.–Dec.

Type locality: "In India Orientalis," *Heyne*.

Distribution: Widespread in hills of peninsular India, 250–1,300 m; Burma, Sri Lanka.

Representative specimens: Bihar: Singbhum, *Haines 459* (CAL). Orissa: Kiapodar, Kalahandi, *Mooney 3172* (DD). Madhya Pradesh: Danga, *Gamble 3935C* (DD); Dahin-Bori road, *Joseph 11,230* (MH). Andhra Pradesh: Adativalasa, Vishakapatnam Distr., *Subbarao 21,618* (MH). Tamil Nadu: Coimbatore, *Fischer 1393* (CAL); Annamalay, *Wight 1174* (CAL, UC); Kumili, Madurai Distr., *Subramanyam 9469* (MH). Kerala: Perambicolum, Cochin, *Meebold 12,359* (CAL). Karnataka: Sampagi, S. Canara, *Barber 2304* (MH); Attapadi Valley, *Fischer 2337* (CAL). Maharashtra: Khandala, *Santapau 8053* (BLAT); Singhgadh, 24 Oct. 1982, *Mukherjee s. n.* (CUH, UC).

11. **Pimpinella sikkimensis** C. B. Clarke in Hook. f., Fl. Brit. Ind. 2: 685. 1879.

Plants erect or scandent, branched, glabrous to hirtellous or scaberulous, 30–120 cm tall from a taproot. Leaves ovate, 4–12 cm long, 5–15 cm broad, ternate-pinnately or bipinnately decompound, the leaflets ovate to lanceolate, spinulose-serrate, incised or pinnately lobed, the terminal often trilobed, 4–15 mm long, 3–12 mm broad; petiole 2–10 cm long, exceeding lamina; cauline leaves reduced upward, ternately or pinnately decompound with oblong ultimate divisions. Peduncles terminal and lateral, stout, 5–15 cm long. Involucre of a few inconspicuous linear bracts ca. 5 mm long, or lacking. Fertile rays 6–16, usually 7, stout, divaricate, angled, unequal, glabrous, 1.5–5 cm long. Umbellets up to 20-flowered, but only 6–8 maturing, the mature pedicels 6–15 mm long, glabrous. Involucel of usually 5 inconspicuous, linear, setaceous bracteoles 1–3 mm long. Flowers greenish-white; petals oblanceolate, subequal; calyx teeth prominent, lanceolate, persistent, to 0.5 mm long; stylopodium stout, conical, about equaling the short styles. Fruit oblong-ovoid, 2.5–6 mm long, 2–3 mm broad, narrowed into a slight neck toward apex, compressed laterally and constricted at commissure, the mericarps subterete in transection, glabrous, the ribs slender, evident but inconspicuous. Vittae 1 or 2 in intervals, 2 on commissure. Commissural face concave. Carpophore entire. July–Nov.

Type locality: "Sikkim; common in damp forests, alt. 4,000–7,500 ft.," *Hooker, Clarke.*

Distribution: Himalaya of Sikkim and West Bengal, 1,000–3,000 m.

Representative specimens: West Bengal: Rungbee, Darjeeling, *Clarke 12,409* (K), *24,818* (K), *24,818D* (CAL); Ghoom Pahar, *Gamble 7278* (K); Darjeeling, *Clarke 9645A* (K), *C* (CAL). Sikkim: w/o further locality, *Hooker* (CAL, K), *King* (CAL).

12. **Pimpinella wallichii** C. B. Clarke in Hook. f., Fl. Brit. Ind. 2: 685. 1879.

Plants slender, erect, corymbosely branched, glabrous, 40–125 cm tall, from a fusiform root. Leaves ovate, 6–15 cm long, pinnately or ternately decompound, the ultimate divisions lanceolate, 1–5 mm long, 1 mm broad. Involucre lacking. Rays 6–12, rigid, very unequal, up to 1.5 cm long. Umbellets 8–12-flowered, the mature pedicels 2–2.5 mm long. Involucel of 2–4 linear bracteoles ca. 2 mm long, or involucel lacking. Flowers not seen; stylopodium inconspicuous. Fruit ellipsoid, 1.5–2 mm in diameter, slightly broader than long, compressed laterally, the mericarps compressed dorsally in transection, glabrous, the ribs prominent. Vittae 2 or 3 in intervals, 4 on commissure. Commissural face plane. Carpophore bipartite. Oct.

Type locality: "NIPAL; *Wallich in Herb. Hook.* SIKKIM; alt. 4,000–6,000 ft., frequent," *Clarke.*

Distribution: Himalaya of Sikkim and Nepal, 1,350–2,350 m.

Representative specimens: Nepal: Nepalia, *Wallich 1830* (K). West Bengal: Darjeeling, *Clarke 9645E* (CAL). Sikkim: Kulhait, *Clarke 25,512C* (K); Heeloo, *Clarke 13,142C*, *D* (CAL), *E* (K); Hee, *Clarke 13,039A* (K).

Note. Clarke's practice of appending letters of the alphabet to his collection numbers, presumably indicating duplicates, has sometimes led to confusion. Thus, while *Clarke 9645A* (K), *C* (CAL), and *13,142* (K) are all *P. sikkimensis*, *9645E* (CAL) and *13,142C*, *D* (CAL), *E* (K) all represent *P. wallichii*.

13. **Pimpinella acuminata** (Edgew.) C. B. Clarke in Hook. f., Fl. Brit. Ind. 2: 686. 1879.

Reutera acuminata Edgew., Trans. Linn. Soc. Bot. 20: 52. 1851.

Pimpinella hazariensis H. Wolff, Feddes Repert. 27: 331. 1930. Based on *Inayat Khan*, "Kashmir: Kund, Hazara," 15 Aug. 1899.

Plants slender, erect, branched, leafy, glabrous or foliage sparsely puberulent, 20–100 cm tall, perennial from a fusiform taproot. Leaves ovate-triangular, 5–20 cm in diameter, 1-2-ternate, -ternate-pinnate, or -pinnate, the leaflets ovate to lanceolate, coarsely serrate to pinnately incised, 1.5–4 cm long, 0.5–2.5 cm broad, puberulent on veins beneath, or glabrate; petiole stout, 5–15 cm long, narrowly sheathing; uppermost cauline leaves pinnate or ternate with narrower divisions, subsessile. Peduncles terminal and lateral, very slender, 3–15 cm long. Involucre of 2–5 filiform, scarious-margined, reflexed bracts 3–20 mm long. Rays 4–10, filiform, spreading, 2–6 cm long. Umbellets 7-12-flowered, the mature pedicels filiform, divaricate, unequal, 8–18 mm long. Involucel of 3–6 filiform, scarious-margined, reflexed bracteoles 2–6 mm long. Flowers yellowish-white; petals oval to obovate, glabrous; calyx obsolete; stylopodium low conical, the styles moderately slender, reflexed. Fruit ovoid to subglobose, 2–3 mm in diameter, compressed laterally, constricted at commissure and subdidymous, the mericarps subterete-pentagonal in transection, glabrous, the ribs filiform, inconspicuous. Vittae 1 or 2 in intervals, 2 on commissure. Commissural face planoconcave. Carpophore bipartite. Aug.–Nov.

Type locality: "Himala, ad alt. ped. 6,000–7,000; Masúri, Chepal, & C.", *Edgeworth 149.*

Distribution: Western Himalaya, from Kumaon to Kashmir and Pakistan, 1,200–2,700 m.

Representative specimens: Jammu & Kashmir: Kishtawar, *Clarke 31, 379C* (CAL). Himachal Pradesh: Manali, *Bor 15,309* (DD); Panwi, Simla Distr., *Brandis 3716* (CAL, DD). Uttar Pradesh: Chapal, *Edgeworth* (K); Masúri, *Edgeworth* (K); Mussoorie, *Saxena 1131* (DD), *Royle 93/61* (DD); Chakrata, *Bhattacharyya 13,061*; Kapkot/Loharket, *T. A. Rao 4177* (BSD); Genwali, Tehri State, *Chand 1201* (UC).

14. **Pimpinella evoluta** (C. B. Clarke) Hiroe, Umbell. World, 831. 1979.

Pimpinella tenera var. *evoluta* C. B. Clarke, Jour. Linn. Soc. Bot. 25: 27. 1899; *Acronema evolutum* (C. B. Clarke) H. Wolff in Engl., Pflanzenreich 90 ($4\frac{228}{}$): 322. 1927.

Plants slender, erect, branched and sparsely leafy above, glabrous, ca. 45 cm tall, from a swollen taproot. Leaves ovate to ovate-lanceolate, 1.5–7.5 cm long, ternate-pinnate, the leaflets ovate to ovate-lanceolate, crenate-serrate and pinnately lobed or incised or trilobed, petiolulate to confluent, 1–2 cm long, 1–1.5 cm broad; petiole very slender, 1–7.5 cm long, narrowly sheathing at base; cauline leaves reduced upward, with narrower divisions. Peduncles very slender, terminal and lateral, 6–12 cm long. Involucre lacking. Rays 4–8, slender, divaricate, subequal, 1.5–4 cm long. Umbellets 6-15-flowered, the mature pedicels capillary, subequal, spreading, 4–6 mm long. Involucel of 3–5 linear bracteoles ca. 2.5 mm long. Flowers not seen; calyx obsolete; stylopodium conical, the styles short, ascending. Fruit ovoid, 3–3.5 mm long, 1.5–2 mm broad, distinctly narrowed at apex, rounded at base, compressed laterally, constricted at commissure and subdidymous, the mericarps subterete, glabrous, the ribs slender. Vittae 2 or 3 in intervals, 2 on commissure. Commissural face plane. Carpophore bipartite. ?–Oct.

Type locality: "Jakhpo, alt. 9,900 feet," Naga Hills, *Clarke 41,277, 41,329, 42,015.*

Distribution: Apparently endemic to the Naga Hills of extreme northeastern India.

Representative specimens: Nagaland: Jakhpo, Naga Hills, *Clarke 41,329E* (CAL), *41,329A, B* (K).

15. Pimpinella tirupatiensis Balakr. & Subram., Bull. Bot. Surv. India 2: 427. 1960.

Plants rather slender but rigid, erect, branched and sparsely leafy above, glabrous, 40–60 cm tall, perennial from a fusiform taproot. Basal and lower cauline leaves ovate, 1.5–4 cm long, 1–3.5 cm broad, simple, obtuse or acute, cordate at base, crenate-serrate with cartilaginous teeth, palmately 9–11-veined, reddish-brown beneath; petiole slender, 2.5–8 cm long, narrowly sheathing; upper cauline leaves petiolate, ternate with linearly lanceolate to linear, acute, lobed or incised segments 1–2.5 cm long, or the uppermost reduced to a narrow sheath. Peduncles slender, mostly opposite leaves. Involucre and involucel of 1 or 2 minute linear bracts or bracteoles, or lacking. Rays 3–6, slender, spreading-ascending, 1–2.5 cm long. Umbellets 5–16-flowered, the mature pedicels slender, 2–4 mm long. Flowers white; petals elliptic to obovate, glabrous; calyx obsolete; stylopodium conical, the styles slender, spreading. Fruit ovoid-globose, 1.5–2 mm in diameter, compressed laterally, constricted at commissure and subdidymous, the mericarps subterete in transection, scaberulous-papillose, the ribs obscure. Vittae 2 or 3 in intervals, 2 on commissure. Commissural face plane. Carpophore entire. Oct.

Type locality: "on the way to Japalathirtha in Tirupati Hills, Chittoor District, Andhra Pradesh," *Subramanyam & Balakrishnan 6889* (CAL, holotype).

Distribution: Apparently confined to the Eastern Ghats of peninsular India, ca. 1,000 m. Only the type collection seen.

16. Pimpinella candolleana Wight & Arn., Prodr. Fl. Ind. Orient. 369. 1834; C. B. Clarke in Hook. f., Fl. Brit. Ind. 2: 687. 1879.

Carum candolleanum Franchet, Bull. Soc. Philom. Paris VIII. 6: 128. 1894, as to name only.

Plants slender, erect, branched and sparsely leafy above, densely shortly pubescent, usually 30–60 cm tall, perennial from a fusiform taproot. Basal leaves subrosulate, ovate-cordate, 2–6 cm long, 1.5–6 cm broad, simple (rarely trilobed), somewhat acute to obtuse at apex, deeply cordate, crenate-serrate with subcartilaginous teeth, palmately 5–9-veined, pubescent on both surfaces but especially beneath; petiole slender, up to 10 cm long, pubescent, sheathing at base; cauline leaves reduced upward, ternate to pinnatifid with oblong, acute, sometimes incised or lobed segments, the uppermost subsessile. Peduncles few, terminal or lateral, elongate. Involucre of 3–8 linear, pubescent bracts up to 10 mm long. Rays 8–20, spreading-ascending, subequal, densely pubescent, ca. 2.5 cm long. Umbellets 10–15-flowered, the central often sterile, the mature pedicels pubescent, unequal, a little contracted in flower but divergent in fruit, 2–4 mm long. Involucel of about 8 linear, pubescent bracteoles 2–4 mm long. Flowers white; petals obovate, hirsute dorsally; calyx obsolete; stylopodium conical, the styles slender, reflexed. Fruit ovoid, ca. 2.5 mm long, 2 mm broad, a little narrowed at apex, compressed laterally, constricted at commissure, the mericarps pentagonal or slightly compressed dorsally in transection, scaberulous-tuberculate, the ribs slender. Vittae 2 or 3 in intervals, 2 on commissure. Commissural surface plane. Carpophore entire or shortly bifid. Mar.–July.

Type locality: Peninsular India, w/o precise locality, *Wight 1202*.

Distribution: Hills of peninsular India, 1,500–2,000 m.

Representative specimens: "Peninsular India," *Wight 1173* (CAL, K). Tamil Nadu. Nilgiris, *Wight* (CAL); Kodaikanal, *Bourne 1098* (CAL). Kerala: Pykara, *King* (CAL); Lockhert Gap, Kottayam Distr., *Sebastine 14,491* (MH).

Note. Buwalda (1936, pp. 188–191) questions the specific distinctness from each other of *P. candolleana*, *P. leschenaultii*, and *P. pulneyensis*, and of these (by implication) from Chinese, southeastern Asian, and Indonesian taxa. We are not convinced that such drastic mergers are required.

17. **Pimpinella stracheyi** C. B. Clarke in Hook. f., Fl. Brit. Ind. 2: 688. 1879.

Plants low, cespitose, freely branched and leafy, grayish-tomentose, 15–40 cm tall, from a slender taproot. Leaves numerous, subrosulate, ovate-cordate, ca. 2 cm long, 1.5 cm broad, simple, serrate, dentate, or incised; petiole elongated, sheathing at base; lower cauline leaves ternate or pinnate, petiolate, up to 8 cm long with 5 ovate leaflets, the uppermost ternate. Peduncles slender but rather short, terminal and lateral. Involucre and involucel lacking. Rays ca. 10, very unequal. Umbellets ca. 10-flowered, the mature pedicels short, subequal. Petals elliptic, pubescent dorsally; calyx obsolete; stylopodium conical; ovary villous. Fruit oblong-elliptic, 2–3 mm long, 1–1.5 mm broad, narrowed at apex and base, compressed laterally, the mericarps subterete in transection, grayish-villous or glabrescent, the ribs slender. Vittae usually 2 in intervals, 2 on commissure. Commissural face plane. Carpophore bilobed.

Type locality: "KUMAON: Dhauli River, Samangentha, alt. 7,500 ft.," Uttar Pradesh, *Strachey & Winterbottom*.

Distribution: Western Himalaya.

Representative specimens: "W. Himalayas," *Hooker & Thomson* (Herb. Ind. Orient., *Pimpinella 3*) (CAL). Himachal Pradesh: Tissa, Chambra Distr., *Nair 32,953* (BSD). Uttar Pradesh: Kumaon, *Strachey & Winterbottom* (CAL).

18. **Pimpinella tomentosa** (Dalzell & Gibson) C. B. Clarke in Hook. f., Fl. Brit. Ind. 2: 689. 1879.

Heracleum tomentosum Dalzell & Gibson, Bombay Fl. 108, 313. 1861.

Plants slender, erect, profusely branched, densely hirtellous 50–200 cm tall, perennial from a taproot. Basal leaves rosulate, 2–3-ternate or -pinnate, the leaflets ovate, spinose-serrate to acutely laciniate, petiolulate; petiole 3–8 cm long, dilated-sheathing at base; upper cauline leaves pinnatifid or reduced to bladeless sheaths. Peduncles mostly leaf opposed. Involucre lacking. Rays 4–10, hirsutulous, 1.5–2.5 cm long. Umbellets 5–15-flowered, the mature pedicels 2.5–4 mm long. Involucel of 2–4 linear bracteoles ca. 2.5 mm long, or involucel lacking. Flowers white; petals ovate; calyx obsolete; stylopodium conical, the styles elongate, erect or diverging. Fruit oblong-ovoid, 3.5–4.5 mm long, 1.5–2 mm broad, narrowed toward apex, rounded to cordate at base, compressed laterally, constricted at commissure, the mericarps pentagonal-subterete in transection, papillose-scaberulous, the ribs obscure. Vittae 2 or 3 in intervals, 2 on commissure. Commissural face plane. Carpophore bipartite. Aug.–Dec.

Type locality: "BOMBAY," *Dalzell*.

Distribution: Western Ghats of peninsular India, ca. 1,500 m.

Representative specimens: Maharashtra: Mahabaleswar, *Sedgewick & Bell 4593* (CAL), 24 Dec. 1982, *Mukherjee s. n.* (CUH, UC); Bombay Presidency, *Gibson* (CAL); Bombay, *Dalzell* (CAL); Concan, *Stocks* (CAL).

19. **Pimpinella wallichiana** (Hohen.) Gandhi in Saldanha & Nicolson, Fl. Hassan Distr. 417. 1976.

Helosciadium wallichianum Miq. ex Hohen., Bot. Zeitg. 7: 775. 1849.

Pimpinella monoica Dalzell, Hook. Jour. Bot. 3: 212. 1851; C. B. Clarke in Hook. f., Fl. Brit. Ind. 2: 687. 1879. Based on Dalzell, "Bombay."

Pimpinella duthiei H. Wolff, Feddes Repert. 27: 330. 1930. Based on *Duthie 10,408*, "India orientalis: Near the Bee Falls, Pacmashi [Panchmari]," Madhya Pradesh (DD, lectotype).

Pimpinella katrajensis Rao & Hemadri, Indian Forester 102: 232. 1976. Based on *Hemadri 108,794A*, "Katraj Ghat near Poona," Maharashtra (CAL, holotype).

Plants slender, erect, much branched above, hirsute to puberulent or glabrate, 6–240 cm tall, biennial (?) from a fusiform root. Basal (and sometimes all) leaves triangular-ovate, ca. 6 cm long, 7 cm broad, simple, cordate to hastate at base, crenate-serrate with white cartilaginous teeth, palmately 3-veined, glabrous above, hirsute on veins beneath; petiole 5–20 cm long, hirsute; cauline leaves usually ternate, petiolate, with ovate-lanceolate divisions, the uppermost with narrower segments or reduced to sheaths. Inflorescence divaricate-paniculate, the peduncles mostly lateral, slender, ca. 5 mm long. Involucre and involucel usually lacking, or of one to several inconspicuous linear bracts or bracteoles. Rays 4–18, filiform, spreading, subequal, glabrous or pubescent, 8–35 mm long. Umbellets 8–20-flowered, all or only the inner maturing, the mature pedicels glabrous, capillary, unequal, 2–6 mm long. Flowers white; petals ovate; calyx obsolete; stylopodium low conical, the styles short, reflexed. Fruit ovoid-globose, 1–1.5 mm in diameter, obtuse at apex, cordate at base, compressed laterally, constricted at commissure and subdidymous, the mericarps subterete in transection, minutely papillose, the ribs slender or obscure. Vittae 2 in intervals, 2 on commissure. Commissural face plane. Carpophore ± bifid. July–Feb.

Type locality: "Pl. e territoris Canara, Maharattarum aust rariori et Malayalim," *Miquel*.

Distribution: Hills of peninsular India, 500–1,500 m.

Representative specimens: Bihar: Jaigi, Palamau, *Gamble 8861* (CAL). Orissa: Jashpur, *Wood 160* (CAL); Jashpur, Sergooja, *Clarke 34,032* (CAL). Andhra Pradesh: Anantagiri, *Subbarao 32,827* (MH); Ramanapenta, Kurnool, *Ellis 32,634* (MH). Tamil Nadu: Madhumalai, Nilghiris, *Gamble 15,960* (CAL); w/o further locality, *Wight 1175* (CAL). Kerala: Kurichiyat, Calicut Distr., *Ellis 20,436* (MH). Karnataka: "prope Mercara," *Miquel* (*Hohenacker*, Pl. Ind. Orient. 633) (BM, K); Bababudan Hills, Mysore Distr., *Talbot 2356* (DD). Maharashtra: Bombay, *Dalzell* (CAL); Concan, *Stocks* (CAL); Yellapur, Dharwar, *Breganza 727* (DD); Katrap Ghat, near Poona, *Bhide 917* (CAL); Pratapgadh Ghat, from Mahabaleswar, 25 Dec. 1982, *Mukherjee s. n.* (CUH, UC).

20. Pimpinella adscendens Dalzell in Hook., Jour. Bot. 2: 261. 1850; C. B. Clarke in Hook. f., Fl. Brit. Ind. 2: 689. 1879.

Plants low, caespitose, with numerous ascending branches, sparsely leafy above, glabrous to hispidulous, 15–60 cm tall, perennial. Leaves clustered toward base, oblong, 6–15 cm long (including petiole), simply pinnate with 4–8 pairs of oval to orbicular, obtuse, cuneate or truncate, coarsely crenate-serrate leaflets 4–15 mm long, the terminal leaflet the largest; cauline leaves reduced upward, the uppermost often more divided and with narrower divisions. Peduncles terminal only. Involucre and involucel usually of 4–8 linearly subulate, ciliolate bracts or bracteoles 2–4 mm long. Rays 4–12, unequal, hispidulous to glabrous, 3–15 mm long. Umbellets 8–16-flowered, the mature pedicels unequal, 1.5–3 mm long. Flowers white; petals obovate, pubescent dorsally; calyx obsolete; stylopodium large, depressed, the styles slender. Fruit ovoid, ca. 1.5 mm long, narrowed toward apex, compressed laterally, constricted at commissure and subdidymous, the mericarps subterete in transection, hispidulous, the ribs obscure. Vittae 2 or 3 in intervals, 2 on commissure. Commissural face plane. Carpophore entire. Oct.–Nov.

Type locality: "In fluminum Concananensium ripis," Bombay.

Distribution: Western Ghats of Peninsular India.

Representative specimens: Maharashtra: Bombay, *Dalzell* (CAL); Malabar, *Stocks* (CAL).

21. Pimpinella urceolata Watt ex M. L. Banerji, Jour. Bombay Nat. Hist. Soc. 50: 89. 1951.

Plants slender, erect, corymbosely branched and sparsely leafy above, pubescent to glabrate, 45–100 cm tall, from a fusiform root. Basal leaves not seen, but probably (like those subtending lateral branches) simple, ovate-cordate, the lower cauline leaves oblong-ovate, 5–15 cm long, simply pinnate, rarely ternate, with 5 (or 3) ovate to elliptic, acute to obtuse, cordate to cuneate, crenate-dentate leaflets 0.5–6 cm in diameter, usually glabrate above, pubescent on veins beneath; petiole slender, 1.5–8 cm long, pubescent, sheathing at base; cauline leaves reduced upward, the uppermost with much narrower divisions. Peduncles terminal and lateral, very slender, pubescent, 0.5–3 cm long. Involucre and involucel usually lacking. Rays 4–6, filiform, spreading-ascending, unequal, pubescent, 3–15 mm long. Umbellets 5–8-flowered, the mature pedicels filiform, pubescent, very unequal, 5 mm long to obsolete. Flowers white; petals ovate to elliptic, pubescent dorsally; calyx obsolete; stylopodium obsolete. Fruit ovoid-globose, 1–1.5 mm in diameter, slightly broader than long, strongly compressed laterally, constricted at commissure and subdidymous, the mericarps pentagonal, pilose-papillose, the ribs obsolete. Vittae 2 or 3 in intervals, 2–4 on commissure. Commissural face plane. Carpophore entire. Sept.–Dec.

Type locality: Sikkim, Rinchingpong and Gassing [not Tonglo, as indicated], *Anderson 623* (CAL, holotype).

Distribution: Darjeeling and Sikkim Himalaya, 600–2,000 m

Representative specimens: West Bengal: Batasi-Manibhanjan, Darjeeling Distr., *Mukherjee 94* (CUH); Little Rangeet-Darjeeling, *Kurz* (CAL). Sikkim: Birhi-Rinchingpong, *Anderson 612* (CAL); w/o further locality, *Kurz* (CAL); Tashiding, *Thomson* (CAL); Gorkhey Village, *Mukherjee 312* (CUH, UC).

Note. The specimens cited by M. L. Banerji in the original description, in my opinion, represent two different taxa. These are remarkably similar in habit but differ substantially in characters of umbel and fruit. *Anderson 623*, although not from Tonglo and not conforming well to the description, must nevertheless be retained as the holotype of *P. urceolata*. A paratype, "Sikkim," *Kurz*, becomes the holotype of *P. tongloensis*. (P. K. M.)

22. Pimpinella bracteata Haines, Jour. Asiat. Soc. Bengal II. 15: 314. 1920.

Plants rather stout, erect, branched above, pubescent, 90–120 cm tall, biennial from a taproot. Basal and lower cauline leaves ovate, 5–25 cm long, simply pinnate with 3–7 ovate to ovate-lanceolate, acute, cordate or caudate, often oblique or falcate, crenate-serrate or -dentate, sometimes lobed, leaflets 1.5–7 cm long, 0.5–4 cm broad, palmately 3-5-nerved or pinnately reticulate, pubescent on both surfaces and on rachis; petiole 4–10 cm long, pubescent, sheathing at base; upper cauline leaves reduced, the uppermost pinnate or ternate with oblong to lanceolate divisions. Peduncles terminal and lateral, pubescent, 3.5–8 cm long. Involucre of 3–6 linear, pubescent bracts 6–11 mm long. Rays 8–12, spreading-ascending, subequal, pubescent, 1–3 cm long. Umbellets 12–20-flowered, the mature pedicels pubescent, unequal (the outer longer than the central), 2–6 mm long. Involucel of 3–6 linear, pubescent bracteoles 2.5–5 mm long. Flowers white; petals elliptic; calyx teeth minute, or calyx obsolete; stylopodium low conical. Fruit ovoid-globose, 1.5–2 mm in diameter, strongly compressed laterally, constricted at commissure, the mericarps compressed laterally, papillose, the ribs slender. Vittae 2 or 3 in intervals, 2 on commissure. Commissural face plane. Carpophore bipartite, the halves adhering to and falling with mericarps. May–Nov.

Type locality: "India. Mountains of Chota Nagpur, 2,000–3,000 ft.!", *Haines*.

Distribution: Hills of peninsular India, from Bihar southward, 900–1,200 m.

Representative specimens: Bihar: Netarhat, *Haines 4742* (K), *4743* (K); Ranchi, *Haines 4742* (CAL). Orissa: Jashpur, *Mooney 1824* (CAL). Andhra Pradesh: Near Araku Valley, Vishakapatnam Distr., *Balakrishnan 10,738* (CAL, MH).

Note. This species may be a connecting link between *P. diversifolia* and *P. wallichiana*. (P. K. M.)

23. Pimpinella diversifolia DC., Prodr. 4: 122. 1830; C. B. Clarke in Hook. f., Fl. Brit. Ind. 2: 688. 1879.

Heracleum diversifolium Wallich, Cat. 574B. 1829, *nomen nudum.*

Pimpinella pubescens Wallich, Cat. 567. 1829, *nomen nudum*; *Helosciadium pubescens* DC., Prodr. 4: 106. 1830. Based on *Wallich* 567, "Kamaon."

Pimpinella diversifolia var. *divisa* C. B. Clarke in Hook. f., Fl. Brit. Ind. 2: 688. 1879. Based on *Thomson* (*Hooker & Thomson*, Herb. Ind. Orient., *Pimpinella 5*), "Kashmir and Kistwar, alt. 7,000–8,000 ft."

Pimpinella diversifolia var. *simplicifolia* Kuntze, Rev. Gen. 1: 269. 1891. Based on *Kuntze*, "Sikkim, 9,000 ft."

Plants slender, erect, corymbosely branched and sparsely leafy above, hirtellous 50–150 cm tall, perennial from a fusiform taproot. Basal leaves simple or ternate, the lamina or leaflets ovate- to orbicular-cordate, 3–10 cm in diameter, acute to acuminate at apex, crenate-dentate to coarsely serrate or incised, hirsutulous on both surfaces, or only on veins, especially beneath; petiole slender, 3–18 cm long, hirsutulous, conspicuously sheathing; lower cauline leaves ternate to pinnate, the leaflets ovate to lanceolate, cordate to cuneate and often oblique at base, the uppermost greatly reduced, with narrower divisions and slightly inflated oblong sheaths. Peduncles slender, hirsutulous, 1–15 cm long. Involucre lacking, or occasionally of 1–3 scarious linear bracts. Rays 5–20, slender, spreading-ascending, unequal, hirsutulous to glabrate, 1–3 cm long. Umbellets 8–16-flowered, the mature pedicels unequal, 1–5 mm long, hirsutulous. Involucel of 3–6 linear bracteoles 1.5 mm long, or rarely lacking. Flowers white, or greenish or yellowish; petals oval to obovate, hirsutulous to glabrous, sometimes a little unequal; calyx obsolete; stylopodium low conical, the styles slender, often reflexed; ovary hirtellous. Fruit ovoid to globose, 1.5–2.5 mm in diameter, often a little broader than long, narrowed at apex, rounded at base, compressed laterally, constricted at commissure and subdidymous, the mericarps subterete in transection, scaberulous-papillose to glabrate, the ribs slender, evident to inconspicuous. Vittae 2–4 in intervals, 2–4 on commissure, or obscure. Commissural face plane. Carpophore entire or ± bifid. July–Nov.

Type locality: "In Sirmore Indiae orientalis," *Wallich 574B*.

Distribution: Throughout the Himalaya, from Assam to Afghanistan and Pakistan, 1,500–3,500 m; Japan, Korea, and China to Vietnam.

Representative specimens: Jammu & Kashmir: Kistwar, *Thomson* (CAL); Astor Valley, *Duthie 12,252* (DD). Himachal Pradesh: Kulu Valley, *Duthie 2449* (DD); Mahasu, Simla, *Gamble 4688C* (CAL); Dalhousie, *Clarke 51* (CAL). Uttar Pradesh: Chakrata, *Mishra 38,062* (BSD); Mussoorie, *Rau 3378* (BSD); Ranikhet, *Duthie 4059a* (DD); Kumaon, *Wallich 567, 574B* (CAL); Rambara, 23 Oct. 1982, *Mukherjee s. n.* (CUH, UC). Nepal: Dhunche, *Mukherjee 240* (CUH). West Bengal: Manebhanjan, Darjeeling Distr., *Mukherjee 3572* (CUH); Tonglo, *King 2464* (CAL). Sikkim: w/o further locality, *Kuntze* (CAL), *Hooker* (CAL); Lachung, *Gammie 702* (CAL). Bhutan: Thimpu, *Abraham 2340* (CAL); Dochula, *Hajra 439* (CAL). Meghalaya: Khasia, *Hooker & Thomson* (CAL), *Simons* (CAL).

Note. Extremely variable as to habit, indumentum, foliage, and fruit, there are pronounced regional differences. Eastern Himalayan plants tend to be taller and more flexuous, less hairy, have smaller basal leaves with more deeply incised margins, and fruit with more prominent ribs and roughened surfaces than do Western Himalayan plants. Plants from Khasia (northeastern India) approach *P. wallichiana* of peninsular India, but differ in possessing hirtellous pedicels, slightly larger, broader, and obtuse fruit, and a more deeply bifid carpophore. One or more of the following variants may be deserving of taxonomic status.

P. diversifolia var. *alpina* C. B. Clarke in Hook. f., Fl. Brit. Ind. 2: 688. 1879 (*P. trifoliata* Wallich, Cat. 565. 1829, *nomen nudum*; *Helosciadium trifoliatum* DC., Prodr. 4: 106. 1830; *P. diversifolia* var. *trifoliata* (DC.) Kuntze, Rev. Gen. 1: 269. 1891), known only from the type collection (Gossainthan, Nepal), is said to differ in having petals merely acute, ovary glabrous, and fruit "yellow-brown, glabrous, much attenuated upwards."

P. tibetanica H. Wolff, Feddes Repert. 27: 319. 1930, based on *Hobson*, "Tibet: Yatung. 27°51′ Lat., 38°35′ Long. or.," 1897, described as having well-developed involucre and involucel and minutely granulate ovary, may be the same. No specimens have been examined.

P. diversifolia var. *stolonifera* Hand.-Mazz., Symbol. Sinicae 7: 714. 1933, based on *Handel-Mazzetti 8283*, "NW. Y[UNNAN]: Im tp. Regenmischwalde und Hochstaudenfluren des birm. Mons. zwischen Mekong und Salwin unter dem Schöndsu-la gegen Londjre," 22 Sept. 1915 and *P. diversifolia* var. *sermentifera* Goel & Bhattacharyya, Jour. Econ. & Taxon. Bot. 6: 213. 1985, based on *Goel 64,313*, "Garhwal, Gangi, alt. 3,000 m. 10.8.78", are the same. They differ from more characteristic forms in emitting from the lower leaf axils slender stolons which root at the apex and nodes. Cannon (1979) reports this from China to Bhutan and Nepal at altitudes of 3,200–3,400 m. The following specimens have been examined: Nepal. Tukucha, Stainton, *Sykes & Williams 8011* (BM); Ratamati, Chakure Lekh, *Stainton, Sykes & Williams 400* (BM).

Pimpinella sp. 1

Plants caulescent, weak and ascending, branching, sparsely hirtellous 15–25 cm tall, perennial from a slightly swollen taproot. Leaves ovate, 3–4 cm in diameter, trifoliate, the leaflets orbicular, 5–15 mm in diameter, petiolulate to sessile, rounded at apex, truncate at base, crenate- to mucronulate-dentate and/or somewhat incised; petiole slender, 3–8 cm long, rather narrowly scarious-sheathing; cauline leaves like the basal, reduced upward, with somewhat dilated sheaths. Peduncles terminal and lateral, filiform, 2–5 cm long, glabrous. Involucre lacking. Rays 3–5, filiform, very unequal, ascending, 1–2 cm long, glabrous. Umbellets 3–8-flowered, the immature pedicels unequal, spreading-ascending, glabrous, 2–4 mm long. Involucel of 1–3 linear bracteoles up to 2 mm long. Flowers white; petals subequal (?), oval, somewhat acute, with a definite inflexed apex; calyx teeth lacking; stylopodium low conical, exceeding the short styles. Very immature fruit suborbicular, ca. 1 mm in diameter. Mature fruit, seed, and carpophore not seen.

Representative specimens: Nepal: Maharigaon, 13,500 ft., 19 July 1952, *Polunin, Sykes & Williams 221* (AAH, BM, E).

Pimpinella sp. 2

Plants rather slender, caulescent, erect but weak, unbranched, very sparsely hirsute or hirtellous, from a slender taproot, 20–30 cm tall. Leaves oblong to ovate, 4–8 cm long, 5–6 cm broad, ternate-pinnate or pinnate-pinnatifid with 5 orbicular leaflets 1–2 cm in diameter, the lower petiolulate, the others subsessile, all ternately incised and dentate, rounded or truncate at base, the margins and rachis ciliolate to hirsutulous, paler beneath; petiole slender, 6–8 cm long, with an oblong scarious sheath; cauline leaves few, like the basal. Peduncles very slender, terminal or in the upper axils, 8–16 cm long, scaberulous at summit. Involucre of a single scarious linear bract up to 4 mm long, or lacking. Rays 4 or 5, filiform, unequal, 1–2 cm long, angled, hirsutulous. Umbellets 4- or 5-flowered, the immature pedicels unequal, spreading-ascending. Involucel of 1–3 linear bracteoles 1–2 mm long. Flowers deep red; petals oval, plane or slightly incurved, obtuse or somewhat acute, with an inflexed apex, apparently subequal and scarcely clawed; calyx lobes evident, ovate; stylopodium rounded conical (hemispherical ?), about equal to the short styles; ovary densely puberulent. Very immature fruit oblong-ovoid, compressed laterally and constricted at the commissure, the ribs obscure, densely puberulent. Vittae, seed face, and carpophore not seen.

Representative specimens: Nepal: Rambrong, Lamjung Himal, 11,500 ft., 29 June 1954, *Stainton, Sykes & Williams 6009* (BM, E).

47. ERIOCYCLA Lindley

In Royle, Ill. Bot. Himal. Mount. 232. 1835.

Pituranthos subgen. *Eriocycla* (Lindley) C. B. Clarke in Hook. f., Fl. Brit. Ind. 2: 680. 1879.
Petrosciadium Edgew., Proc. Linn. Soc. 1: 252. 1846.
Crematosciadium Rech. f. in Rech. f. & Riedl, Biol. Skr. 13(4): 47. 1963.

Slender, erect, caulescent, sparsely branching, pubescent, sometimes woody-based, perennial herbs from a taproot. Leaves petiolate, membranaceous, pinnate or bipinnate with oblong to ovate, entire to pinnatifid divisions; petiole slender, sheathing; cauline leaves usually strongly reduced and stems appearing naked above. Inflorescence of lax terminal and lateral compound umbels. Involucre and involucel of several narrow or broader, pubescent bracts or bracteoles, or involucre occasionally lacking. Rays few, subequal, spreading-ascending. Umbellets few to several flowered, the flowers pedicellate. Flowers white or yellowish-white; petals obovate, with a narrower inflexed apex, usually pubescent dorsally; calyx obsolete, or calyx teeth minute, triangular; stylopodium conical or low conical, with a conspicuous undulate disk, the styles short to slender, erect to sharply divaricate; ovary pubescent. Pedicels slender. Fruit oblong to oblong-ovoid, compressed laterally, constricted at commissure, slightly narrowed at apex, rounded at base, the mericarps pentagonal-subterete in transection, hairy, the ribs prominent to obscure, often hidden by hair. Commissural face plane. Vittae large and solitary in intervals, 2 on commissure, or each interval with 1 or 2 small additional vittae. Carpophore bipartite.

Type species: *Eriocycla nuda* Lindley.

Distribution: A genus of 6–9 species, extending from China to the Western Himalaya and southwestern Asia. Some species are very similar, at least in general aspect, to the Mediterranean-African genus *Deverra* DC. (*Pituranthos* D. Viviani, 1824 = *Pityranthus* H. Wolff, not Martius, 1814.)

KEY TO SPECIES

1. Plants low and caespitose, with many short, leafy stems; bracteoles broad, overlapping laterally .... 1. *E. caespitosa.*
1. Plants with a single stem at base and long virgate branches, almost naked to sparsely leafy above; bracteoles narrow, not overlapping.
    2. Leaves bipinnate, the divisions mostly obtuse; fruit oblong-elliptic or oblong-ovoid, 3–6 mm long.
        3. Plants stout and broomlike, nearly naked above, 30–100 cm tall; involucre of 2 or 3 bracts 1–5 mm long .... 2. *E. nuda.*
        3. Plants slender, sparsely leafy above, 15–45 cm; involucre of 4–8 bracts 4–8 mm long .... 3. *E. thomsonii.*
    2. Leaves mostly simply pinnate (but leaflets sometimes pinnatifid), the divisions usually acute; fruit ovoid, 1.5–2.5 mm long. .... 4. *E. stewartii.*

1. **Eriocycla caespitosa** (Edgew.) H. Wolff in Engl., Pflanzenreich 90 ($4^{\underline{228}}$): 316, 366. 1927.

Petrosciadium caespitosum Edgew., Proc. Linn. Soc. 1: 252. 1846; *Pimpinella caespitosa* (Edgew.) C. B. Clarke in Hook. f., Fl. Brit. Ind. 2: 689. 1879.
Crematosciadium leucocorymbe Rech. f. in Rech. f. & Riedl, Biol. Skr. 13(4): 47, figs. 17, 18. 1963. Based on *Neubauer 749*, "NE: Nuristan: Zwischen Ort Waigel und Nischai, aus senkrechter Felswand aus Ritzen herabhängend," 12 Aug. 1951.

Plants low, caespitose with many short, few branched, leafy stems arising from a slender, woody, branched, sheath covered caudex 10–25 cm tall, villous to tomentulose throughout. Leaves oblong, 3–8 cm long, 2–3 cm broad, simply pinnate, the leaflets ovate to orbicular, 3–15 mm in diameter, crenate-serrate to shallowly lobed, villous on both surfaces and on rachis; petiole slender, 3–6 cm long, villous, narrowly sheathing; cauline leaves similar, gradually reduced upward. Peduncles terminal and lateral, villous, 1–5 cm long. Involucre and involucel of 4–6 lanceolate to linearly lanceolate, acute, villous to tomentose bracts or bracteoles 2–5 mm long. Rays 4–7, slender, spreading, subequal, villous, 1–1.5 cm long. Umbellets 6–12-flowered, not all maturing, the mature pedicels slender, unequal, tomentose, 2–5 mm long. Petals obovate, tomentose dorsally; calyx teeth obsolete, or minute and hidden by pubescence; stylopodium low conical, the styles slender, recurved; ovary tomentose. Fruit oblong, narrowed at apex, 2.5–3.5 mm long, 2–2.5 mm broad, the mericarps pentagonal in transection, tomentose to glabrate, the ribs hidden by pubescence. Vittae solitary in intervals, 2 on commissure. Commissural face plane. July–Oct.

Type locality: "Himala, ad alt. ped. 10,000–11,000, infra Niti," Western Himalaya, *Edgeworth*.

Distribution: Western Himalaya, 3,000–4,000 m; Afghanistan.

Representative specimens: Himachal Pradesh: Kunawar, *Thomson* (CAL). Uttar Pradesh: Ganges Valley, *Duthie 1333* (DD); Damodar Valley, *Duthie 1132* (CAL, DD).

2. **Eriocycla nuda** Lindley in Royle, Ill. Bot. Himal. Mount. 232. 1835.

Pituranthos nuda (Lindley) C. B. Clarke in Hook. f., Fl. Brit. Ind. 2: 680. 1879.

Plants stout, much branched, nearly leafless above, pubescent, 30–100 cm tall. Leaves oblong, 3–5 cm long, 1.5–2.5 cm broad, bipinnate, the divisions ovate to oblong, dentate or lobed, 3–8 mm long, 1–2 mm broad; petiole 3–5 cm long, sheathing; upper cauline leaves greatly reduced, often to bladeless sheaths. Peduncles 3–20 cm long, pubescent. Involucre of 2 or 3 linear bracts 1–5 mm long. Rays 3–10, unequal, 1–7 cm long, pubescent. Involucel of 4–10 linear bracteoles 1.5–2 mm long. Pedicels 8–12, pilose, 1–4 mm long. Petals light yellow, ovate-elliptic, pubescent; calyx teeth small, triangular; stylopodium conical, the styles to 1 mm long, divergent. Fruit oblong-ovate, compressed laterally, 1.5–2 mm long, 0.5–1 mm broad, narrowed toward apex, the mericarps subterete in transection, pubescent, the ribs obtuse (hidden by hairs). Vittae solitary in intervals, 2 on commissure. Commissural face plane.

Type locality: "Rarung in Kunawur," Himachal Pradesh, *Royle*.

Distribution: Western Himalaya above 3,000 m; Chitral (?).

Representative specimens: Himachal Pradesh: Kunawar, *Thomson s. n.* (CAL), *Royle 93/69* (DD).

3. **Eriocycla thomsonii** (C. B. Clarke) H. Wolff in Engl., Pflanzenreich 90 ($4^{\underline{228}}$): 106. 1927.

Pituranthos thomsonii C. B. Clarke in Hook. f., Fl. Brit. Ind. 2: 680. 1879.

Plants slender, erect, much branched and sparingly leafy above, grayish-villous to tomentulose-pubescent, 15–45 cm tall, with persistent leaf sheaths. Leaves oblong, to 9 cm long, bipinnate, the divisions ovate to oblong, dentate to pinnatifid, up to 2 cm long; petiole 5–6 cm long, sheathing; upper cauline leaves greatly reduced, often ternate. Peduncles terminal and lateral, elongate to obsolete. Involucre of 4–8 linear bracts 4–8 mm long. Rays 3–8, villous or tomentulous, 1–3.5 cm long. Involucel of 4–10 villous bracteoles. Pedicels tomentulous. Petals obovate, hairy dorsally; calyx teeth evident, villous; stylopodium conical, the styles slender, deflexed. Fruit oblong-elliptic, strongly compressed laterally and constricted at commissure, 3.5–4 mm long, 1.5–2 mm broad, rounded at both ends, the mericarps subterete in transection, villous to glabrate, the ribs obtuse, the lateral slightly displaced onto commissure. Vittae solitary in intervals, 2 on commissure. Commissural surface planoconcave.

Type locality: "WEST TIBET; Piti, alt. 10,000–12,000 ft.," *Thomson* (*Hooker & Thomson*, Herb. Ind. Orient., *Eriocycla ? 2*).

Distribution: Western Himalaya, from Tibet to Chitral, Pakistan.

Specimen examined: Himachal Pradesh. Piti [Spiti?] Valley, *Thomson* (CAL).

4. **Eriocycla stewartii** (Dunn) H. Wolff in Engl., Pflanzenreich 90 ($4^{\underline{228}}$): 107. 1927.

Pituranthos stewartii Dunn, Kew Bull. 1924: 385. 1924; *Pimpinella stewartii* (Dunn) Nasir, Fl. W. Pakistan 20: 66. 1972; *Trachyspermum stewartii* (Dunn) Hedge, Lamond & Rech. f., Fl. Iranica 162: 338. 1987.

Plants slender, erect, branched and sparsely leafy above, villous to glabrate, 35–100 cm tall, somewhat woody at base, from an elongated taproot. Leaves oblong to oval, 3.5–12 cm long, 1.5–3 cm broad, 1–2-pinnate, the divisions lanceolate to obovate, 5–20 mm long, 2–15 mm broad, acute, cuneate at base, pinnately or ternately lobed to entire; petiole slender, 2–8 cm long, narrowly sheathing; cauline leaves greatly reduced upward, the uppermost linear and entire. Peduncles terminal and lateral, slender, 3–13 cm long. Involucre and involucel of 2–8 linear to linearly-lanceolate, scarious, villous bracts or bracteoles 1–5 mm long. Rays 3–12, slender, subequal, villous, spreading-ascending, 0.5–2 cm long. Umbellets 12–18-flowered, the mature pedicels slender, unequal, villous, 1–5 mm long. Flowers white; petals oval to obovate, hairy dorsally; calyx obsolete; stylopodium conical, the styles slender, reflexed; ovary densely villous to tomentulose. Fruit ovoid, 1.5–2.5 mm long, 1.5–2 mm broad, constricted into a very short neck at apex, rounded at base, the mericarps pentagonal-subterete in transection, densely hirtellous, the ribs obtuse, largely concealed by pubescence. Vittae 1–3 in intervals, 2–4 on commissure. Commissural surface plane. July–Nov.

Type locality: Murree Hills, *R. R. Stewart 4109* (K, holotype).

Distribution: Western Himalaya, Kashmir, 4,000–5,000 m.

Representative specimen: Jammu & Kashmir. Abbottabad, *Duthie 7482* (DD).

Note. Conflicting evidence as to the number of fruit vittae has led to placement of this species in different genera, but by habits at least, it would appear to belong here.

48. TRACHYSPERMUM Link

Enum. Hort. Berol. 1: 267. 1821, *nomen conserv.*

Ptychotis sect. *Trachyspermum* (Link) DC., Prodr. 4: 108. 1830; *Carum* sect. *Trachyspermum* (Link) Benth. in Benth. & Hook. f., Gen. Pl. 1: 891. 1867.
Ammios Moench, Meth. 99. 1794.
Carum sensu C. B. Clarke in Hook. f., Fl. Brit. Ind. 2: 680. 1879, *pro parte*, non Linn. 1753.

Slender and erect or low and caespitose, branching, glabrous to hirtellous, sometimes glaucous, annual or perennial herbs from a taproot. Leaves petiolate, membranaceous, simply ternate or pinnate to pinnately decompound, the leaflets serrate, oblong to oblong-oval, or the ultimate divisions entire, linearly filiform; petiole slender, sheathing; cauline leaves reduced upward. Inflorescence of lax terminal and lateral umbels, the peduncles elongate. Involucre and involucel of several narrow, entire bracts or bracteoles, or rarely lacking. Rays few, slender, subequal or unequal, spreading-ascending. Umbellets several to many flowered, the flowers pedicellate. Flowers white, rarely pink; petals oval to obovate, emarginate, with a narrower inflexed apex; calyx teeth evident, subulate, or calyx obsolete; stylopodium conical or low conical, with a prominent crenulate-margined disk, the styles slender, recurved. Pedicels slender, unequal. Fruit lanceolate-ovoid to ovoid-globose, strongly compressed laterally, constricted at commissure and subdidymous, usually sharply constricted into a very short neck at apex, rounded at base, the mericarps pentagonal-subterete in transection, hirsute to hirsutulous (rarely glabrous) or variously hispid or scaberulous-papillose, the ribs filiform, prominent to obscure. Vittae 1 or more in intervals, 2 on commissure. Commissural face plane. Carpophore bipartite, bifid, or entire.

Type species: *Sison ammi* Linn. [*Trachyspermum ammi* (Linn.) Sprague].

Distribution: An ill-defined pimpinelloid genus of 6–10 species of southern Asia, two of them widely cultivated as condiments and naturalized.

KEY TO SPECIES

1. Basal leaves pinnately or ternately decompound with linear or filiform ultimate divisions 1 mm or less broad.
    2. Plants hirtellous at least in inflorescence; petals oval or elliptic, hirtellous dorsally; calyx usually obsolete; fruit densely hirtellous.
        3. Plants hirsutulous throughout; rays usually 15–20 (–40 ?); fruit 1 mm long, 1.5 mm broad .... 1. *T. anethifolium.*
        3. Plants glabrate below; rays usually 3–7; fruit 2–3 mm long, 2 mm broad .... 2. *T. falconeri.*
    2. Plants glaucous, glabrous to minutely puberulent throughout; petals obcordate, hairy only on midvein; calyx teeth usually evident; fruit scaberulous-papillose. .... 3. *T. ammi.*
1. Basal leaves 1–2-pinnate, -ternate, or -ternate-pinnate, with oblong to obovate leaflets.

4. Annual; stems profusely branched and leafy throughout .... 4. *T. roxburghianum.*

4. Perennial; stems sparsely branched and sparsely leafy above .... 5. *T. khasianum.*

1. Trachyspermum anethifolium (D. Don) H. Wolff in Engl., Pflanzenreich 90 ($4^{\underline{228}}$): 90. 1927.

Pimpinella anethifolia D. Don, Prodr. Fl. Nepal. 184. 1825; *Athamanta anethifolia* Wallich, Cat. 569. 1829, *nomen nudum*; *Ptychotis anethifolia* (D. Don) DC., Prodr. 4: 108. 1830; *Carum anethifolium* (D. Don) C. B. Clarke in Hook. f., Fl. Brit. Ind. 2: 683. 1879.

Plants slender, erect, branched above, finely grayish-hirtellous, 30–120 cm tall, from a fusiform root. Leaves ovate, 5–8 cm long, pinnately decompound, the ultimate divisions linearly lanceolate, 3–6 mm long, 1 mm broad; petiole slender, 2–12 cm long; cauline leaves reduced upward, ternate-pinnate, ternate, or entire, the uppermost reduced to bladeless sheaths. Peduncles elongate, 2–18 cm long. Involucre and involucel of 4–8 linear, pubescent bracts or bracteoles 3–5 mm long. Rays usually 15–20, slender, hirtellous, spreading-ascending, unequal, 1–2 cm long. Umbellets 8–30-flowered, the mature pedicels slender, subequal, hirsutulous, 3–6 mm long. Petals elliptic, acute, pubescent dorsally; calyx obsolete; stylopodium low conical, the styles slender, reflexed. Fruit ovoid, ca. 1 mm long, 1.5 mm broad, the mericarps pentagonal-subterete in transection, densely hirsutulous, the ribs acute, the lateral slightly displaced onto commissure. Vittae solitary in intervals, 2 on commissure. Carpophore entire.

Type locality: Narainhatty, Nepal, *Buchanan-Hamilton* (*Wallich 569*).

Distribution: Himalaya, from Nepal to Garhwal, Uttar Pradesh, 1,200–3,000 m.

Representative specimens: Uttar Pradesh: Makteswar Hill near Almorah, *Duthie 2947* (DD); Gurna, Pithoragarh, *Bhattacharyya 15,174* (BSD) Garhwal, *Durga native 1170* (CAL, DD). Nepal: Narainhatty, *Buchanan-Hamilton* (*Wallich 569*) (CAL).

2. Trachyspermum falconeri (C. B. Clarke) H. Wolff in Engl., Pflanzenreich 90 ($4^{\underline{228}}$): 90. 1927.

Carum falconeri C. B. Clarke in Hook. f., Fl. Brit. Ind. 2: 683. 1879.

Plants slender, erect, sparsely branched and nearly leafless above, 50–100 cm tall, the inflorescence hirtellous, otherwise glabrous, from a fusiform root. Leaves ovate (?), 8–15 cm in diameter, ternately or pinnately decompound, the ultimate divisions linear, 2–7.5 cm long, 1 mm or less broad; petiole slender, 5–15 cm long; cauline leaves strongly reduced upward. Peduncles terminal and lateral, 4–6 cm long. Involucre and involucel of 4–6 linearly lanceolate or -oblong, scarious-margined, ciliolate, ultimately reflexed bracts or bracteoles 2–8 mm long. Rays 3–7, filiform, glabrate, spreading-ascending, unequal, 1–4 cm long. Umbellets 6–15-flowered, the mature pedicels slender, unequal, spreading-ascending, hirtellous 2–7 mm long. Petals oval, hirsutulous dorsally; calyx obsolete; stylopodium low conical, the styles slender, recurved; ovary densely hirsutulous. Fruit ovoid, 2–3 mm long, 2 mm broad, slightly narrowed toward apex, rounded at base, the mericarps pentagonal-subterete, densely hirsutulous, the ribs inconspicuous. Vittae solitary in intervals, 2 on commissure. Carpophore bipartite.

Type locality: w/o precise locality, *Hooker & Thomson* (Herb. Ind. Orient., *Ptychotis 8*).

Distribution: Western Himalaya, Garhwal, 1,600-2,300 m.

Representative specimens: Locality uncertain, *Hooker & Thomson* (CAL, K). Uttar Pradesh: Mussoorie, *Bacon* (K), *Duthie 23,018* (DD); Gurhwal, *Strachey & Winterbottom 8* (K), *Falconer* (CAL, K); Januma, Tehri, *Koelz 21,844* (DD, UC); Kalamuni Pass, *Gamble 15,051* (CAL, DD); Bunuj, *Edgeworth 150* (K).

3. **Trachyspermum ammi** (Linn.) Sprague, Kew Bull. 1929: 228. 1929.

Sison ammi Linn., Sp. Pl. 252. 1753.

Ammi copticum Linn., Mant. 1: 56. 1767; *Daucus copticus* (Linn.) Pers., Syn. Pl. 1: 307. 1805; *Bunium copticum* (Linn.) Sprengel, Umbell. Prodr. 28. 1813; *Trachyspermum copticum* (Linn.) Link, Enum. Hort. Berol. 1: 267. 1821; *Ptychotis coptica* (Linn.) DC., Mém. Soc. Phys. Hist. Nat. Genève 4: 496. 1829; *Carum copticum* (Linn.) Hiern in Oliver, Fl. Trop. Afr. 3: 12. 1871; C. B. Clarke in Hook. f., Fl. Brit. Ind. 2: 682. 1879. Based on *Forskahl*, "Habitat in AEgypto."

Ligusticum ajawain Roxb. in Fleming, Asiat. Res. 11: 171. 1810; *L. ajouan* Roxb., Hort. Bengal 21: 1814, *nomen nudum*; *Ptychotis ajowan* (Roxb.) DC., Mém Soc. Phys. Hist. Nat. Genève 4: 497. 1829; *Athamanta ajouan* (Roxb.) Wallich, Cat. 1829, *nomen nudum*. Based on *Roxburgh* (?), India (no specific locality indicated).

Plants low and bushy to slender and erect, profusely branching, glaucous, glabrous or minutely puberulent, annual, 15–90 cm tall, from a slender taproot. Leaves oblong to oval, 2–8 cm long, 2–6 cm broad, pinnately decompound with linearly filiform ultimate divisions 2–25 mm long, 0.5 mm or less broad; petiole slender, 1-5 cm long, narrowly sheathing; cauline leaves similar, petiolate, but gradually reduced upward. Peduncles slender, 1-11 cm long. Involucre of 4–10 linear, entire or occasionally trifid linear bracts 2–10(–15) mm long. Rays usually 6–20(–25), slender, glabrous or puberulent, spreading-ascending, unequal, 0.5–2.5(–6 ?) cm long. Umbellets 10–20-flowered, the mature pedicels slender, unequal, 1–6 mm long. Involucel of 4–8 linear or oblong, usually entire, ciliolate bracteoles 1–5 mm long. Petals obcordate, sparsely hairy on midvein dorsally; calyx teeth evident, ovate, or calyx obsolete; stylopodium conical, the styles slender, recurved. Fruit ovoid-globose, 1.5–2.5 mm long, 1.5–2 mm broad, strongly narrowed at apex, cordate at base, the mericarps pentagonal in transection, scaberulous-papillose, the ribs slender, inconspicuous. Vittae solitary in intervals, 2 on commissure. Carpophore bipartite. Jan.–Nov.

Type locality: "Habitat in Apulia, AEgypto," no collector indicated.

Distribution: Cultivated throughout southern and southwestern Asia and North Africa, adventive; presumably indigenous to India.

Representative specimens: Punjab: Hansi, *Nair 25,985* (BSD). Uttar Pradesh: Dehra Dun, *Babu 35,213* (BSD); Pithoragarh, *Bhattacharyya 15,173* (BSD). West Bengal: Uttarpara, *Mukherjee s. n.* (CUH). Tripura: Agartala, *Debberman 1187* (CAL). Andhra Pradesh: Tawang, *Mukherjee s. n.* (CUH); Vijayapuri, *Sebastine 9841* (MH). Tamil Nadu: w/o further locality, *Wight 1161* (CAL).

4. **Trachyspermum roxburghianum** (DC.) Craib, Fl. Siam. Enum. 1: 788. 1931.

Apium involucratum Roxb. in Fleming, Asiat. Res. 11: 157. 1810; *Pimpinella involucrata* (Roxb.) Wight & Arn., Prodr. Fl. Ind. Orient. 369. 1834; *Ptychotis involucrata* (Roxb.) Lindley in Royle, Ill. Bot. Himal. Mount. 1: 229. 1839; *Trachyspermum involucratum* (Roxb.) H. Wolff in Engl., Pflanzenreich 90 ($4\underline{^{228}}$): 89. 1927, non Maire 1922.
Athamanta roxburghiana Wallich, Cat. 571. 1829, *nomen nudum*; *Ptychotis roxburghiana* DC., Prodr. 4: 109. 1830; *Carum roxburghianum* (DC.) Kurz, Jour. Asiat. Soc. Bengal II. 46: 114. 1877; C. B. Clarke in Hook. f., Fl. Brit. Ind. 2: 682. 1879; *Trachyspermum roxburghianum* (DC.) H. Wolff in Engl., Pflanzenreich 90 ($4\underline{^{228}}$): 57. 129. 1927, sphalm. Based on *Apium involucratum* Roxb.
Pimpinella lateriflora Dalzell in Dalzell & Gibson, Bombay Fl. 106. 1861, non Link 1821; *P. dalzellii* P. K. Mukherjee, Indian Forester 95: 567. 1969. Based on *Dalzell*, "Ravines in the Deccan."
Carum stictocarpum C. B. Clarke in Hook. f., Fl. Brit. Ind. 2: 681. 1879; *Trachyspermum stictocarpum* (C. B. Clarke) H. Wolff in Engl., Pflanzenreich 90 ($4\underline{^{228}}$): 89. 1927. Based on *Stocks & Law*, "Concan."
Carum stictocarpum var. *hebecarpa* C. B. Clarke in Hook. f., Fl. Brit. Ind. 2: 682. 1879; *Trachyspermum stictocarpum* var. *hebecarpum* (C. B. Clarke) H. Wolff in Engl., Pflanzenreich 90 ($4\underline{^{228}}$): 89. 1927. Based on *Stocks & Law*, "Concan."
Carum involucratum var. *anomalum* Kuntze, Rev. Gen. 1: 269. 1891; *Trachyspermum involucratum* var. *anomalum* (Kuntze) H. Wolff in Engl., Pflanzenreich 90 ($4\underline{^{228}}$): 90. 1927. Based on *Kuntze*, "West Ghats."

Plants slender, erect, much branched, hirtellous at least in inflorescence, puberulent, or glabrate, 20–100 cm tall, annual from a taproot. Leaves triangular-ovate, 3–8 cm long, 2–12 cm broad, bipinnate or ternate-pinnate, with oblong to obovate, cuneate, lobed to pinnatifid divisions, the segments narrowly oblong, 5–20 mm long, 2–3 mm broad; petiole slender, 1–2 cm long, with a scarious-margined sheath at base; cauline leaves reduced upward, the uppermost pinnate with narrowly oblong or linearly lanceolate divisions. Peduncles terminal and lateral, 5–9 cm long. Involucre and involucel of 3–8 lanceolate or linearly-subulate, scarious, ciliate bracts or bracteoles, usually 3–5 mm long. Rays 4–12, filiform, spreading-ascending, unequal, glabrous to hirsutulous, 1–3 cm long. Umbellets 12-20-flowered, the mature pedicels unequal, hirsutulous, 1–5 mm long. Flowers white or pink; petals elliptic to obovate, pubescent dorsally; calyx obsolete; stylopodium low conical, the styles slender, reflexed. Fruit ovoid, 1.5–3 mm long, 1.5–2 mm broad, contracted into a very short neck at apex, compressed laterally, constricted at commissure, the mericarps pentagonal-subterete in transection, densely hirsutulous to glabrous, the ribs prominent, acute. Vittae 1 or more in intervals, 2 on commissure. Commissural face plane. Carpophore bipartite. Jan.–May.

Type locality: Cultivated, Bengal, no collector indicated.

Distribution: Apparently native to the Western Ghats of Peninsular India, but cultivated as a spice throughout India and southeastern Asia and Indonesia.

Representative specimens: "N. W. India," *Royle* (CAL). Uttar Pradesh: Dehra Dun, *Duthie* (DD); Naini Tal, *Reid* (DD). West Bengal: Raneegunge, *Kurz* (CAL). Sikkim: w/o further locality, *Hooker* (CAL). Tripura: Agartala, *Debbarman 350* (CAL). Orissa: Sambalpur, *Mooney 3173* (K). Madhya Pradesh: Panchmarhi, *Haines 212P* (K); Rao

Ghat, Bastar Distr., *Balakrishnan & Henry 12,146* (MH); Jubbalpur, *Beddome 10* (CAL). Tamil Nadu: Courtallam, *Wallich 7217* (CAL). Karnataka: Bababoodan Hill, *Talbot 2356* (CAL). Maharashtra: Concan, *Stocks & Law* (CAL); Bombay, *Dalzell* (CAL); Singhgadh, 24 Oct. 1982, *Mukherjee s. n.* (CUH, UC); Pratapgadh Ghat, 25 Dec. 1982, *Mukherjee. s. n.* (CUH, UC). Rajasthan: Mount Abu, *Duthie 6650* (DD, K).

Note. As indicated by Clarke (1879, p. 682), *Carum stictocarpum* and *C. roxburghianum* represent wild and cultivated forms respectively, of the same taxon, which differ primarily in size and pubescence of fruit, but are connected by many intermediates. Discrepancies in reported number of vittae in the fruit—apparently a variable feature—have led to confusion in proper generic placement.

In excluding taxa from *Apium*, Wolff (1927, p. 57) made the new combination "*Trachyspermum roxburghianum* (Wall.) Wolff," but later in the same volume (p. 89), considered *Athamanta roxburghiana* Wallich to be a synonym of *Apium involucratum* Roxb., and created the new combination *Trachyspermum involucratum* (Roxb.) Wolff, itself a later homonym of *T. involucratum* Maire, 1922. Since Wolff did not accept *Trachyspermum roxburghianum* in treating that genus, the binomial *T. roxburghianum* (DC.) Craib, 1931, is accepted here for this much named taxon.

5. **Trachyspermum khasianum** (C. B. Clarke) H. Wolff in Engl., Pflanzenreich 90 ($4^{\underline{228}}$): 91. 1927.

Carum khasianum C. B. Clarke in Hook. f., Fl. Brit. Ind. 2: 682. 1879.

Plants slender, erect, corymbosely few branched above, 60–120 cm tall, hirtellous throughout or glabrate below, perennial from a fusiform root. Leaves ovate, 5–15 cm in diameter, simply ternate or pinnate, the leaflets oblong to oblong-obovate, serrate, 2–6 cm long, 0.5–1.5 cm broad; petiole slender, hirsutulous, 4–7 cm long, abruptly sheathing at base; cauline leaves strongly reduced upward, with linearly lanceolate divisions. Peduncles terminal and lateral, slender, hirsutulous, 2–10 cm long. Involucre and involucel of 2–5 linear, ± hirsutulous bracts or bracteoles 2–5(–15) mm long. Rays 4–10, slender, hirsutulous, unequal, spreading-ascending, 1–3 cm long. Umbellets 8–15-flowered, the mature pedicels slender, hirsutulous, unequal, 1–5 mm long. Petals oval, hairy dorsally; calyx teeth minute, ovate, or calyx obsolete; stylopodium conical, the styles slender, recurved. Fruit ovoid-globose, 1.5–2.5 mm in diameter, constricted into a very short neck at apex, rounded at base, the mericarps pentagonal and slightly compressed dorsally in transection, sparsely hispidulous, the ribs prominent, acute. Vittae solitary in intervals, 2 on commissure. Carpophore ± bifid. Aug.–Oct.

Type locality: Sillet, Bangladesh, *Wallich 7218*.

Distribution: Northeastern India (apparently confined to Khasia Hills), 1,000–3,000 m; Bangladesh.

Representative specimens: Meghalaya: Khasia, Nonkrem, *Hooker & Thomson* (CAL); Maokadokadok, *Clarke 45,364* (CAL); Nongstan, *Panigrahi 16,579* (ASSAM); Mawrngkneng, Khasia Hills, *Koelz 28,931* (UC). Bangladesh: Sillet, *Wallich 7218* (CAL); East Bengal, *Griffith 2600* (CAL).

49. SESELI Linn.

Sp. Pl. 259. 1753; C. B. Clarke in Hook. f., Fl. Brit. Ind. 2: 693. 1879.

Seseli subgen. *Euseseli* Drude in Engl. & Prantl, Pflanzenfam. 38: 20. 1898.
Libanotis Hill, Brit. Herb. 420. 1756; *Seseli* subgen. *Libanotis* (Hill) Gren. & Godron, Fl. Franc. 1: 707. 1848.
Scaphespermum Edgew., Trans. Linn. Soc. 20: 58. 1851.

Stout to slender, caulescent to acaulescent, erect, or rarely decumbent, usually branching, glabrous to pubescent, perennial, annual, or biennial herbs arising from a fibrous rootcrown surmounting a taproot. Leaves petiolate, membranaceous, the basal often rosulate, pinnate, or ternate to pinnately decompound, the leaflets or ultimate divisions ovate to filiform; petiole slender, sheathing at base; cauline leaves reduced upward and often conspicuously sheathing, or lacking. Inflorescence of lax to condensed terminal or lateral compound umbels, the peduncles usually elongate. Involucre of several narrow bracts, or lacking. Rays numerous, spreading or spreading-ascending. Umbellets many flowered, the flowers pedicellate. Involucel of several narrow, glabrous or pubescent bracteoles. Flowers white or sometimes pinkish or yellow; petals ovate to obcordate, with a narrower inflexed apex; calyx teeth prominent to evident, subulate to obtuse, often deciduous, or occasionally lacking; stylopodium conical to depressed, the styles short to slender, recurved to erect. Pedicels slender to abortive, spreading-ascending. Fruit oblong to ovoid, rounded at both ends, subterete and broad at commissure, the mericarps subterete to compressed dorsally in transection, pubescent to glabrous, the ribs prominent, acute or obtuse, sometimes corky thickened, subequal, unwinged to narrowly winged. Vittae usually solitary in intervals, 2–4 on commissure. Commissural face plane. Carpophore entire to bipartite.

Type species: *Seseli tortuosum* Linn.

Distribution: A Eurasian genus of perhaps 70–80 species, three of which are reported to occur in India. Some authors prefer to retain *Libanotis* as a separate genus.

KEY TO SPECIES

1. Plants perennial, erect; calyx teeth evident, subulate; carpophore bipartite; Himalayan.
    2. Stems nearly leafless above; foliage membranaceous, pubescent; involucre usually present; involucel bracteoles 5–8; fruit ovoid. .... 1. *S. trilobum.*
    2. Stems leafy above; foliage subcoriaceous, glabrous; involucre lacking; involucel bracteoles 8–12; fruit oblong. .... 2. *S. cyclolobum.*
1. Plants annual, diffuse; calyx teeth obsolete or minute; carpophore entire; plains. ... . 3. *S. diffusum.*

1. Seseli trilobum (Edgew.) C. B. Clarke in Hook. f., Fl. Brit. Ind. 2: 694. 1879.

Scaphespermum trilobum Edgew., Trans. Linn. Soc. 20: 58. 1851.

Plants perennial, slender, erect, nearly leafless above, pubescent throughout, 60–100 cm tall. Leaves oblong-ovate, 2–3 cm long, bipinnate or biternate, the leaflets ovate, crenate-serrate to lobed, minutely pubescent above, villous to tomentose beneath, 0.5–3 cm long, 0.5–2 cm broad; petiole 7–15 cm long; cauline leaves few, ternate-pinnate. Peduncles stout, terminal or opposite leaves, 3–12 cm long. Involucre of 3–5 linearly lanceolate, hirsute, deciduous bracts 5–12 cm long, rarely lacking. Rays 5–15, villous, 2–4 cm long. Umbellets 10–20-flowered (one sessile), the pedicels 4–5 mm long, pubescent. Involucel of 5–8 villous, linear bracteoles ca. 5 mm long, persistent. Petals ovate, pubescent; calyx teeth evident, subulate, deciduous. Fruit ovoid, slightly narrowed toward apex, 3–3.5 mm long, ca. 2 mm broad, the mericarps compressed dorsally, densely villous, the ribs prominent, acute, unwinged. Commissural face slightly concave (?). Carpophore bipartite. Jul.

Type locality: "Himala, in glareosis, alt. ped. 10,000–11,000; prope Niti," *Edgeworth.*

Distribution: Western Himalaya.

Representative specimens: Uttar Pradesh: Ganges Valley, *Duthie 1133* (CAL); Niti, *Edgeworth 172* (K).

2. **Seseli cyclolobum** (Gilli) Pimenov & Sdobnina,
Bot. Zhurn. 60: 1120. 1975.

Libanotis cycloloba Gilli, Feddes Repert. 61: 196. 1959.
Seseli sibiricum sensu C. B. Clarke in Hook. f., Fl. Brit. Ind. 2: 693. 1879, non *Athamanta sibirica* Linn. 1753.

Plants perennial, stout, erect, villous in inflorescence, the foliage glabrous, 60–100 cm tall from a thick, elongated, woody root. Leaves subcoriaceous, oblong to oblong-ovate, 10–40 cm long, 5–20 cm broad, bipinnate, the leaflets oblong to ovate, 1–4 cm long, 0.5–3 cm broad, petiolulate to sessile, spinulose-dentate or -lobed, conspicuously venose; petiole slender, 8–12 cm long; cauline leaves reduced upward, mostly pinnate and wholly sheathing. Peduncles terminal and lateral, 5–15 cm long. Involucre lacking. Rays 20–30, slender, spreading-ascending, subequal, 2–5 cm long, pubescent to glabrate. Umbellets 25–35-flowered, the mature pedicels slender, unequal, spreading-ascending, pubescent, 2–6 mm long. Involucel of 8–12 villous, linear to lanceolate bracteoles 3–6 mm long, slightly connate at base. Flowers white; petals obovate, pubescent; calyx teeth linear-subulate, pubescent, persistent; stylopodium conical, shorter than styles, the ovaries densely villous. Fruit oblong-cylindrical, 3–4 mm long, 1–2 mm broad, slightly compressed laterally, the mericarps subterete in transection, villous, the ribs evident, acute, unwinged. Carpophore bipartite. July–Aug.

Type locality: "Ost-Afghanistan: Gebirge oberhalb des zwischen Gulbahar und Surobi gelegenen Nedschrab-(Nidjrau-) Tales," 3,100 m, *Gilli*.

Distribution: Western Himalaya from Uttar Pradesh to Pakistan and Afghanistan.

Representative specimens: Jammu & Kashmir: w/o further locality, *Thomson* (CAL). Himachal Pradesh: Pangee, Lahul, *Lace 2010* (CAL). Uttar Pradesh: Shelang Glacier moraine, Kumaon, *T. A. Rao 6918* (BSD).

3. **Seseli diffusum** (Roxb. ex Smith) Santapau & Wagh.,
Bull. Bot. Surv. Ind. 5: 108. 1963.

Ligusticum diffusum Roxb. ex Smith in Rees, Cyclop. 21: no. 11. 1812; *Athamanta diffusa* (Roxb. ex Smith) Wallich, Cat. 570. 1829, *nomen nudum*; *Cnidium diffusum* (Roxb. ex Smith) DC., Prodr. 4: 153. 1830.

Seseli indicum Wight & Arn., Prodr. Fl. Ind. Orient. 371. 1834; C. B. Clarke in Hook. f., Fl. Brit. Ind. 2: 693. 1879. Based on *Wight 1200*, no locality indicated.

Plants annual, diffuse, much branched, slightly villous, 8–45 cm tall. Leaves oblong, 4–8 cm long, 1.5–2.5 cm broad, bipinnate, the leaflets ovate-cuneate, pinnately lobed to pinnatifid, 3–7 mm long, 2–5 mm broad, the lobes obtuse, apiculate; petiole 2.5–6 cm long; cauline leaves similar, reduced upward, but mostly petiolate. Peduncles opposite leaves, 1.5–8 cm long. Involucre and involucel of 5–8 conspicuous, linear-lanceolate, broadly scarious-margined, ciliate bracts or bracteoles 3–8 mm long, the former reflexed in fruit. Rays 8–25, spreading-ascending, subequal, 1–2.5 cm long. Umbellets ca. 15-flowered, the mature pedicels 3–5 mm long, subequal. Flowers pinkish; petals broadly ovate; calyx teeth minute, triangular, or obsolete; stylopodium low conical, shorter than the slender, recurved styles. Fruit broadly ovoid, 1.5–2.5 mm long, 2–3 mm broad, slightly compressed laterally, the mericarps pentagonal or slightly compressed dorsally in transection, stellate-hispid to glabrous, the ribs prominent, acute, unwinged. Carpophore entire. Feb.–Mar.

Type locality: "Vicinity of Calcutta," *Roxburgh* (?).

Distribution: Plains of India.

Representative specimens: "East India," *Roxburgh* (K). Punjab: Banda, *Bell 104* (CAL). Uttar Pradesh: Oudh, *Inayat Khan 21,659* (CAL). Bihar: Gorakhpur, *Harsukh 21,659b* (CAL). West Bengal: Ballygunge, *Clarke 33,346A* (CAL); Sarsuna, *Mukherjee s. n.* (CUH). Madhya Pradesh: Jubbalpore, *Beddome* (CAL); Nagpur, *Qurashi 631* (CAL).

GROUP V

1. Flowers white, red, purple, etc., but never yellow; vittae 1-5 in intervals, 2–8 on commissure.
   2. Plants distinctly caulescent, usually with cauline leaves, and bearing only evidently compound umbels.
      3. Bracts and bracteoles prominent to obsolete, but usually not conspicuously scarious.
         4. Bracteoles flaccid, not strongly imbricated laterally and not closely embracing umbellet; rays usually numerous and slender.
            5. Calyx teeth usually minute or obsolete; involucre usually inconspicuous or lacking.
               6. Plants perennial; fruit ribs thin winged; vittae commonly 2 or 3 in dorsal intervals. ....50. *Ligusticum*.
               6. Plants annual; fruit ribs corky thickened; vittae solitary in intervals. ....51. *Cnidium*.
            5. Calyx teeth usually prominent; involucre usually prominent; vittae commonly solitary in dorsal intervals. ....52. *Selinum*.
         4. Bracteoles rigid, strongly imbricated laterally and closely embracing umbellet; rays few and stout. ....53. *Pleurospermopsis*.
      3. Bracts and bracteoles prominent and conspicuously scarious or scarious-margined. ....54. *Pleurospermum*.
   2. Plants acaulescent or subcaulescent, the rays appearing like peduncles bearing simple umbels.
      7. Involucre obsolete; fruit ribs corky winged. ....52. *Selinum*.
      7. Involucre usually conspicuous, of foliaceous or scarious, often dissected bracts; fruit ribs or wings thin, never corky.
         8. Fruit lightly compressed either laterally or dorsally, or subterete.
            9. Involucel of usually conspicuous, ± scarious bracteoles; calyx teeth small or lacking. ....54. *Pleurospermum*.
            9. Involucel of narrow, foliaceous bracteoles; calyx teeth evident, persistent. ....37. *Keraymonia*.
         8. Fruit strongly compressed or flattened dorsally.
            10. Rays few (2–15); carpophore bipartite. .... 55. *Cortia*.
            10. Rays numerous (ca. 30); carpophore obsolete. ....56. *Cortiella*.
1. Flowers definitely yellow; vittae numerous and continuous around seed cavity; fruit wings conspicuously corrugate-plicate. ....57. *Prangos*.

50. **LIGUSTICUM** Linn.

Sp. Pl. 250. 1753; C. B. Clarke in Hook. f., Fl. Brit. Ind. 2: 698. 1879.

Ligusticum subgen. *Eu-Ligusticum* Drude, Engl. & Prantl, Pflanzenfam. *38*: 211. 1898.
Coristospermum Bertol., Fl. Ital. 3: 466. 1837; *Ligusticum* subgen. *Coristospermum* (Bertol.) Leute, Ann. Naturhist. Mus. Wien 74: 458. 1970.
Haloscias Fries, Summa Veg. Scand. 1: 180. 1846; *Ligusticum* subgen. *Haloscias* (Fries) Drude in Engl. & Prantl, Pflanzenfam. *38*: 212. 1898.
Dystaenia Kitagawa, Bot. Mag. Tokyo 51: 805. 1937.

Slender to moderately stout, erect, caulescent, branching, scaberulous or variously pubescent to glabrous perennial herbs, arising from a caudex usually surmounting taproots and bearing fibrous leaf sheaths. Leaves petiolate, membranaceous to subcoriaceous, 1–2-pinnate or ternate, with definite leaflets, or pinnately or ternate-pinnately decompound with short, narrow ultimate divisions; petiole slender to stout, sheathing at base; cauline leaves similar, reduced upward, at least the uppermost sessile, sometimes with a prominent sheath. Inflorescence of lax compound umbels, the peduncles terminal and lateral, or terminal only, often elongate, or the terminal umbel subsessile. Involucre of 1 to several narrow, usually inconspicuous and often deciduous, narrow, entire bracts, or often lacking. Rays usually numerous, usually slender, unequal or equal, often scaberulous or hirtellous, usually spreading-ascending. Involucel of few to numerous linear or filiform, entire bracteoles, longer or shorter than flowering umbellets, or less commonly lacking. Umbellets many-flowered, usually not all maturing, the flowers pedicellate. Flowers white or pinkish; petals spatulate to obovate, with a narrower inflexed apex; calyx teeth evident but minute, or calyx obsolete; stylopodium conical, the styles short or slender, Pedicels slender, usually unequal, spreading-ascending. Fruit oblong to elliptic or orbicular, lightly compressed laterally to subterete, to strongly compressed dorsally, the mericarps compressed or flattened dorsally, glabrous, the ribs subequally or unequally thin winged, or merely prominent, filiform. Vittae usually 2 or more (rarely only 1) in intervals, 2–8 on commissure. Commissural face plane to slightly concave, usually grooved beneath intervals. Carpophore bipartite.

Type species: *Ligusticum scothicum* Linn.

Distribution: A circumboreal genus of perhaps two dozen species, with the removal of *Pachypleurum* Ledeb. (Shishkin, 1950), *Halosciastrum* Pimenov & Tikhomirov (1968), *Ligusticopsis* Leute (1969, 1970), *Paraligusticum* Tikhomirov (1973), *Tilingia* Regel & Tiling (Gorovoy, 1978), and *Cyclorhiza* Sheh & Shan (1980). The Southern Hemisphere species once attributed to *Ligusticum* have long since been referred to other genera, viz. *Anisotome* Hook. f., *Apium* Linn., and *Dasispermum* Necker ex Raf. (*Heteroptilis* E. Meyer ex Meisner).

Note. The boundaries and content of the genus, however, are still somewhat problematical. A particularly difficult boundary problem is that concerning the Indo-Himalayan species treated here under *Selinum* and by Leute (1969–1970) under *Cortia* and *Ligusticum*. Nasir (1972) says of *Ligusticum*: "It is very near *Selinum* but differs in the absence of prominent calyx teeth, more prominent dorsal and intermediate ridges and in having more than one vitta in the fruit furrows" (p. 120). However, *S. papyraceum* C. B. Clarke violates the first criterion, *S. candollii* DC. the second, and several species at least occasionally have multiple vittae. Hiroe & Constance (1958) said of *Ligusticum* (in which they included *Tilingia* Regel & Tiling): "The really distinctive generic character appears to be the possession of a fruit that is compressed laterally, but with its carpels more or less compressed dorsally, and all the ribs subequally prominent to narrowly winged" (p. 72). This definition also is inapplicable to Indian species, which may generally be recognized by their larger leaflets, entire bracts and bracteoles (although this does not distinguish them from some species of *Selinum*), and more numerous vallecular vittae.

The Indo-Himalayan species here referred to *Selinum* and *Ligusticum*, appear to be more similar to each other than to the type species of their respective genera.

KEY TO SPECIES

1. Plants strongly caulescent, 50–160 cm tall; involucre inconspicuous or lacking.
    2. Fruit subterete, lightly compressed dorsally, the ribs subequally winged.
        3. Stem slightly angled; leaves mostly simply pinnate; pedicels 3–8 mm long; Himalaya. .... 1. *L. thomsonii.*
        3. Stem sulcate; leaves bipinnate; pedicels 25–35 mm long; peninsular India. . ... 2. *L. alboalatum.*
    2. Fruit strongly compressed or flattened dorsally, the ribs unequally winged, the lateral broader than dorsal or intermediate wings (if any).
        4. Plants usually scaberulous or hirtellous, conspicuously hirsutulous at summit of peduncle; bracteoles ciliolate. .... 3. *L. marginatum.*
        4. Plants essentially glabrous, not hirtellous at summit of peduncle; bracteoles glabrous. .... 4. *L. eletum.*
1. Plants subcaulescent, low, only 10–40 cm tall; involucre of 5–8 bracts 4–10 mm long. .... 5. *L. afghanicum.*

1. Ligusticum thomsonii C. B. Clarke in Hook. f., Fl. Brit. Ind. 2: 698. 1879.

Ligusticum thomsonii var. *evolutior* C. B. Clarke in Hook. f., Fl. Brit. Ind. 2: 698. 1879. Based on *Clarke* (?), "North Kashmir, alt. 6,000–9,000 ft."

Plants rather stout, erect, sparsely branched above, scaberulous to hirtellous, rarely glabrate, 50–150 cm tall from a fibrous base surmounting fascicled roots, the stem striate and somewhat angled. Leaves oblong to oblong-oval, 10–15 cm long, 3–8 cm broad, usually simply pinnate, the leaflets ovate to orbicular, 1–5 cm long, 1–3 cm broad, sessile, pinnately lobed to pinnatifid, mucronate-dentate; petiole slender, 5–20 cm long, sheathing at base; cauline leaves similar, the uppermost sessile with prominent oblong sheaths and reduced laminae. Peduncles terminal and lateral, 8–22 cm long, hirsutulous at summit. Involucre of 1 to several inconspicuous, caducous, linear-filiform, hirsutulous bracts 3–12 mm long, or involucre lacking. Rays 15–35, slender, angled, hirsutulous, unequal, spreading-ascending to ascending-divaricate, 1.5–3.5 cm long. Umbellets 15–25-flowered, not all maturing, the mature pedicels rather slender, angled, hirsutulous, spreading-ascending, unequal, 3–8 mm long. Involucel of 8–12 linear to linearly lanceolate, hirsutulous, narrowly scarious-margined bracteoles 3–10 mm long. Petals oval to obovate; calyx teeth minute, ovate, or calyx obsolete; stylopodium conical, the styles longer than stylopodium, reflexed. Fruit oval, scarcely compressed dorsally, 3.5–5 mm long, 2–3 mm broad, rounded at apex and base, the mericarps compressed dorsally, glabrous, the ribs subequally thin winged. Vittae 2 or 3 in intervals, 6–8 on commissure. Commissural face plane, slightly grooved beneath intervals. July–Sept.

Type locality: "WEST TIBET, alt. 11,000–13,000 ft.," *Thomson* (*Hooker & Thomson*, Herb. Ind. Orient., *Oreocome 2*). [Zausker, W. Tibet, *Thomson* (BM, K)].

Distribution: Himalaya, western Tibet to Kashmir, Afghanistan, and Pakistan, 2,000–4,500 m.

Representative specimens: Jammu & Kashmir: Sind Valley, *Clarke 30,942C* (BM), *32,942B* (CAL); above Doien, *Duthie 12,453* (DD); 8 miles S. of Karaghal, *Koelz 9229* (UC).

2. **Ligusticum alboalatum** Haines, Jour. Asiat. Soc. Bengal II. 15: 314. 1920.

Plants stout, erect, branched above, hirtellous, 60–150 cm tall, from a fibrous caudex surmounting a fusiform root, the stem fistulose, striate, and somewhat sulcate. Leaves oblong-lanceolate, 10–25 cm long, up to 20 cm broad, bipinnate, the leaflets ovate, 2–8 cm long, pinnately incised to pinnatifid, spinulose-serrate, the ultimate divisions linearly-lanceolate; petiole slender, cauline leaves reduced upward, 1-2-pinnate, the divisions narrow, the sheaths dilated. Peduncles terminal and lateral, elongate, hirsutulous at summit. Involucre of 1 or 2 inconspicuous linear bracts, or lacking. Rays 12–25, hirsutulous, unequal, spreading-ascending, 2–5 cm long. Umbellets 20–30-flowered, the mature pedicels 25–35 mm long. Involucel of 8–12, narrowly linear, unequal bracteoles. Petals obovate; calyx obsolete. Fruit elliptic, compressed dorsally, 4–5 mm long, 2.5-3.5 mm broad, the mericarps compressed dorsally, glabrous, the ribs subequally thin winged. Vittae 1-3 in intervals, 4–8 on commissure. Commissural face slightly concave. May–June.

Type locality: "Mountains of Chota Nagpur, elev. 2,500–3,000 ft., near streams," Bihar, *Haines*.

Distribution: Hills of Bihar, Peninsular India, 1,000–12,000 m.

Representative specimens: Bihar. Ranchi, *Haines* (CAL); Serguja, *Mooney 1307* (DD); Netarhat, *Haines 4319, 4319a* (K).

3. **Ligusticum marginatum** C. B. Clarke in Hook. f., Fl. Brit. Ind. 2: 698. 1879.

Plants rather stout, erect, sparsely branched above, scaberulous to hirtellous, rarely glabrate, 50–160 cm tall, the stem terete, ribbed but not angled. Leaves oblong to ovate, 15–40 cm long, 5–10 cm broad, simply pinnate to bipinnate, the leaflets ovate-lanceolate, 3–6 cm long, 1.5–4 cm broad, pinnately incised to pinnatifid and mucronulate-serrate, with shortly lanceolate to oblong ultimate divisions, scaberulous to hirsutulous; petiole slender, 15–40 cm long, dilated and scarious-sheathing at base; cauline leaves reduced upward, the upper sessile with slightly inflated, oblong sheaths. Peduncles terminal and lateral, 8–26 cm long, hirsutulous at summit. Involucre lacking, or occasionally of 1 or 2 linear or filiform, entire, hirsutulous, scarious-margined bracts 5-15 mm long. Rays 20–30, very slender, angled, hirsutulous, unequal, spreading-ascending to ascending-divaricate, 2.5–6 cm long. Umbellets 15–30-flowered, not all maturing, the mature pedicels slender, angled, hirsutulous, unequal, spreading-ascending or ascending-divaricate, 2–10(–15) mm long. Involucel of 6–12 linear-filiform, ciliolate, narrowly scarious-margined entire bracteoles 3–7 mm long, mostly shorter than flowering umbellets. Petals oval to obovate; calyx teeth minute, ovate, or calyx obsolete; stylopodium conical, about half as long as the reflexed styles. Fruit oval to orbicular, flattened dorsally, 3–5 mm long, 2.5–3 mm broad, rounded at apex, truncate at base, the mericarps flattened dorsally in transection, glabrous, the ribs unequally winged, the lateral much narrower than body but twice as broad as dorsal and intermediate ribs. Vittae 2 or 3 in intervals, 4–8 on commissure. Commissural face plane, grooved under intervals. Aug.–Oct.

Type locality: "North-west HIMALAYA; alt. 7,000–10,000 ft.; Simla and Naini Tal," *Thomson.*

Distribution: Western Himalaya, Garhwal and Kumaon to Pakistan, 2,000–4,000 m.

Representative specimens: "Himal. Bor. Occ.," *Thomson* (CAL). Jammu & Kashmir: Kargeil Valley, *Duthie 13,042* (CAL, DD). Himachal Pradesh: Simla, *Thomson* (K); Kardang, Lahul, *Mukherjee 331* (CUH, UC). Uttar Pradesh: Mahasu, *Meebold 4957* (CAL).

4. **Ligusticum elatum** (Edgew.) C. B. Clarke in Hook. f., Fl. Brit. Ind. 2: 698. 1879.

Cortia elata Edgew., Trans. Linn. Soc. 20: 55. 1851.

Plants rather stout, erect, sparsely branched above, essentially glabrous throughout, 50–150 cm tall, from a fibrous base, the stem terete, ribbed. Leaves ovate to ovate-lanceolate, 15–35 cm long, 15–20 cm broad, bipinnate, the leaflets ovate to ovate-lanceolate, 2.5–12 cm long, 1–8 cm broad, pinnately incised to pinnatifid with mucronulate-serrate, lanceolate ultimate divisions; petiole slender, 8–20 cm long, broadly sheathing below; cauline leaves gradually reduced upward, the uppermost with conspicuous sheaths and abortive laminae. Peduncles terminal and lateral, slender, 8–20 cm long, glabrous at summit. Involucre of 1 to several inconspicuous, glabrous, linear, entire bracts, or involucre lacking. Rays 10–30(–40 ?), slender, angled, nearly glabrous, slightly unequal, spreading-ascending, 2–4 cm long. Umbellets 15–30-flowered, not all perfect, the mature pedicels slender, angled and subwinged, 3–8 mm long. Involucel of 5–10 linear to filiform, entire, glabrous bracteoles 3–8 mm long. Petals a little unequal, oval to obovate; calyx obsolete; stylopodium conical, about equal to the reflexed styles. Fruit oblong to elliptic, compressed dorsally, 3–6 mm long, 1.5–3 mm broad, rounded at apex and base, the mericarps flattened dorsally in transection, glabrous, the ribs unequally winged, the lateral twice as broad as the dorsal. Vittae 2–3(–5 ?) in intervals, 4–6(–8 ?) on commissure, slightly grooved beneath intervals. Commissural face nearly plane. Aug.–Oct.

Type locality: "Himala, in apricis graminosis, alt. ped. 7,000–9,000. Mahaseo, etc.," *Edgeworth.*

Distribution: Western Himalaya, from Garhwal to Pakistan and Afghanistan, 1,800–3,000 m.

Representative specimens: Jammu & Kashmir: Kagan Valley, *Inayat Khan* 19,605 (DD). Himachal Pradesh: Lahul, *Jaeschke* 109 (CAL); Mahasu: Simla, *Edgeworth* 175 (K). Uttar Pradesh: Mussoorie, *Saxena* 921 (DD), *s. n.* (DD).

Note: We could not decide its synonymy with *Levisticum argutum* Lindley, in Royle, Ill. Bot. Himal. Mount. 232. 1835 (Based on: *Royle* "Mussoorie.") as suggested by Edgeworth (1851) and Clarke (1879).

5. **Ligusticum afghanicum** Rech. f. in Rech. f. & Riedl, Biol. Skr. 14(4): 94. 1963.

Cortia schmidii Nasir, Fl. W. Pakistan 20: 115. 1972; *Ligusticum schmidii* (Nasir) Farille & Malla in Farille, Cauwet-Marc & Malla, Candollea 40(2): 551. 1985. Based on *Schmid 2365*, "Chitral, Upper Yarkhum Vy," Pakistan.

Plants low, subcaulescent, essentially glabrous except for the scaberulous inflorescence, the flowering stems numerous, simple, almost naked, spreading to erect, 10–40 cm tall, arising from a fibrous caudex surmounting a long taproot. Leaves rosulate or clustered at base, oblong-lanceolate to oblong, 3–6 cm long, 0.8–1.5 cm broad, pinnate with 3–5 pairs of ovate, sessile leaflets 2–10 mm long, 2–6 mm broad, obtuse at apex, cuneate to truncate at base, pinnately incised to pinnatifid; petiole slender, 2–5 cm long, dilated and scarious-sheathing below; cauline leaves only 1 or 2, like the basal. Peduncles slender, terminal, 6–30 cm long. Involucre of 5–8 linear, mostly entire bracts 4–10 mm long, scarious-ciliolate. Rays 10–30, stout, unequal, 4–10 mm long, ± winged or angled, scaberulous. Involucel dimidiate, of 6–8 linearly lanceolate, mostly entire bracteoles 3–6 mm long, about equal to the flowers. Umbellets 15–20-flowered, the pedicels short, unequal, spreading-ascending, 1–3 mm long. Flowers white or purplish; calyx lobes minute, lanceolate, unequal; petals oblanceolate to obovate, with a narrower inflexed apex; stylopodium low conical, much shorter than the flattened, spreading styles; ovaries glabrous. Fruit oblong-oval, compressed dorsally, 3–3.5 mm long, ca. 2 mm broad, rounded at apex and base, the mericarps compressed dorsally, glabrous, the ribs unequally winged, the lateral much broader than the dorsal. Vittae 1 or 2 in intervals, 4 on commissure. Commissural face plane.

Type locality: "AFGHANISTAN: Nuristan: Sanglich, Minjan Prov., 1,300 ft., among stones and gravel, flrs. green-white," 25 Jul. 1937, *Koelz 12,633* (W, holotype; MICH, UC, US, isotypes).

Distribution: Western Himalaya, from Kashmir to Afghanistan and Pakistan.

Representative specimen: Jammu & Kashmir: Bara Deosai, SW of Satpara Pass, 13,000–13,500 ft., 6 Aug. 1955, *Webster & Nasir 6376* (UC).

51. **CNIDIUM** Cusson
Mém. Soc. Méd. Paris 280. 1782.

Ligusticum sect. *Cnidium* Calest. Webbia 1: 211. 1905.
Pinasgelon Raf. Good Book 51. 1840.

Slender, erect, branching, glabrous to variously pubescent perennial, biennial, or annual herbs from a rootstock or taproot. Leaves petiolate, membranaceous, 2–3-pinnate or pinnately decompound with small, narrow divisions; petiole broadly sheathing; cauline leaves subsessile and usually wholly sheathing. Inflorescence of lax terminal and lateral compound umbels, the peduncles short to elongate. Involucre and involucel of several to numerous, mostly narrow bracts or bracteoles, or the former sometimes lacking. Rays few to numerous, subequal to unequal, spreading-ascending. Umbellets few to rather many flowered, the flowers pedicellate. Flowers white, or rarely pinkish or purplish; petals elliptic to obcordate with a short, narrower inflexed apex; calyx teeth minute or obsolete; stylopodium low conical to conical, the styles slender, reflexed. Pedicels slender, spreading-ascending. Fruit oblong-ovoid to subglobose, rounded at apex and base, slightly compressed laterally or subterete, the mericarps slightly compressed dorsally in transection, glabrous, the ribs corky, prominent to broadly winged, the wings subequal or the lateral slightly broader, broader to narrower than intervals. Vittae large, solitary in intervals, 2 or 4 on commissure. Commissural face nearly plane. Carpophore bipartite.

Type species: *Selinum monnieri* Linn. [*Cnidium monnieri* (Linn.) Cusson].

Distribution: A Eurasian genus of probably fewer than a dozen species. Leute (1970) referred the genus to *Selinum*, in recognition of the close resemblance of *Cnidium* spp. to *S. carvifolium* (Linn.) Linn., type species of *Selinum*. Although we recognize the confused status of *Selinum*, we tentatively keep the two genera distinct, as was done by Shishkin (1950) and Tutin (1968).

1. **Cnidium monnieri** (Linn.) Cusson, Mém. Soc. Méd. Paris 280. 1782.

Selinum monnieri Linn., Cent. Pl. 1: 9. 1755; *Cicuta monnieri* (Linn.) Crantz, Class. Umbell. 98. 1767; *Pinasgelon monnieri* (Linn.) Raf., Good Book 52. 1840; *Ligusticum monnieri* (Linn.) Calest., Webbia 1: 211. 1905.
Ammi daucifolium Buchanan-Hamilton in Wallich, Cat. 7209. 1829, *nomen nudum*; *Seseli daucifolium* C. B. Clarke in Hook. f., Fl. Brit. Ind. 2: 693. 1879. Based on "*Wallich, etc.*, EAST BENGAL; from BHOTAN and ASSAM to DACCA and CHITTAGONG."

Plants annual, branched, glabrous to variously pubescent or scaberulous, 20–80 cm tall from a slender vertical taproot. Leaves oblong to ovate, 6–15 cm long, 8–12 cm broad, pinnately decompound, the ultimate divisions linear to lanceolate, acute, 2–10 mm long, 1–3 mm broad; petiole 1–2 cm long; cauline leaves numerous, like the basal but reduced and subsessile upward. Peduncles short and stout, 3–7 mm long. Involucre of 8–10 linearly subulate, acuminate, ciliolate bracts 2–10 mm long. Rays 8–30, scabrous, unequal, 0.5–2 cm long, becoming incurved. Involucel of 5–12 linearly lanceolate to filiform, ciliolate bracteoles 2–4 mm long. Umbellets 10–25-flowered, the pedicels 3–5 mm long. Flowers white or pinkish; calyx obsolete. Fruit ovoid, 2.5–3 mm long, 1.5–2 mm broad, the ribs conspicuously and subequally corky winged, the ribs broader than intervals. Apr.–July.

Type locality: "In Gallia australi," no collector indicated.

Distribution: East Asia; sporadically adventive elsewhere.

Representative specimens: Meghalaya: Khasia, *Mann* (CAL). Assam: Sibsagar, *Clarke 38,033B* (CAL); Assam, *Jenkins* (CAL); Mangaldai, *Chatterjee* (CAL).

52. **SELINUM** Linn.

Sp. Pl. ed. 2. 350. 1762, *nomen conserv.*; C. B. Clarke in Hook. f., Fl. Brit. Ind. 2: 699. 1879.

Oreocome Edgew. Proc. Linn. Soc. 1: 252. 1846.

Slender to stout, erect, caulescent, branching, or rarely low, dwarfed and acaulescent, scaberulous or variously scaberulous or pubescent to glabrous perennial herbs arising from a caudex bearing fibrous leaf bases, or these lacking, surmounting a taproot. Leaves petiolate, membranaceous, pinnate or bipinnate to pinnately or ternate-pinnately decompound, with definite leaflets, or short, narrow, ultimate divisions; petiole slender to stout, sheathing at base; cauline leaves similar, reduced upward, at least the uppermost sessile, often with prominently dilated sheaths. Inflorescence of lax terminal and lateral compound umbels, the peduncles slender or stout, often elongate, or the terminal umbel short pedunculate to sessile. Involucre of 1 to several conspicuous entire, lobed, or pinnatifid, pubescent or glabrous, sometimes scarious-margined bracts, often reflexed, or bracts caducous, or involucre obsolete or lacking. Rays few to numerous, usually slender, unequal or subequal, scaberulous to hirtellous, often angled, spreading-ascending to ascending-divaricate. Involucel of bracteoles like the bracts, shorter or occasionally longer than flowering umbellet. Umbellets few to many flowered, usually not all maturing, the flowers pedicellate. Flowers white or rarely purplish; petals oblanceolate to obovate, with a narrower inflexed apex, rarely clawed, often a little unequal; calyx teeth prominent, linear or linearly lanceolate, or merely evident, or calyx obsolete; stylopodium conical, the styles slender to short, usually reflexed. Pedicels slender to stout, usually unequal, scaberulous to hirtellous. Fruit oblong-oval to subquadrate or orbicular, slightly compressed to flattened dorsally, rounded, truncate or cordate at apex and base, the mericarps compressed to flattened dorsally, glabrous, the ribs subequally thin winged, or the lateral wings broader than the dorsal. Vittae solitary in the intervals, or commonly 2(1–4?) in lateral intervals, 2–6 (and visible) on commissure, or totally obscure. Commissural face plane to slightly concave, usually grooved under intervals. Carpophore bipartite, the halves capillary, or carpophore rarely obsolete.

Type species: *Seseli carvifolia* Linn. [*Selinum carvifolium* (Linn.) Linn.].

Note. The genus *Selinum* as treated here is almost certainly *not* a "natural" aggregation of congeneric species. There is little if any agreement on the scope or limits of *Selinum* and as to its place in the family, except that it must include the type species, *S. carvifolium*. Leute (1969–1970, pp. 80–81, 496–497) has summed up the vicissitudes the genus has undergone. Most recent authors (Thellung, Shishkin, Tutin, Nasir) have followed Candolle and Drude in maintaining *Selinum* and *Cnidium* Cusson as distinct genera. Leute, however, follows Bentham and Hooker in combining them under *Selinum*.

The Indo-Himalayan species treated here, however, Leute excludes from *Selinum* entirely and refers to *Cortia*, pointing out the many characters they share with *Cortia depressa* and *C. hookeri*. Hedge & Rechinger f. in Fl. Iranica (1987) follow Leute's example. Franchet (1894) transferred two species to *Ligusticum*, another genus with which they seem to have much in common. They are retained in *Selinum* largely for convenience and because their proper position does not seem to have been unequivocally established.

Whatever qualms may be expressed about their generic affinities, *S. cortioides*, *S. vaginatum*, and *S. papyraceum* are generally recognized as valid species, but there agreement ends. Clarke (1879) accepted *S. candollii* but combined the present *S. wallichianum* and *S. elatum* under *S. tenuifolium*. Hiroe's treatment (1958) was substantially the same. Nasir (1972) recognized *S. candollii*, *S. wallichianum* (*S. tenuifolium*), and *S. filicifolium* as distinct taxa. Cannon (1979) combined *S. wallichianum* with *S. candollii*, and *S. filicifolium* with *S. tenuifolium*, emphasizing their close relationship to each other and with *S. striatum*, and astutely observed that "a complete revision of the genus is needed for clarification" (p. 189). Chang (1985) combines *S. wallichianum* and *S. tenuifolium* under *S. candollii*. Because all four of the species described by Edgeworth under the generic name *Oreocome* are regarded here as synonyms of either *S. candollii* or *S. wallichianum*, *O. stelliphora* Cauwet-Marc & Farille is also treated here, but with no firm decision as to its natural affinities.

KEY TO SPECIES

1. Plants low and dwarfed, acaulescent; carpophore obsolete. ....1. *S. cortioides*.
1. Plants stout or slender, erect, caulescent, branching; carpophore well developed.
   2. Both dorsal and commissural vittae obsolete; bracteoles usually pinnatifid. ....2. *S. vaginatum*.
   2. Both dorsal and commissural vittae evident; bracteoles usually entire, less commonly lobed, not pinnatifid.
      3. Cauline petioles conspicuously dilated; rays 10–30; bracteoles 2–15 mm long, little if any surpassing umbellet.
         4. Calyx obsolete; caudex lacking fibrous leaf bases. ....3. *S. papyraceum*.
         4. Calyx prominent; caudex conspicuously fibrous.
            5. Fruits subterete, only the mericarps flattened dorsally; ribs subequally winged.
               6. Stems angled; rays 8–16; calyx persistent, the teeth equal to stylopodium; seed face concave with central ridge. . 4. *S. striatum*.
               6. Stems striate or ribbed but not angled; rays 20–30; calyx persistent, but teeth shorter than stylopodium; seed face concave. 5. *S. candollii*.
            5. Fruits and mericarps both flattened dorsally; ribs unequally winged.
               7. Dorsal and intermediate fruit ribs unwinged; ultimate leaf divisions linearly lanceolate. ....6. *S. wallichianum*.
               7. Dorsal and intermediate ribs narrowly winged; ultimate leaf divisions broadly elliptic, toothed. ....7. *S. elatum*.
      3. Cauline petioles very narrow, scarcely dilated; rays ca. 8; bracteoles 10–20 mm long, conspicuously surpassing umbellet. ....8. *Oreocome stelliphora*.

1. Selinum cortioides Norman, Jour. Bot. 75: 95. 1937.

Cortia hookeri C. B. Clarke in Hook. f., Fl. Brit. Ind. 2: 702. 1879, *pro parte*.

Plants low, dwarfed, acaulescent, ± tomentose, from an elongated taproot. Leaves rosulate, oblong, 8–12 cm long (including petiole), 2–3 cm broad, simply pinnate, the 4 or 5 pairs of leaflets 1–1.5 cm long, 7–12 mm broad, pinnatifid, the ultimate divisions linear, acute, ca. 5 mm long, 1 mm broad; petiole tomentose. Inflorescence of a sessile or shortly pedunculate (0–3.5 cm) terminal umbel and several pedunculate lateral umbels. Involucre obsolete. Rays of terminal umbels numerous, fistulose, unequal, tomentose, 1.5–6 cm long, shorter than foliage leaves; rays of lateral umbels ca. 10, 4–9 mm long. Flowers of terminal umbel perfect, those of lateral umbels staminate only. Involucel of several linearly lanceolate, entire, narrowly scarious-margined bracteoles 2.5–5 mm long. Petals ovate to elliptic; calyx teeth small but evident, lanceolate; stylopodium conical, the styles slender, reflexed. Fruit suborbicular, strongly compressed dorsally, ca. 3 mm long, 2.5 mm broad, truncate at apex, cordate at base (sometimes also at apex), the mericarps flattened dorsally in transection, glabrous, the ribs unequally corky winged, the lateral broader than the narrowly winged or sometimes unwinged dorsal ribs. Vittae large, solitary in intervals, 2 on commissure. Commissural face plane or slightly concave. Carpophore obsolete. Aug.–Sept.

Type locality: "SIKKIM: Phalloot, 13,000 ft.," *Kurz* (K, holotype) (Norman 1937).

Distribution: Sikkim Himalaya, 3,000–5,000 m.

Representative specimens: w/o definite locality, *Thomson* (CAL). West Bengal: Singaleleh, *Kurz* (CAL). Sikkim: Tankara, *Gammie 562* (CAL); Phalloot, *Kurz* (CAL); Yakla, *Clarke 9911* (K); Llonak, *Cave 41* (K).

2. **Selinum vaginatum** (Edgew.) C. B. Clarke in Hook. f., Fl. Brit. Ind. 2: 700. 1879.

Cortia vaginata Edgew., Trans. Linn. Soc. 20: 55. 1851.

Plants slender to stout, erect, sparsely branched above, finely hirtellous at least in inflorescence, (6)20–120 cm tall, from a strongly fibrous caudex surmounting a taproot, the stem ribbed but scarcely angled. Leaves oblong to ovate-lanceolate, 5–30 cm long, 3–14 cm broad, bipinnate, the leaflets lanceolate to ovate-lanceolate, 5–30 mm long, 3–10 mm broad, sessile, spinulose-serrate and incised to pinnatifid, pubescent on rachis and nerves or glabrate, conspicuously venose; petiole stout, 2–10 cm long, broadly scarious-sheathing below; cauline leaves reduced upward, sessile, with inflated, elliptic, conspicuously purple-veined sheaths. Peduncles terminal and lateral, stout, 4–18 cm long, hirsutulous at summit. Involucre of 1 to several narrow, usually pinnatifid, scarious-margined, hirsutulous bracts 1–3 cm long, or lacking. Rays 10–30, rather stout, angled, hirsutulous, unequal, spreading-ascending, 2–5 cm long. Umbellets 15–25-flowered, not all maturing, the mature pedicels slender, angled, hirsutulous, unequal, spreading-ascending, 3–8 mm long. Involucel of 8–12 lanceolate to obovate, usually pinnatisect, ciliate, scarious-margined, reflexed bracteoles 5–15 mm long, exceeding flowering umbellets. Petals a little unequal, oval to obovate, narrowed but scarcely clawed at base, glabrous; calyx teeth prominent, linear, ca. 1 mm long; stylopodium conical, the styles slender, reflexed. Fruit oblong-elliptic to subquadrate, compressed dorsally, 4–5 mm long, 2–2.5 mm broad, the mericarps flattened dorsally, glabrous, the ribs unequally thin winged, the lateral about as broad as body, the dorsal half as broad. Vittae obsolete. Commissural face plane, grooved under intervals. July–Oct.

Type locality: "Himala, ad alt. ped. 9,000–11,000, in apricis graminosis. Chúr, etc., Kamalori, Hattu, Garhwál," *Edgeworth 178* (K, holotype) (Nasir 1972).

Distribution: Western Himalaya, Garhwal to Kashmir and Pakistan, 2,000–4,500 m.

Representative specimens: Jammu & Kashmir: Kalahoi, *Duthie 13,503* (CAL); Lidder Valley, *Inayat Khan 25,627* (DD); Pahlgam, *R. R. Stewart 21,723* (UC), *21,823* (UC); Sonamarg, *R. R. Stewart 22,384* (UC). Himachal Pradesh: Lahul, Koksar, *Nair 17,025* (BSD); Dalhousie, *Clarke 40* (CAL); Choor, Simla, *Raizada 13,486* (DD). Uttar Pradesh: Rambara, Garhwal, *Rau 8728* (CAL); Hemkund, *Bhattacharyya 29,527* (BSD); Kedarnath, Mandakini River, 26 Oct. 1982, *Mukherjee s. n.* (CUH, UC).

3. **Selinum papyraceum** C. B. Clarke in Hook. f., Fl. Brit. Ind. 2: 701. 1879.

Cortia papyracea (C. B. Clarke) Leute, Ann. Naturhist. Mus. Wien 73: 84. 1969.

Plants stout, erect, branched above, glabrous except for the scaberulous inflorescence, 20–100 cm tall, from a woody, nonfibrous base. Leaves ovate, 3–10 cm long, 3–8 cm broad, bipinnate, the leaflets ovate-lanceolate to ovate, 5–50 mm long, 3–20 mm broad, pinnatifid to pinnatisect, the ultimate divisions linearly lanceolate to oval; petiole slender, 3–12 cm long, rather broadly sheathing; cauline leaves gradually reduced upward, with conspicuous oblong sheaths. Peduncles terminal and lateral, stout, 8–25 cm long, scaberulous at summit. Involucre of 1 or 2 filiform bracts, or lacking. Rays 8–20, slender, angled, scaberulous, subequal, ascending-divaricate, 1–3 cm long. Umbellets 15–30-flowered, not all maturing, the mature pedicels slender, angled, scaberulous, unequal, spreading-ascending, 2–8 mm long. Involucel of 6–10 linear or filiform, entire, scaberulous-ciliolate but not white-margined bracteoles 4–6 mm long, often exceeding flowering umbellets. Flowers white or purplish; petals ovate; calyx obsolete; stylopodium conical, a little shorter than the reflexed styles. Fruit oblong-elliptic, compressed dorsally, 3–4 mm long, 1.5–2.5 mm broad, the mericarps flattened dorsally, glabrous, the ribs all thin winged, the lateral broader than the dorsal. Vittae solitary (occasionally 2) in intervals, 2 or 4 on commissure. Commissural face plane or shallowly concave. Carpophore bipartite. July–Sept.

Type locality: "From BALTISTHAN and KASHMIR to SIKKIM, alt. 8,000–12,000 ft.; frequent," [Pahlgam], *Clarke 31,047* (K, lectotype) (Nasir, 1972).

Distribution: Western Himalaya, from Himachal Pradesh and Kashmir to Baltistan, 2,500–4,000 m; Sikkim.

Representative specimens: Jammu & Kashmir: Pahlgam, *Clarke 31,067B* (CAL); Baltal, *Thomson* (K); Tilail, *Clarke 30,664* (K). Himachal Pradesh: Lahul, *Jaeschke 64* (K); Kardang, Lahul, *Mukherjee 332* (CUH, UC); Keylong, *Mukherjee s. n.* (CUH, UC). Uttar Pradesh: Kuanri Pass, *Giles 701* (K). Sikkim: Ilumbo, *Clarke 25,621* (K).

4. **Selinum striatum** (DC.) C. B. Clarke in Hook. f., Fl. Brit. Ind. 2: 699. 1879.

Laserpitium striatum Wallich, Cat. 583. 1829, *nomen nudum*; *Ligusticum striatum* DC., Prodr. 4: 158. 1830; *L. wallichii* Franchet, Bull. Soc. Philom. Paris VIII. 6: 1894; *Cortia striata* (DC.) Leute, Ann. Naturhist. Mus. Wien 73: 85. 1969.

Pimpinella bengalensis H. Wolff, Feddes Repert. 17: 170. 1921. Based on *Griffith 2615/1*, "Bengali orientalis sine loco indic." [actually Khasia, NE India] (CAL, isotype).

Plants slender, erect, sparsely branched above, glabrous except for the scaberulous inflorescence, 75–120 cm tall, from a slender fibrous caudex, the stem angled. Leaves ovate-lanceolate, 4–8 cm long, 2–4 cm broad, pinnately decompound, the ultimate divisions linearly lanceolate, mucronate, 1–5 mm long, 1–1.5 mm broad; petiole slender, 2–10 cm long, oblong-sheathing below; cauline leaves few, reduced upward, the uppermost sessile with a prominent oblong sheath. Peduncles terminal and lateral, rather stout, elongate, 7–20 cm long, scabrous at summit. Involucre of 4–10 linear to filiform, entire, reflexed bracts 2–10(–15) mm long. Rays 8–16, slender, angled, scabrous-papillose, unequal, ascending, 1.5–2.5 cm long. Umbellets 10–20-flowered, not all maturing, the mature pedicels slender, angled, scaberulous, unequal, spreading-ascending, 1–8 mm long. Involucel of 6–10 linearly lanceolate, entire, reflexed bracteoles 2–5(–8) mm long. Petals oval to obovate, narrowed into claw at base; calyx teeth prominent, linear, 0.5–0.8 mm long, persistent; stylopodium conical, the styles about equal to stylopodium, reflexed. Fruit oblong-oval, subterete, 4–6 mm long, 1.5–2.5 mm broad, rounded at apex and base, the mericarps pentagonal, compressed dorsally in transection, glabrous, the ribs all subequally narrow winged. Vittae large, solitary in intervals, occasionally 2 smaller ones in lateral valleculae, 2–4 on commissure. Commissural face plane to slightly concave, grooved under vittae. July–Oct.

Type locality: Nepal, *Wallich 583* (CAL, isotype).

Distribution: Nepal; Khasia and Naga hills of northeastern India, 1,500–2,100 m.

Representative specimens: Nepal: Syabrubenshi, *Mukherjee 246* (CUH, UC). Meghalaya: Shillong, *Clarke 5714* (CAL); Khasia, *Hooker & Thomson* (CAL); Mawphlang, Khasia Hills, *Chand 2352* (MICH, UC), *7932* (MICH, UC), *Koelz 31,153* (MICH, UC), *31,284* (MICH, UC).

5. **Selinum candollii** DC., Prodr. 4: 165. 1830; C. B. Clarke in Hook. f., Fl. Brit. Ind. 2: 700. 1879.

Angelica ? candollii Wallich, Cat. 582. 1829, *nomen nudum*; *Oreocome candolliana* (DC.) Edgew., Trans. Linn. Soc. 20: 55. 1851; *Cortia candollii* (DC.) Leute, Ann. Naturhist. Mus. Wien 73: 83. 1969.

Plants rather stout, erect, branched above, hirtellous at least in inflorescence to glabrate, 1–3 m tall, the stem fistulose, striate. Leaves ovate, 10–25 cm long, 8–20 cm broad, tripinnately or ternate-pinnately decompound, the ultimate divisions short, lanceolate to oblong, up to 1 mm broad; petiole dilated and broadly sheathing, 5–12 cm long; cauline leaves like the basal but reduced upward, sessile with prominent sheaths. Peduncles terminal and lateral, stout, 5–15 cm long, hirsutulous at summit. Involucre of 5–12 linear or lanceolate, entire, scarious-margined bracts 1.5–2.5 cm long, often deciduous. Rays 20–30, slender, hirsutulous, unequal, ascending, 3–9 cm long. Umbellets ca. 15-flowered, the mature pedicels slender, hirsutulous, unequal, 4–15 mm long. Involucel of several linear or lanceolate, entire, scarious-margined bracteoles 5–10 mm long. Petals a little unequal, obovate, emarginate; calyx teeth linearly lanceolate; stylopodium conical, the styles slender, recurved. Fruit elliptic or subquadrate, compressed dorsally, 4–8 mm long, 2–4 mm broad, the mericarps flattened dorsally, glabrous, the ribs unequally thin winged, the lateral narrower than the body but usually broader than the often unequal dorsal ribs. Vittae usually solitary in dorsal intervals, 1 or 2 in lateral intervals, 2–4 on commissure. Commissural face plane to slightly concave. July–Oct.

Type locality: "in montibus Nepaliae, ad Kamaon," [Nepalia, 1821, Kumaon, *Blinkworth*], *Wallich 582*.

Distribution: Himalaya, from Nepal to Kashmir, 3,000–4,000 m.

Representative specimens: "N. W. India," *Royle* (CAL). Jammu & Kashmir: w/o precise locality, *Thomson* (CAL). Uttar Pradesh: Brahmatal, *Bhattacharyya 31,055* (BSD); Kumaon, *Wallich 582B* (CAL). Nepal: Langtang, *Mukherjee 281* (CUH, UC); Sheopore, *Wallich 582A* (CAL).

6. **Selinum wallichianum** (DC.) Raizada & Saxena, Indian Forester 92: 232. 1966.

Selinum tenuifolium Wallich, Cat. 579. 1829, *nomen nudum*; C. B. Clarke in Hook. f., Fl. Brit. Ind. 2: 700. 1869; *Peucedanum wallichianum* DC., Prodr. 4: 181. 1830; *Ligusticum tenuifolium* (Clarke) Franchet, Bull. Soc. Philom. Paris VIII. 6: 136. 1894; *Cortia wallichiana* (DC.) Leute, Ann. Naturhist. Mús. Wien 73: 83. 1969.

Laserpitium coniifolium Wallich, Cat. 584. 1829, *nomen nudum*; *Ligusticum coniifolium* DC., Prodr. 4: 158. 1830. Based on *Wallich 584*, "in Nepaliae montibus altis" [Nepalia, Kamaon, 1821, *Blinkworth*].

Pleurospermum cicutarium Lindley in Royle, III. Bot. Himal. Mount. 233. 1839; *Oreocome cicutaria* (Lindley) Edgew. ex Leute, Ann. Naturhist. Mus. Wien 73: 1869, *nomen in synon*. Based on *Royle*, "Choor and Simla."

Oreocome filicifolia Edgew., Proc. Linn. Soc. 1: 253. 1846; *Selinum tenuifolium* var. *filicifolia* (Edgew.) C. B. Clarke in Hook. f., Fl. Brit. Ind. 2: 700. 1879; *S. filicifolium* (Edgew.) Nasir, Fl. W. Pakistan 20: 119. 1972. Based on *Edgeworth*, "Himala, in sylvis, alt. ped. 9,000–10,000. Kamalori, Hattu, etc.

Plants stout, erect, sparsely branched above, hirtellous to scaberulous, 20–150 cm tall, from a strongly fibrous caudex, the stem ribbed but scarcely angled. Leaves triangular-ovate, 15–35 cm in diameter, finely pinnately or ternate-pinnately decompound, the ultimate divisions short, linear, 1–3 mm long, up to 1 mm broad, venose; petiole stout, 7–20 cm long, broadly sheathing below; cauline leaves mostly ternately decompound, the upper sessile with dilated, oblanceolate, purple-veined sheaths. Peduncles terminal and lateral, stout, 3–18 cm long, hirtellous at summit. Involucre of 5–10 narrow, pinnatifid or entire, hirsutulous, scarious-margined bracts up to 3 cm long, or involucre lacking. Rays 20–30, slender, angled, scaberulous to hirsutulous, subequal, spreading-ascending to ascending-divaricate, 3–6 cm long. Umbellets 20–30-flowered, not all maturing, the mature pedicels slender, angled, hirsutulous or scaberulous, unequal, spreading-ascending, 3–15 mm long. Involucel of 6–12 linear to lanceolate, entire or occasionally few lobed distally, hirsutulous, scarious-margined bracteoles 5–15 mm long, the longer often exceeding flowering umbellet. Petals a little unequal, oblanceolate to obovate, scarcely clawed, glabrous; calyx teeth prominent, linear, white-margined, hirsutulous, 0.5–1.5 mm long; stylopodium conical, about equal to the short, reflexed styles. Fruit oblong-elliptic to oval, flattened dorsally, 4–6 mm long, 2–3 mm broad, the mericarps flattened dorsally, glabrous, the dorsal ribs filiform, the lateral thin winged, the wing about half as broad as body. Vittae mostly solitary in dorsal intervals, 2(1–4?) in intermediate intervals, 2–6 on commissure. Commissural face plane, grooved under intervals. Carpophore halves capillary. July–Oct.

Type locality: "In India or. ad Sirinagur, in Nepaliae monte Emodo ad Kamaon et Gossain-Than," *Wallich 579*.

Distribution: Himalaya, Sikkim, and Nepal to Kashmir and Pakistan, 2,500–5,000 m.

Representative specimens: Jammu & Kashmir: Pir Pass, *T. A. Rao* 7689 (BSD). Himachal Pradesh: Chitkul, *Nair 34,357* (BSD); Rothung, *Brandis 3738* (CAL); Choor, *Royle 93/22* (DD). Uttar Pradesh: Mussoorie, *Edgeworth 176* (K), *Duthie 1650* (DD); Naini Tal, *Gillett, Pangtey & Gupta 2615* (CUH, UC); Kumaon, *Wallich 5797* (CAL). Nepal: Gossainthan, *Wallich 579A* (CAL). West Bengal: Tonglo, *Thomson* (CAL); Darjeeling, *Clarke 13,283B* (CAL); Ghoom to Bhanjan, *Mukherjee 3530* (CUH, UC). Sikkim: Jongri, *Anderson 618* (CAL); Lachung, *Gammie 361* (CAL).

7. **Selinum elatum** (Edgew.) Hiroe, Umbell. Asia 154. 1958.

Oreocome elata Edgew., Proc. Linn. Soc. 1: 253. 1846; *Selinum tenuifolium* var. *elata* (Edgew.) C. B. Clarke in Hook. f., Fl. Brit. Ind. 2: 700. 1879.

Plants stout, erect, branched, puberulent, 1–3 m tall, the stem striate and fistulose. Leaves oblong-ovate, 30–45 cm long, 15–25 cm broad, pinnately decompound, the ultimate divisions broadly elliptic to ovate-lanceolate, incised-dentate, sessile, 5–20 mm long, 3–10 mm broad, puberulent on veins; petiole elongate, inflated-sheathing; cauline leaves reduced upward, the uppermost consisting of bladeless sheaths. Peduncles terminal and lateral, stout, elongate, 5–10 cm long. Involucre of a few linear deciduous bracts. Rays numerous (to 25), unequal, 3–6 cm long. Umbellets many flowered, the mature pedicels unequal, 3–5 mm long. Involucel of several linearly subulate, entire, puberulent and ciliolate bracteoles 3–4 mm long. Petals obovate, emarginate; calyx teeth small; stylopodium conical, the styles short, reflexed. Fruit elliptic-subquadrate, flattened dorsally, 4–5 mm long, 2–2.5 mm broad, the dorsal ribs narrowly winged, the lateral broadly thin winged. Vittae 1–3 in intervals, 4–6 on commissure. Commissural face plane, grooved under intervals. Oct.

Type locality: "*Hab*. Himala, in sylvis, alt. ped. 8,000–10,000. Mahaseo, Chúr," *Edgeworth*.

Distribution: Western Himalaya, from Uttar Pradesh to Kashmir, 3,800 m.

Representative specimens: Uttar Pradesh: Valley of Flowers, Garhwal, *Bhattacharyya 24,385* (BSD). Himachal Pradesh: Mahasu, *Edgeworth 174* (K); Mahasu; Simla, *Edgeworth* (K).

Note. Although Hiroe (1958) made the proper nomenclatural combination, the specimens he cited are all referable to one or another species of *Ligusticum*.

8. **Oreocome stelliphora** Cauwet-Marc & Farille in Farille, Cauwet-Marc & Malla, Candollea 40(2): 549. 1985.

Plants slender, erect, simple or branched above, glabrous or a little scaberulous in the inflorescence, 20–50 (±) cm tall, from a slender taproot. Leaves ovate-lanceolate, 20–25 cm long, 6–12 cm broad, bipinnate, acuminate, the leaflets linear to ovate, entire to pinnatifid, the ultimate lobes or divisions, linear, acute or acuminate, 3–40 mm long, 1–5 mm broad; petiole slender, 4–10 cm long, very narrowly sheathing; cauline leaves like the basal, longly-acuminate, the sheaths very narrow. Peduncles terminal and often also lateral, 6–15 cm long, scaberulous at summit. Involucre of a single foliaceous bract, linear and entire or pinnate with linear divisions, 5-7 cm long, equal to or exceeding the umbel. Rays slender, 4–15, unequal, 3–15 cm long, scaberulous, spreading-ascending. Umbellets 20–40-flowered, the immature pedicels 2–5 mm long, scaberulous. Involucel of about 8 elongated linearly-filiform, ciliolate bracteoles 1-2 cm long, exceeding the umbellets. Flowers "pink"; petals obovate, obtuse but with a small, acute incurved apex, conspicuously clawed, slightly unequal; calyx teeth evident, very unequal; stylopodium low conical, about equal to the divergent styles. Immature fruit oblong, ca. 4 mm long, 1.5–2 mm broad, a little narrowed at apex, truncate at base, the mericarps subterete in transection, glabrous, the ribs prominent or narrowly winged, subequal. Vittae 2 or 3 in intervals, 4–6 on commissure. Commissural face plane. Carpophore bipartite. Aug.–Sept.

Type locality: "C. Népal: N. Annapurna, high valley of Marsyangdi, upstream from Manang, 3,800 m," 18 Sep. 1981, *Farille 81,467*.

Distribution: Himalaya of central Nepal to Uttar Pradesh, 3,700–3,850 m.

Representative specimens: Nepal: Taglung (S. of Tukucha), Kali Gandaki, *Stainton, Sykes & Williams 7951* (BM, E); Annapurna, Muktinath Range, *Farille 81,421* (E).

Selinum dissectum (DC.) C. B. Clarke in Hook. f.,
Fl. Brit. Ind. 2: 701. 1879.

Selinum ? dissectum Wallich, Cat. 581. 1829, *nomen nudum*; *Peucedanum dissectum* DC., Prodr. 4: 178. 1830. Based on *Wallich 581*, "In Emodo seu Himalaya Nepalensi ad Gossain-Than."

Mentioned, although scarcely accepted by Clarke (1879), who noted that the appropriate sheet in the Wallichian herbarium was an incongruous assemblage of two or three different elements, this epithet was not included by Hiroe (1958), Leute (1969–1970), or Cannon (1979). The type number at CAL is a mixture of *S. striatum* and *S. wallichianum*. However, neither Candolle's description nor the microfiche of the Prodromus holotype permits reliable identification. The epithet probably should be abandoned as a source of continuing confusion.

53. PLEUROSPERMOPSIS Norman
Jour. Bot. 76: 200. 1938.

Stout, erect, caulescent, usually branching above, a little scaberulous or puberulent in inflorescence, otherwise glabrous, biennial or perennial from a stout taproot. Leaves petiolate, membranaceous, pinnate with ovate leaflets; petiole dilated and broadly sheathing at base; cauline leaves like the basal. Inflorescence of a large terminal and several small lateral compound (or occasionally simple) umbels arising in the upper axils, the terminal peduncle short and stout, continuing the main stem. Involucre of several conspicuous, foliaceous, nonscarious, apically incised or pinnatifid bracts shorter than the rays. Rays few, stout, unequal, spreading-ascending, slightly dilated at summit. Involucel of numerous persistent, rigid, foliaceous, nonscarious, apically trifid to deeply lacerate-dentate and strongly imbricated bracteoles embracing and exceeding the umbellet. Umbellets numerous flowered, the flowers pedicellate. Flowers purple; petals oblong, with a short incurved acute apex; calyx teeth small, ovate; stylopodium depressed conical, the styles slightly longer. Pedicels short, subequal. Fruit oblong, lightly compressed laterally, the mericarps subterete, glabrous, the ribs equally narrowly winged. Vittae 1 or 2 in intervals, 4 on commissure. Commissural face concave. Carpophore bipartite.

Type species: *Pleurospermum sikkimense* C. B. Clarke [*Pleurospermopsis sikkimensis* (C. B. Clarke) Norman].

Distribution: A monotypic genus of the Eastern Himalaya. The author (Norman) regarded his new genus as intermediate between *Ligusticum* and *Pleurospermum*.

1. Pleurospermopsis sikkimensis (C. B. Clarke) Norman,
Jour. Bot. 76: 200. 1938.

Pleurospermum sikkimense C. B. Clarke in Hook. f., Fl. Brit. Ind. 2: 702. 1879.

Plants stout, erect, the stem fistulose, simple or few branched, a little scaberulous or puberulent at base of umbels and umbellets, 10–55 cm tall. Leaves narrowly oblong, 5–20 cm long, 2–5 cm broad, simply pinnate, the leaflets ovate to orbicular, 5–25 mm in diameter, subsessile, somewhat acute, truncate to cuneate at base, coarsely mucronate-dentate; petiole slender, 2.5–14 cm long. Terminal peduncle 5–15 cm long, the lateral peduncles much more slender and raylike. Involucre of usually 5 oblanceolate to obovate, prominently sheathing foliaceous bracts 2–5 cm long. Rays 4–7, stout, unequal, 3–10 cm long. Umbellets 40–60-flowered, the mature pedicels stout, angled, 3–5 mm long. Involucel of numerous, obovate, foliaceous, distally deeply incised bracteoles 1–2 cm long, exceeding umbellets. Flowers apparently all fertile, appearing black in dried specimens. Fruit oblong to oblong-oval, 3–4.5 mm long, 1.5–2 mm broad, the prominent light-colored ribs narrower than the brownish intervals. Oct.

Type locality: "SIKKIM, alt. 13,000–16,000 ft.," *Hooker, Clarke* [Jongri, *Clarke 26,207* (K, lectotype)] (Norman 1938).

Distribution: Himalaya of Sikkim and Nepal, 4,000–5,000 m; Tibet.

Representative specimens: Nepal: Topke Gola, *Beer, Lancaster & Morris 10,783* (BM). Sikkim: Jongri, *Clarke 26,207E* (CAL), *26,164C* (K), *26,179A* (K), *Hooker* (CAL, K), *Anderson 620* (BM, CAL); Yakla, *Clarke 10,190* (CAL); Changu, *Ribu & Rhomoo 4372* (CAL).

54. PLEUROSPERMUM Hoffm.

Gen. Umbell. 8, 29. 1814; C. B. Clarke in Hook. f., Fl. Brit. Ind. 2: 702. 1879.

Hymenolaena DC. Prodr. 4: 244. 1830; *Pleurospermum* subgen. *Hymenolaena* (DC.) Drude in Engl. & Prantl, Pflanzenfam. 38: 171. 1898.

Hymenidium Lindley in Royle, Ill. Fl. Himal. Mount. 233. 1835; *Pleurospermum* subgen. *Hymenidium* (Lindley) Drude in Engl. & Prantl, Pflanzenfam. 38: 172. 1898.

Pterocyclus Klotzsch in Klotzsch & Garcke, Bot. Ergebn. Reise Prinz Waldemar 150. 1862.

Aulacospermum sect. *Hymenolaenopsis* Kljuykov, Pimenov & Tichomirov, Bjull. Moskovsk. Obsc. Isp. Prir., Otd. Biol. 81: 65. 1976.

Stout to very slender, erect, caulescent, simple or sparsely branching, or nearly acaulescent and the rays appearing like peduncles bearing simple umbels, glabrous or scaberulous or puberulent to sparsely, pubescent perennial or biennial herbs from a taproot. Leaves petiolate, membranaceous, ternate or pinnate or bipinnate with large distinct leaflets, to pinnately or ternate-pinnately decompound with linearly filiform ultimate divisions, or leaves rarely simple; petiole sheathing, often broadly so; cauline leaves reduced upward, with usually dilated sheaths. Inflorescence of terminal or terminal and lateral compound umbels, the peduncles elongate to abortive. Involucre and involucel of few to numerous, usually conspicuous, scarious- or white-scarious-margined or ± foliaceous entire or variously dissected bracts and bracteoles, the latter often exceeding umbellets, or an involucre rarely lacking. Rays few to numerous, subequal to unequal, spreading-ascending. Umbellets few to numerous flowered, the flowers pedicellate. Flowers white, rose, brownish, or purple; petals oblong to obovate, acute or obtuse and nearly plane, or with a very short, narrower inflexed apex, often clawed, the outer sometimes slightly radiant; calyx teeth small but evident, or calyx obsolete; stylopodium conical to depressed conical, the styles usually slender, divergent or recurved. Pedicels slender to stout, subequal to unequal, spreading-ascending. Fruit oblong to ellipsoid or obovoid, rounded or slightly narrowed toward apex, rounded to subcordate at base, lightly compressed either laterally or dorsally or subterete, the mericarps correspondingly compressed in transection, glabrous, the epicarp sometimes loose and spongy, smooth or papillate, the ribs equally or unequally winged or subwinged, sometimes corrugated. Vittae 1–3 in intervals, 2–6 on commissure. Commissural face sulcate to concave and the seed lunate, or plane. Carpophore bipartite.

Type species: *Ligusticum austriacum* Linn. [*Pleurospermum austriacum* (Linn.) Hoffm.].

Distribution: A Eurasian genus of some 50 species, extending from central Europe to Siberia and Japan, but especially well developed throughout the Himalaya and in western China.

KEY TO SPECIES

1. Leaves divided into definite large lanceolate to obovate leaflets 0.5–8 cm in diameter.

2. Involucral bracts greenish, usually pinnatisect and nearly equaling rays. ....1. *P. benthamii*.
2. Involucral bracts conspicuously scarious or white-margined, mostly entire to crenulate and shorter than rays.
   3. Cauline leaf sheaths very strongly inflated, orbicular; mature pedicels 10–20 mm long; fruit 10–20 mm long. ....2. *P. angelicoides*.
   3. Cauline leaf sheaths slightly or only moderately inflated and much narrower; mature pedicels 2–12 mm long; fruit 3–9 mm long.
      4. Bracts and bracteoles spreading or ascending; fruit mericarps smooth.
         5. Leaves ovate; bracteoles linear to ovate-lanceolate; fruit 7–9 mm long, or if smaller, the bracteoles linear.
            6. Bracteoles narrow and narrowly or obscurely white-margined; umbellets 8–16-flowered; pedicels 5–10 mm long.
               7. Fruit compressed laterally; leaflets simple or trilobed, 3–8 cm long; stem fistulose. ....3. *P. rotundatum*.
               7. Fruit compressed dorsally; leaflets incisely lobed, 8–20 mm long; stems not fistulose. ....4. *P. pumilum*.
            6. Bracteoles broader and broadly white-margined; umbellets 15–30-flowered; pedicels 2–6 mm long; fruit compressed dorsally. ....5. *P. dentatum*.
         5. Leaves oblong; bracteoles broadly lanceolate to obovate; fruit 3–6 mm long.
            8. Bracteoles longly acuminate, scaberulous; vittae 2 or 3 in intervals. ....6. *P. apiolens*.
            8. Bracteoles obtuse, sometimes apiculate; vittae solitary in intervals.
               9. Involucral bracts lanceolate-acuminate; rays 5–8; fruit compressed dorsally. ....7. *P. erosum*.
               9. Involucral bracts oblong to oval; rays 10–20; fruit compressed laterally. ....8. *P. candollii*.
      4. Bracts and bracteoles reflexed; mericarps papillose and wings ± transversely plicate. ....9. *P. stylosum*.

1. Leaves decompound and without definite leaflets, or the small leaflets deeply incised to pinnatifid with small ultimate divisions.
   10. Involucral bracts decidedly shorter than mature rays; cauline leaf sheaths no more than moderately dilated; vittae usually solitary.
       11. Epicarp of mature fruit neither lax nor lustrous.
           12. Involucral bracts narrow, linear to lanceolate; umbellets many flowered.
               13. Plants scaberulous in inflorescence, otherwise glabrous; fruit wings glabrous. ....10. *P. hookeri*.
               13. Plants setulose-pilose throughout; fruit wings scaberulous. ....11. *P. pilosum*.
           12. Involucral bracts broad, oblong to obovate; umbellets 15–25-flowered.
               14. Herbage scaberulous; seed face plane. ....12. *P. brunonis*.
               14. Herbage glabrous; seed face grooved. ....13. *P. densiflorum*.

11. Epicarp of mature fruit lax and lustrous.
    15. Bracteoles mostly pinnatifid; fruit ribs triangular or narrowly thick winged.
        16. Plants glabrous; leaves narrowly oblong. ....14. *P. crassicaule.*
        16. Plants usually ± scaberulous; leaves oblong to oval. 15. *P. stellatum.*
    15. Bracteoles mostly entire; fruit ribs broad and slender winged. ....16. *P. lindleyanum.*

10. Involucral bracts equaling or exceeding rays; cauline leaves with greatly dilated sheaths; vittae several in intervals.
    17. Plants ± scaberulous or puberulent; flowers greenish-brown. .. 17. *P. album.*
    17. Plants glabrous; flowers white or purplish, or yellow.
        18. Leaves bipinnate with palmatisect leaflets and somewhat broader ultimate divisions; bracts entire; calyx teeth prominent. .... 18. *P. corydalifolium.*
        18. Leaves ternately or ternate-pinnately decompound with narrow ultimate divisions; bracts with a dissected foliaceous apex, or this sometimes lacking; calyx teeth minute or lacking.
            19. Ultimate divisions linear to lanceolate, 1–3 mm long; rays 20–50, 3–5 cm long; flowers white or purplish. ....19. *P. amabile.*
            19. Ultimate leaf divisions cuneate-obovate, 4–6 mm long; rays ca. 10, ca. 1 cm long; flowers yellow. ....20. *P. wolffianum.*

Note. A number of authors, viz. Shishkin (1950 [1973]) and Rechinger f. in Flora Iranica (1987), have followed Candolle in recognizing *Pleurospermum* Hoffm. and *Hymenolaena* DC. as two distinct genera. We, however, follow Clarke (1879), Nasir (1972), Fu & Ho (1979), and Farille (1985) in combining most of the species into a single genus. Farille (1985) has proposed a total of seven new or reconstituted subgenera within *Pleurospermum*, but we do not believe these were validly published.

We find Rechinger's comment in Flora Iranica (1987, p. 172) pertinent: "The demarcation line between *Pleurospermum* and *Hymenolaena* is far from clear. The former genus has reputedly a very loose mesocarp with a prominent air space between exocarp and endocarp, whereas the latter does not. Whether this character holds true for all species is uncertain and, in any case, would be hard to confirm in the absence of mature fruit in some species." Until some of the uncertainties can be settled, we prefer to retain *Pleurospermum* intact, except for the exclusion of *Pleurospermopsis* Norman.

1. Pleurospermum benthamii (DC.) C. B. Clarke in Hook. f., Fl. Brit. Ind. 2: 703. 1879.

Ligusticum benthamii Wallich, Cat. 542. 1829, *nomen nudum*; *Hymenolaena benthamii* DC., Prodr. 4: 246. 1830.

Plants low, caulescent, leafy, the stem very stout, fistulose, simple or with a whorl of fertile branches below the terminal umbel, scaberulous on leaf veins and in inflorescence, 60–150 cm tall, from a massive swollen taproot. Leaves ovate, 5–15 cm in diameter, pinnate or ternate-pinnate, the leaflets ovate, 2–7 cm long, 1–4 cm broad, deeply pinnately incised to pinnatifid and coarsely serrate; petiole slender, 3–15 cm long, broadly dilated and sheathing at base; cauline leaves rather numerous, pinnate or pinnately incised with decurrent, acuminate lobes, the sheaths narrow, winged. Terminal peduncles stout,

3–10 cm long, the lateral more slender. Involucre of 5–8 oblong to oblanceolate, apically pinnatisect, or rarely entire, greenish bracts 2–8 cm long, nearly equal to rays. Rays of terminal umbel 15–30, slender, unequal, spreading-ascending, 5–10 cm long. Umbellets 20–40-flowered, the mature pedicels slender, unequal, spreading-ascending, 10–30 mm long. Involucel of 8–10 oblanceolate to obovate, greenish, narrowly scarious-margined, scaberulous, apically pinnatisect, or rarely entire bracteoles 10–20 mm long, exceeding flowering umbellets. Flowers pinkish-lavender; petals ovate-lanceolate to ovate, with a scarcely inflexed acute tip; calyx teeth obsolete; stylopodium large, conical, a little shorter than the slender styles. Fruit ovoid, slightly compressed dorsally, 6–10 mm long, 2–5 mm broad, the mericarps slightly compressed dorsally in transection, glabrous but roughened, the ribs prominent, narrowly winged, the lateral slightly broader than the dorsal. Vittae large, solitary in intervals, 2 on commissure. Commissural face concave. June–Aug.

Type locality: "ad Gossain-Than Nepalensium," *Wallich 542* (CAL, K-*pars*, isotypes).

Distribution: Eastern Himalaya from Bhutan to Nepal, 3,000–5,100 m; Tibet.

Representative specimens: Nepal: Bhut-Kehs, *Dhwoi 7* (BM); Rambrong, Lamjung Himalaya, *Stainton, Sykes & Williams 6094* (BM); Tolo Gompa Khola, *Nakao 195* (UC-*pars*). Sikkim: Changu, *W. W. Smith 3586* (CAL); Jongri, *King's collector* (CAL). Bhutan: Pangotang, Tsampa, *Ludlow, Sherriff & Hicks 19,155* (BM).

Note. Farille (1985) reports *P. davidii* Franchet "du Népal au Bhutan," and distinguishes it from the closely related *P. benthamii* as being monocarpic. Cannon (1979) suggests that the former "may be better considered as a subspecies of *P. benthamii*." We have seen no material of *P. davidii* from our area.

2. **Pleurospermum angelicoides** (DC.) C. B. Clarke in Hook. f., Fl. Brit. Ind. 2: 703. 1879.

Ligusticum angelicoides Wallich, Cat. 548. 1829, *nomen nudum*; *Hymenolaena angelicoides* DC., Prodr. 4: 245. 1830; *Pterocyclus angelicoides* (DC.) Klotzsch in Klotzsch & Garcke, Bot. Ergebn. Reise Prinz Waldemar, 150. 1862.

Plants erect, caulescent, the stem stout, fistulose, branching above, the foliage slightly scaberulous, otherwise glabrous, 90–200 cm tall. Leaves ovate-triangular, 20–45 cm long, ternate-pinnate or bipinnate, the leaflets oblong-ovate to ovate, acute or acuminate, coarsely dentate or serrate, 5–15 cm long, 3–7 cm broad; petiole stout, broadly dilated into scarious sheath at base; cauline leaves ternate or ternate-pinnate, sessile, with strongly dilated obovate sheaths. Peduncles stout, terminal and lateral, 15–40 cm long. Involucre of 5–8 linear to lanceolate, entire or foliaceous and distally incised bracts 2.5–5 cm long. Rays 15–30, spreading-ascending or ascending, subequal, 2.5–12 cm long. Umbellets 15–30-flowered, the mature pedicels stout, subequal, angled, spreading-ascending or ascending, 10–20 mm long. Involucel of 8–12 linear to lanceolate, broadly scarious-white-margined, often denticulate or ciliolate bracteoles 5–20 mm long, about equal to the umbellets. Flowers greenish or purplish; petals obovate with a slightly inflexed somewhat acute apex; calyx teeth minute, ovate; stylopodium obtusely conical, about equal to the short reflexed styles. Fruit oblong, subterete, 10–20 mm long, 2–5 mm broad, the mericarps compressed dorsally in transection, glabrous, the ribs narrowly winged, the lateral slightly broader than the dorsal. Vittae solitary in intervals, 2 on commissure. Commissural face concave. June–Sept.

Type locality: "in Nepaliae montibus altissimis ad Kamaon et Gossain-Than," *Wallich 548*.

Distribution: Himalaya from Bhutan to Kashmir, 2,500–4,500 m; Assam, Tibet, and Yunnan.

Representative specimens: Jammu & Kashmir: Kashmir, *Duthie 13,099* (CAL); Gulmarg, *Duthie 11,395* (DD). Uttar Pradesh: Kumaon, *Wallich 548A* (CAL); Kumaon, Ralam River, *Strachey & Winterbottom 1* (K); Kunari Pass, *Duthie 4068* (DD). Nepal: Gossainthan, *Wallich 548B* (CAL); Tangba, *Dhwoj 0220* (BM); above Ranmagaon, *Stainton, Sykes & Williams 3337* (BM); Gheding, *Dhwoj 0319* (BM); Tinjura Danda, *Williams & Stainton 8406* (BM). West Bengal: Phulloot, *Thomson* (CAL). Sikkim: Jongri, *Prain's collector* (CAL).

3. **Pleurospermum rotundatum** (DC.) C. B. Clarke in Hook. f., Fl. Brit. Ind. 2: 703. 1879.

Ligusticum rotundatum Wallich, Cat. 549. 1829, *nomen nudum*; *Hymenolaena rotundata* DC., Prodr. 4: 245. 1830.

Plants slender, erect, caulescent, the stem rather stout, fistulose, unbranched below inflorescence, hirtellous at base of umbel, the foliage hirsutulous or scaberulous on veins, or glabrate, 60–80 cm tall, fibrous at base, from a taproot. Leaves ovate, 12–20 cm in diameter, pinnate or 1-2-ternate, the petiolulate leaflets ovate-lanceolate to orbicular, 3–8 cm in diameter, simple or trilobed and crenate-dentate; petiole 15–25 cm long, broadly dilated into an orbicular scarious sheath 2–4 cm long; cauline leaves few, the lower ternate or tripartite, sessile, the uppermost sheaths broadly dilated and bladeless. Involucre of 1 or 2 scarious, sheathlike bracts 5–20 mm long, or lacking. Peduncles stout, the terminal 5–13.5 cm long, accompanied by 1 to several smaller peduncles from the upper leaf axils. Rays 8–15, slender, unequal, spreading-ascending, 2.5–4.5 cm long. Umbellets 8–16-flowered, mostly fertile, the mature pedicels stout, unequal, angled, spreading-ascending, 5–10 mm long. Involucel of 5–8 lanceolate, acuminate, entire bracteoles 4–11 mm long, shorter than mature umbellets. Flowers yellowish-green or purplish; petals obovate, acutish but scarcely inflexed, clawed; calyx teeth minute, ovate, or obsolete; stylopodium obtusely conical, about equal to the short, reflexed styles. Fruit oblong, compressed laterally, 7–8 mm long, 3–4 mm broad, the mericarps subterete in transection, glabrous, the ribs subequally narrowly winged. Vittae 1 or 2 in intervals, usually 4 (?) on commissure. Commissural face slightly grooved. Aug.–Sept.

Type locality: "in Nepalia ad Gossain-Than et Himalaya," *Wallich 549* (CAL, K, isotypes).

Distribution: Himalaya of eastern and central Nepal, 1,700–4,000 m.

Representative specimens: Nepal: Seti Khola, *Stainton, Sykes & Williams 6567* (BM); Nardi Khola, *Stainton, Sykes & Williams 8511* (BM); Chiyak, *Dhwoj 33* (BM).

Note. Although Clarke (1879, p. 703) questioned the generic placement of this species because of the obsolescent involucre and (relatively) inconspicuous involucel, carpological characters appear to confirm its position here.

4. Pleurospermum pumilum (DC.) C. B. Clarke in Hook. f., Fl. Brit. Ind. 2: 704. 1879.

Ligusticum pumilum Wallich, Cat. 550. 1829, *nomen nudum*; *Hymenolaena pumila* DC., Prodr. 4: 245. 1830.

Plants slender to moderately stout, glabrous, to 1 m tall. Leaves oblong, 1-2-pinnate, the leaflets ovate, 8–20 mm long, 6–15 mm broad, pinnately incised or lobed and irregularly mucronate-dentate; petiole equal to or exceeding the blade, broadly sheathing at base; cauline leaves few, reduced upward, with prominent oblong sheaths. Peduncles rather slender, 10–15 cm long. Involucre of 1–3 linear, entire or trifid bracts 1–3 cm long, or lacking. Rays 8–15, angled, unequal, 1–3.5 cm long. Umbellets 8–12-flowered, the mature pedicels unequal, 4–8 mm long, exceeding bracteoles. Involucel of 3-8 linear, entire, obscurely white-scarious-margined bracteoles 3–8 mm long. Flowers polygamous, few fertile; calyx teeth small, triangular; stylopodium low conical, about equal to the very short reflexed styles. Fruit broadly ovoid, subterete, 4–6 mm long, 3–4 mm broad, shiny, slightly narrowed at apex, cordate at base, the mericarps compressed dorsally in transection, the ribs hollow, prominently and subequally winged. Vittae large, solitary in intervals, 2 on commissure. Commissural face sulcate. Oct.

Type locality: "ad Gossain-Than Nepalensium," *Wallich 550* (CAL, isotype).

Distribution: Eastern Himalaya, West Bengal to Nepal, 3,500–4,000 m.

Representative specimens: Nepal: Langtang, *Mukherjee 277* (CUH). West Bengal: Phaloot, *Thomson* (CAL); Phalut Top, *Mukherjee 307* (CUH, UC). Sikkim: Islumbo, *Clarke 25,616* (K); w/o precise locality, *Kurz* (CAL).

5. Pleurospermum dentatum (DC.) C. B. Clarke in Hook. f., Fl. Brit. Ind. 2: 704. 1879.

Ligusticum dentatum Wallich, Cat. 547. 1829, *nomen nudum*; *Hymenolaena dentata* DC., Prodr. 4: 245. 1830.

Plants tall, caulescent, the stem slender, fistulose, simple or few branched above, conspicuously scaberulous in inflorescence, sparsely so in foliage, 30–120 cm tall, from a slender taproot. Leaves oblong-ovate to triangular-ovate, 3–25(–40) cm long, 3–15 cm broad, 1-2-pinnate or ternate-pinnate, the leaflets ovate, acute or acuminate, the larger petiolulate, 1–6 cm long, 0.5–5 cm broad, irregularly mucronulate-serrate and often incised; petiole slender, 5–20 cm long, narrowly sheathing at base; cauline leaves like the basal, reduced upward with dilated oblong sheaths. Peduncles slender, terminal and lateral, 3–20 cm long, densely scaberulous below umbel. Involucre of 1–5 linear to lanceolate, broadly white-scarious-margined, entire or denticulate, minutely scaberulous bracts 1–3 cm long. Rays 5–15, slender, scaberulous, unequal, spreading-ascending, 2–5 cm long. Umbellets 15–30-flowered, the mature pedicels stout, unequal, scaberulous, spreading-ascending, 2–6 mm long. Involucel of 8–12, narrowly lanceolate to ovate, broadly white-scarious-margined, with a darker center prolonged into an acute or acuminate apex, entire, denticulate or erose bracteoles 6–18 mm long, exceeding or at least equal to the umbellets. Flowers white or purplish; petals oval to obovate, with a narrower inflexed apex; calyx obsolete; stylopodium obtusely conical, much shorter than the slender styles. Fruit ovoid, compressed dorsally, 7–9 mm long, 4–5 mm broad, the mericarps compressed dorsally in transection, glabrous, the ribs broadly and subequally winged or the lateral ones slightly broader. Vittae 1 or 2 in intervals, 4–6 on commissure. Commissural face shallowly concave or plane. Aug.–Oct.

Type locality: "in Nepalis ad Kamaon et Gossain-Than," *Wallich 547* (*pars*).

Distribution: Himalaya, from Sikkim to Kumaon, 3,000–4,300 m.

Representative specimens: Uttar Pradesh: Kuari Pass, *Duthie 4066* (BM, CAL, DD); Kumaon, *Arora 49,863* (BSD, CAL). Nepal: Bheding, *Dhwoj 0273* (BM); Lamrak, *Dhwoj 178* (BM). Sikkim: Karikola, *Hooker* (*Hooker & Thomson*, Herb. Ind. Orient., *Hymenolaena dentata*) (K); Jongri, *Clarke 26,217B* (CAL), *26,217D* (BM); Yeumtong, *Gammie 975* (CAL).

6. **Pleurospermum apiolens** C. B. Clarke in Hook. f., Fl. Brit. Ind. 2: 705. 1879.

Pleurospermum apiolens var. *nipaulensis* Farille & Malla in Farille, Cauwet-Marc & Malla, Candollea 40(2): 524. 1985. Based on *Farille 81,697*, "E. Népal: Magar do Dandala, 3,500 m," 17 Oct. 1981.

Plants low and shortly caulescent to tall and caulescent, the stem stout or slender, fistulose, simple or few branched, celery-scented, scaberulous in inflorescence and foliage, or the latter glabrous, 0.8–4.5 cm tall, from a fibrous base surmounting a branched taproot. Leaves oblong, 6–24 cm long, 3–10 cm broad, 1- (rarely 2-) pinnate, the leaflets ovate to orbicular, 1–5 cm in diameter, shortly petiolulate to sessile, mucronate-dentate to pinnatifid; petiole stout, 2–16 cm long, dilated at base into an oblong, usually scaberulous sheath 2–5 cm long; cauline leaves few, like the basal but reduced upward, with a prominent oblong sheath. Peduncles stout, terminal, often with some smaller lateral ones arising from upper leaf axils, 2–20 cm long, scaberulous at summit. Involucre of 4–6 oblong, denticulate, scarious bracts 1–3 cm long. Rays 8–22, slender, spreading-ascending, scaberulous, 1.5–5 cm long. Umbellets 20–30-flowered, mostly maturing, the mature pedicels slender, unequal, angled, spreading-ascending, 4–6 mm long. Involucel of 6–10 lanceolate to obovate, broadly scarious- and denticulate-margined, longly acuminate bracteoles 5–12 mm long, scaberulous, equal to or exceeding umbellets. Flowers white; petals oval to obovate with a narrower slightly inflexed apex; calyx teeth obsolete; stylopodium obtusely conical, exceeding the very short reflexed styles. Fruit oblong-oval, compressed dorsally, 4–6 mm long, 2–3 mm broad, the mericarps compressed dorsally in transection, glabrous, the ribs narrowly winged, subequal. Vittae solitary in intervals, 2 on commissure. Commissural face plane.

Type locality: "Interior valleys of SIKKIM, alt. 11,000–14,000 ft.," *Hooker* (*Hooker & Thomson*, Herb. Ind. Orient., *Hymenolaena 10*).

Distribution: Himalaya of Sikkim and Nepal, 3,000–5,000 m.

Representative specimens: Nepal: Pongsing, *Dhwoj 98* (BM); Lanjung Himalaya, *Stainton, Sykes & Willams 8636* (AAH, BM); Laurivinayak, *Shrestha & Shakya 3800* (BM). Sikkim: w/o further locality, *Hooker* (CAL); Yeumtong, *Hooker* (*Hooker & Thomson*, Herb. Ind. Orient., *Hymenolaena 8*) (K); Lachung, *Gammie 954* (CAL), *955* (CAL); Jongri to Olaktong, *Anderson 628* (CAL).

7. **Pleurospermum erosum** (DC.) P. K. Mukherjee, Bull. Bot. Surv. India 24(1–4): 44. 1983.

Hymenolaena dentata var. *erosa* DC., Prodr. 4: 245. 1830; *Pleurospermum dentatum* var. *erosum* (DC.) C. B. Clarke in Hook. f., Fl. Brit. Ind. 2: 704. 1879.

Plants tall, caulescent, the stem terete, striate, few branched, 15–35 cm tall. Leaves oblong, 10–15 cm long, 3–6 cm broad, pinnate, the leaflets ovate-lanceolate, acuminate, sessile or subsessile, irregularly incised-serrate, the lowest leaflets often trilobed, 1.5–3 cm long, 1–2.5 cm broad; petiole longer than lamina, sheathing at base. Peduncles to 10 cm long. Involucre of 5 lanceolate, acuminate, white-margined, crenulate bracts 1–2 cm long. Rays 5–8, subequal, 2–3 cm long. Umbellets many flowered. Involucel of 8–12 oblong-obovate, apiculate, white-margined, erose-serrate bracteoles 8–12 mm long, exceeding umbellets. Calyx teeth obsolete; stylopodium low conical. Fruit elliptic, compressed dorsally, 3–5 mm long, ca. 2.5 mm broad, the ribs equally winged. Vittae solitary in intervals, 4 on commissure. Carpophore bipartite.

Type locality: "in Nepalia ad Kamaon," *Wallich 547* (pars) (CAL, K, isotypes).

Distribution: Known only from Kumaon, Uttar Pradesh.

Representative specimens: Uttar Pradesh. "Nepalia," *Wallich 547A* (CAL), *547/2 bis* (CAL); above Garbyang in Kali Valley, *Duthie 2959* (CAL).

8. **Pleurospermum candollii** (DC.) C. B. Clarke in Hook. f., Fl. Brit. Ind. 2: 703. 1879.

Ligusticum candollii Wallich, Cat. 544. 1829, *nomen nudum*; *Hymenolaena candollii* DC., Prodr. 4: 245. 1830.

Hymenolaena latifolia Lindley in Royle, Ill. Bot. Himal. Mount. 233. 1839. Based on *Royle*, "Peer Punjal."

Plants low, short caulescent, the stem stout, fistulose, mostly simple, somewhat scaberulous at base of umbel, otherwise glabrous, 8–45 cm tall, from a stout taproot. Leaves oblong, 3–15 cm long, 1.5–6 cm broad, pinnate, the leaflets ovate to orbicular, 0.5–2 cm in diameter, dentate and ternately or pinnately lobed; petiole broadly dilated into an ovate-lanceolate scarious sheath 2–5 cm long; cauline leaves few, like the basal. Peduncles stout, terminal and scarcely distinguishable from the stem, up to 5 cm long to obsolete, or occasionally accompanied by 1 to several slender, raylike lateral peduncles. Involucre poorly organized, of 2–6 oblong to oval, broadly scarious-margined entire or denticulate sheathlike bracts 1–3 cm long, often borne on the rays. Rays usually 10–20, stout to slender, ascending or spreading-ascending, 3–12 cm long. Umbellets 20–35-flowered, not all maturing, the mature pedicels stout, unequal, winged, spreading-ascending, 5–8 mm long. Involucel of 8–12 oblanceolate to obovate, broadly scarious-margined, obtuse, entire bracteoles 8–15 mm long, 4–10 mm broad, exceeding umbellets. Flowers white; petals ovate to obovate, with a very slightly inflexed acute apex; calyx teeth minute, oval; stylopodium obtusely conical, much shorter than the flattened, reflexed styles. Fruit oblong-oval, compressed laterally, 4–6 mm long, 2–3 mm broad, the mericarps subterete in transection, glabrous, the ribs narrowly winged, subequal or the lateral slightly broader. Vittae 2 or 3 in intervals, ca. 6 on commissure. Commissural face deeply concave. June–Sept.

Type locality: "in Nepaliae Himalaya et Buddrinath templum," *Wallich 544* (CAL, isotype).

Distribution: West Himalaya, from Kumaon to Kashmir and Pakistan, 3,000–5,000 m; China and Tibet (?).

Representative specimens: Jammu & Kashmir: Kishanganga Valley, *Keshavanand 317* (DD); Kunware, *Royle 93/10* (DD); Karpuchu Valley, Baltistan, *Duthie 11,954* (CAL). Himachal Pradesh: Sach Pass, *Lace 1472* (CAL); Lahul, *Bor 15,403* (DD). Uttar Pradesh: Tehri Garhwal, *Duthie 1157* (DD).

9. **Pleurospermum stylosum** C. B. Clarke in Hook. f., Fl. Brit. Ind. 2: 704. 1879.

Aulacospermum stylosum (C. B. Clarke) Rech. f. & Riedl, Biol. Skr. 13(4): 28. 1963.

Pleurospermum pulchrum Aitch. & Hemsley, Jour. Linn. Soc. Bot. 18: 13. 1880; *Aulacospermum pulchrum* (Aitch. & Hemsley) Rech. f. & Riedl, Biol. Skr. 13(4): 32. 1963. Based on *Aitchison 743*, *744*, "Kuram Valley, etc., Afghanistan."

Plants stout, erect, caulescent, branching, scaberulous or puberulent at least in inflorescence, 20–150 cm tall, the stem conspicuously fibrous at base, from a thick taproot. Leaves oblong-ovate to triangular-ovate, 3–8 cm long, 2–6 cm broad, bipinnate or ternate-pinnate, the ultimate divisions lanceolate to oblong or oval, 0.5–2 cm long, 2–5 mm broad, confluent or winged at base, sharply mucronate-serrate and pinnatifid, usually scaberulous; petiole slender, 3–8 cm long, inconspicuously dilated at base; cauline leaves numerous, like the basal, but reduced upward. Peduncles terminal and lateral, the former much larger than the latter, 4–18 cm long, scaberulous. Involucre of 5–8 lanceolate to oblong, foliaceous, mostly ternate or pinnatifid, narrowly white-margined, usually puberulent bracts 1-3 cm long, becoming reflexed. Rays 15-35, rather stout, 3–10 cm long, spreading-ascending, scaberulous. Umbellets 20–30-flowered, not all maturing, the mature pedicels slender, subequal, angled, scaberulous, spreading-ascending, 7–12 mm long. Involucel of 6–10 lanceolate to ovate-lanceolate, broadly scarious-margined, acute or acuminate, entire or apically incised, reflexed bracteoles 5–12 mm long, 2–5 mm broad, exceeding umbellets. Flowers white or rose-tinged; petals obovate, with a slightly inflexed acute triangular apex; calyx teeth evident, ovate; stylopodium conical, about equal to the short, recurved styles. Fruit oblong-ovoid to ellipsoid, compressed laterally, 4–6 mm long, 2–4 mm broad, the mericarps compressed laterally in transection, papillose, the pericarp loose, the ribs rather narrowly and subequally winged, the wings undulate or somewhat transversely plicate. Vittae solitary in intervals, 2 on commissure. Commissural face shallowly concave. June–Oct.

Type locality: "KASHMIR: *Falconer*. LAHUL; alt. 11,000–12,000 ft., *Rev. H. Jaeschke*" [Kashmir, *Falconer 3150* (K, lectotype)] (Rech. f. & Riedl 1963; Nasir 1972).

Distribution: West Himalaya, Lahul Valley to Pakistan and Afghanistan, 3,500–4,000 m.

Representative specimens: Jammu & Kashmir: Ladak, *R.R. Stewart 21,066* (DD). Himachal Pradesh: Leylang, Lahul, *Bor 14,875* (DD); Lahul, *Jaeschke* (CAL); Chang la Gali, Murree Hill, *Saxton 2375* (CAL); Biling Lampe, Lahul, *Bor 12,673* (DD).

10. **Pleurospermum hookeri** C. B. Clarke in Hook. f., Fl. Brit. Ind. 2: 705. 1879.

Aulacospermum hookeri (C. B. Clarke) Farille & Malla in Farille, Cauwet-Marc & Malla, Candollea 40(2): 525. 1985.

Plants low or tall, caulescent, the stem slender, fistulose, mostly few branched, scaberulous in inflorescence, 10–50 cm tall, from a somewhat fibrous base surmounting a slender taproot. Leaves oblong-oval to ovate, 5–30 cm long, 2–12 cm broad, bi-pinnate or ternate-1-2-pinnate, the leaflets oval to orbicular, 3–10 mm in diameter, pinnately incised or deeply pinnatifid; petiole slender, 2–15 cm long, dilated into an oblong sheath; cauline leaves few, like the basal, but sessile and with a conspicuously dilated oblong to oval, purplish sheath. Peduncles terminal and lateral, rather stout, 3–23 cm long, scaberulous beneath umbel. Involucre of usually 5 linear to lanceolate, narrowly white-scarious-margined bracts with a darker center exserted as an entire acuminate tip or more rarely foliaceous and pinnatifid apically, 1–3 cm long. Rays 4–10, slender to rather stout, spreading-ascending, angled, a little scaberulous or glabrate, 2–8 cm long. Umbellets 25–50-flowered, the mature pedicels stout, unequal, angled, spreading-ascending, 3–8 mm long. Involucel of 8–15 linear to ovate, narrowly denticulate-ciliolate, scarious-margined bracteoles 5-15 mm long, the darker center prolonged into an acuminate apex, usually exceeding umbellets. Flowers white or brownish-pink; petals oval to orbicular, with a narrower inflexed apex; calyx teeth evident, ovate; stylopodium obtusely conical, shorter than the slender styles. Fruit oval, compressed dorsally, 2.5–3.5 mm long, 2–2.5 mm broad, the mericarps dorsally compressed in transection, glabrous, the ribs moderately and subequally winged. Vittae 1 or 2 in intervals, 2 on commissure. Commissural face concave. July–Sept.

Type locality: "SIKKIM, alt. 10,000–16,000 ft.," *Hooker* (*Hooker & Thomson*, Herb. Ind. Orient., *Hymenolaena 8, Oreocome 1*).

Distribution: Himalaya from Bhutan to Nepal, 3,500–5,500 m; Tibet.

Representative specimens: Nepal: Namdo (N. of Mustano), *Stainton, Sykes & Williams 8854* (BM); Sangdan (N. of Tuicucha), *Stainton, Sykes & Williams 21,854* (BM); Chaulagiri Himal, *Wald 13* (COLO, UC); Mt. Everest Expedition, *Wollaston 102* (K). Sikkim: Karakola, *Hooker* (CAL, K); Lachen, *Ribu & Rhomoo 3026* (CAL); Llonak, *Smith & Cave 1869* (CAL); Tankra Mountain, *Gammie 387* (CAL). Bhutan: Upper Mangde Chu, *Ludlow, Sheriff & Hicks 16,806* (BM).

Note. *Pleurospermum hookeri* var. *thomsoni* C. B. Clarke, described from "WEST TIBET," is apparently conspecific but consisted of plants with less well-developed involucral bracts. The plant figured by Nasir (1972, fig. 43, I–J), although superficially very similar, was described as *Ligusticum afghanicum* Rech. f. (Rechinger f. & Riedl, 1963, fig. 58).

11. **Pleurospermum pilosum** C. B. Clarke ex H. Wolff, Feddes Repert. 27: 117. 1929.

Plants setulose-pilose throughout, to 20 cm tall, the stems few, branched from base, the branches nearly naked. Leaves triangular-ovate, 10 cm long, ca. 5 cm broad, tripinnatisect, the leaflets oblong-oval, to 10 mm long, 5–6 mm broad, deeply pinnatisect, the ultimate divisions lanceolate, abruptly acute, to 3 mm long; petiole slender, to 10 cm long, dilated and broadly sheathing; cauline leaves few, like the basal. Peduncles stout, elongated. Involucre of a few narrow, white-margined, apically incised or foliaceous bracts less than half length of rays. Rays 6–12, 5 cm long, subequal, thickish, spreading-ascending. Involucel of about 10 very shortly petiolulate, broadly cuneate-obovate, apically lacerate, green, broadly white-margined bracteoles, the midvein and rarely also the lateral shortly exserted, exceeding young fruit. Umbellets many flowered, the immature pedicels about equal to the flowers. Flowers numerous; petals obcordate, with an inflexed apex; calyx lobes evident, ovate; stylopodium semiglobose, half as long as the divaricate styles. Fruit oblong-ovoid, the ribs winged, scaberulous. Vittae solitary in intervals, 2 on commissure.

Type locality: "Himalaya, Chumbi: Ka-poofo," *King's collector 70*, 28 Jul. 1884 (K, holotype) (Wolff, 1929).

Distribution: Eastern Himalaya.

Representative specimen: Bhutan: Padina Tso near Thampa La, *Ludlow, Sheriff & Hicks 17,175* (BM).

12. **Pleurospermum brunonis** (DC.) C. B. Clarke in Hook. f., Fl. Brit. Ind. 2: 706. 1879.

Ligusticum brunonis Wallich, Cat. 545. 1829, *nomen nudum*; *Hymenolaena brunonis* DC., Prodr. 4: 245. 1830; *Hymenidium brunonis* (DC.) Lindley in Royle, Ill. Bot. Himal. Mount. 233. 1835.

Plants low, short caulescent, the stem slender, simple or usually branched near base, somewhat scaberulous in inflorescence and foliage, 10–35 cm tall, from a stout, often branching taproot. Leaves oval to triangular-ovate, 3–10 cm in diameter, 1–3-pinnate or ternate-1–3-pinnate, or pinnately or ternate-pinnately decompound, the leaflets oval to orbicular, 3–5 mm in diameter, pinnately incised or pinnatifid, or the ultimate divisions linearly lanceolate, 1–3 mm long; petiole slender, 2–15 cm long, narrowly oblong-sheathing; cauline leaves like the basal, with somewhat dilated sheath. Peduncles slender to stout, 2–8 cm long, scaberulous beneath umbel. Involucre of 5–8 oblong-oval to obovate, broadly white-scarious-margined bracts 1–4 cm long, either obtuse and entire to denticulate, or usually prolonged into a foliaceous pinnatifid apex, shorter than the rays. Rays 3–13, slender to stout, unequal, spreading-ascending, 1–5 cm long. Umbellets 15–25-flowered, the mature pedicels angled, spreading-ascending, 3–5 mm long. Involucel of 6–10 oblanceolate to obovate, broadly white-scarious-margined, entire to denticulate, obtuse bracteoles 8–12 mm long, or the green central portion prolonged into an acuminate tip, or less frequently a ± pinnatifid foliaceous apex, equal to or usually exceeding umbellets. Flowers white; petals oblanceolate to obovate with a narrower inflexed apex, green with a white margin; calyx obsolete; stylopodium conical, shorter than the slender recurved styles. Fruit oblong-ovoid, compressed dorsally, 3–5 mm long, 1.5–2.5 mm broad, the mericarps compressed dorsally in transection, glabrous, the ribs narrowly and subequally winged, or the lateral a little broader. Vittae solitary in intervals, 2–6 on commissure. Commissural face plane to ± sulcate (?). July–Oct.

Type locality: "in montibus Nepaliae ad Himalaya, Kamaon, Gossain-Than," *Govan, Wallich 545*.

Distribution: Central and Western Himalaya, from Nepal to Kashmir and Pakistan, 2,700–4,500 m.

Representative specimens: Jammu & Kashmir: Kanwar, *Thomson* (*Hooker & Thomson*, Herb. Ind. Orient.) (CAL, K); Kashmir, *Falconer 502* (CAL, K); Pir Panjal, *Clarke 27,143* (K); Pahlgam, Lidder Valley, *R. R. Stewart 21,833* (UC). Himachal Pradesh: Laka, Dharamsala, *Clarke 23,838C* (CAL), *28,843* (BM), *Edgeworth* (CAL); Chor, Simla Hill States, *Raizada 13,471* (UC); Simla, *Drummond 1610* (K, UC); N. W. India, *Royle* (CAL). Uttar Pradesh: Kedarkanta, *Duthie 1133* (CAL). Nepal: Lulo Khola, *Polunin, Sykes & Williams 3510* (BM); Gossainthan, *Wallich 545a* (CAL).

13. **Pleurospermum densiflorum** (Lindley) C. B. Clarke in Hook. f., Fl. Brit. Ind. 2: 706. 1879.

Hymenidium ? densiflorum Lindley in Royle, Ill. Bot. Himal. Mount. 233. 1835.
Hymenidium suaveolens Klotzsch in Klotzsch & Garcke, Bot. Ergebn. Reise Prinz Waldem., 148. 1862. Based on *Hoffmeister*, "Himalaya."

Plants slender, caulescent, glabrous, 10–35 cm tall, the stem terete, striate, sparsely branched. Leaves triangular-ovate, membranaceous, 5–10 cm long, 6–10 cm broad, ternately decompound, the ultimate divisions linearly lanceolate, ternately or pinnately lobed, 2–7 mm long, 1–2 mm broad; petiole 2–6 cm long. Involucre of 5–8 oblong to obovate, green-veined, broadly white-scarious-margined (at apex) pinnatifid or pinnate bracts 2–3 cm long. Rays 8–12, slender, unequal, ca. 4 cm long. Umbellets several flowered, the mature pedicels equal to the fruit. Involucel of usually 8 elliptic-obovate, green-veined, broadly white-scarious-margined bracteoles 1–1.5 cm long, the margins crispulate-serrate or entire, apiculate at apex. Petals oblong-obovate, with a narrowly inflexed apex; calyx teeth obsolete. Fruit ellipsoid, subquadrate, compressed dorsally, 3–5 mm long, 2–2.5 mm broad, the mericarps compressed dorsally in transection, the ribs subequally winged. Vittae 1 in intervals, 4–6 on commissure. Commissural face grooved. Aug.–Sept.

Type locality: "Cashmere, Choor, and Kedarkanta," *Royle*.

Distribution: Northwestern Himalaya, from Kumaon to Kashmir, 3,500–5,500 m.

Representative specimens: Jammu & Kashmir: Kunwar. *Royle 93/89* (DD). Himachal Pradesh: Rohtang slope, *Bhattacharyya 40,480* (BSD); Chitora, *Koelz 22,017* (DD). Uttar Pradesh: Kedarkanta, *N.P. Singh 1133* (DD); Ganges Valley, *Duthie 1637* (DD); Valley of Flowers, *Bhattacharyya 24,384* (BSD).

14. **Pleurospermum crassicaule** H. Wolff, Feddes Repert. 21: 241. 1925.

Physospermopsis lalabhduriana Farille & Malla in Farille, Cauwet-Marc & Malla, Candollea 40(2): 516. 1985. Based on *Farille 81,185*, "W. Népal: Dhorpatan," 3,400–3,600 m, 18 Aug. 1981.

Plants caespitose, caulescent, simple or sparsely branched, glabrous, 5–10 cm tall, from a fibrous caudex surmounting a slender taproot. Leaves narrowly oblong, 5–15 cm long, 1–3 cm broad, subbipinnatisect, the leaflets broad, deeply lobed or pinnatifid, the ultimate divisions small and narrow; petiole short, flattened and moderately dilated, longer than or equal to the blade; cauline leaves like the basal. Peduncles terminal, usually with 1 or more smaller lateral flowering branches. Involucre of ca. 5 oblanceolate to obovate foliaceous, broadly white-margined, apically pinnatisect bracts. Rays 5–10, 1–3 cm long at anthesis, spreading-ascending. Umbellets 10–30-flowered, the pedicels 1–5 mm long. Involucel of several oboval bracteoles like the bracts. Flowers gray-green; petals ovate, obtuse, clawed; calyx teeth indistinct; stylopodium low conical, forming a broad flat disk at anthesis, the styles short, divergent. Fruit (immature) rhomboidal, the ribs thick, rugulose, the pericarp lax, reticulate. Vittae 1 (or 2) in intervals, 2 on commissure. Commissural face weakly concave. Mature fruit not seen. Aug.

Type locality: "Eastern Tibet," [locality not indicated], July 1894, *Soulié 2185*.

Distribution: Eastern Tibet to western Nepal, 3,400–3,600 m.

Representative specimen: Nepal: Dhorpatan, *Farille 81,185* (E).

15. **Pleurospermum stellatum** (D. Don) C. B. Clarke in Hook. f., Fl. Brit. Ind. 2: 705. 1879.

Selinum stellatum D. Don, Prodr. Fl. Nepal. 185. 1825; *Hymenolaena stellata* (D. Don) Lindley in Royle, Ill. Bot. Himal. Mount. 239. 1835.
Ligusticum govanianum Wallich, Cat. 546. 1829, *nomen nudum*; *Hymenolaena govaniana* DC., Prodr. 4: 246. 1830; *Pleurospermum govanianum* (DC.) C. B. Clarke in Hook. f., Fl. Brit. Ind. 2: 702. 1879. Based on *Govan* (*Wallich 546*), "in montibus nivosis Sirmore" (CAL, isotype).

Plants low and nearly acaulescent to tall and prominently caulescent, the stem abortive to elongate, very stout to slender, fistulose, simple or with a lateral branch in each of the upper leaf axils, scaberulous at base of umbels and umbellets, or glabrous throughout, up to 50 cm tall, from a fibrous base surmounting a stout, often branched taproot. Leaves oblong to oval, 2–10 cm long, 1–3 cm broad, 1–2-pinnate, the leaflets oval to orbicular, 3–20 mm in diameter, pinnately or ternately incised to pinnatifid, sessile, with mucronulate divisions; petiole slender, dilated into an oblong sheath, 1–8 cm long; cauline leaves, if present, like the basal. Peduncles stout and terminal, often accompanied by one or more slender ones, up to 10 cm tall, or dwarfed to nearly aborted, and the individual rays appearing to be peduncles bearing simple umbels. Involucre of 2–8 oblong to ovate or orbicular, foliaceous, pinnatifid bracts 1–6 cm long, broadly white-scarious-margined throughout, or only at base, or not at all and nearly indistinguishable from foliage leaves, shorter than umbels. Rays 2–20, rather slender, ascending or spreading-ascending, 2–20 cm long. Umbellets 15–40-flowered, the mature pedicels unequal, angled, spreading-ascending, 1–5 mm long. Involucel of 8–12 oblong to obovate, pinnatifid (or some only trifid or even entire), broadly white-scarious-margined bracteoles 8–15 mm long, exceeding umbellets. Flowers white; petals obovate, with a narrower inflexed apex; calyx teeth minute, ovate; stylopodium depressed conical, a little shorter than the recurved styles. Fruit oblong-ovoid, subterete, 4–6 mm long, 2.5–3 mm broad, the mericarps subterete, glabrous, the ribs broadly and subequally winged, the pericarp loose, papillate, often glistening. Vittae solitary in intervals, 2 on commissure. Commissural face sulcate. June–Aug.

Type locality: "Nepalia," *Buchanan-Hamilton* (CAL, isotype).

Distribution: Himalaya, from Nepal to Kashmir and Pakistan, 3,500–5,500 m.

Representative specimens: Jammu & Kashmir: Ladak, *Stoliczka 17,384* (BM); Karakoram, *Russell 1485* (BM); Peer Panjal, *Royle 93/9* (DD). Himachal Pradesh: Chitkal Pass, Simla, *Parmanand 969* (UC); *1080* (UC), *1102* (UC); Sarchu Lahul, *Koelz 6677* (BM); Chundra Valley, *Stoliczka* (CAL); Losar, Spiti, *Bhattacharyya 49,406* (CAL); Bara lacha, Lahul, *Bor 13,980* (DD). Uttar Pradesh: Bamsar Pass, Tehri Garhwal, *Duthie 1160*

(DD); Almora, *Champion* (DD); Cheena Hill, Kumaon, *Vicary 404* (CAL); Kedarkanta, *Royle 93/11* (DD). N. W. India, *Royle* (CAL, K).

Note. Nasir (1957) was correct in merging *Pleurospermum govanianum* and *P. stellatum*, but the epithet *stellatum* is the older name and must be adopted for the resultant taxon.

16. **Pleurospermum lindleyanum** (Klotzsch & Garcke) B. Fedtsch., Rastit. Turkest. 604. 1915.

Hymenolaena lindleyana Klotzsch & Garcke, Bot. Ergebn. Reise Prinz Waldem. 150. 1862; *Pleurospermum stellatum* var. *lindleyana* (Klotzsch & Garcke) C. B. Clarke in Hook. f., Fl. Brit. Ind. 2: 705. 1879.

Plants low, short caulescent, the stem rather stout, simple or with 1 or 2 slender lateral branches, glabrous (not scaberulous) throughout, 2–10 cm tall, from a taproot. Leaves oblong to oval, 2–6 cm long, 1.5–2 cm broad, pinnate or ternate-pinnate, the leaflets oval to orbicular, 5–10 mm in diameter, pinnately lobed with rounded lobes; petiole slender, 2–6 cm long, narrowly sheathing; cauline leaves like the basal, broadly dilated into scarious, oval, purple-veined sheaths. Peduncles short, nearly indistinguishable from the stem, 2–5 cm long. Involucre of 2 or more broadly scarious-sheathing (or sometimes consisting of sheath, only) foliaceous bracts scarcely distinguishable from cauline leaves, shorter than umbels. Rays 3–23, rather slender, unequal, spreading-ascending, 3–9 cm long. Umbellets 20–25-flowered, the mature pedicels angled, 2–5 mm long. Involucel of 8–10 oblong to obovate, entire (rarely trifid or even pinnatifid), broadly white-scarious-margined bracteoles 5–8 mm long, equal to or exceeding umbellets. Flowers white; petals oblanceolate to spatulate, clawed, with a narrower inflexed apex; calyx lobes minute, ovate; stylopodium obtusely conical, a little shorter than styles. Fruit ellipsoid, slightly compressed laterally, 4–6 mm long, 3–4 mm broad, the mericarps subterete, glabrous, the ribs broadly and subequally winged, the pericarp loose, papillate, glistening. Vittae solitary in intervals, 2 on commissure. Commissural face sulcate. Aug.–Sept.

Type locality: "Himalaya," *Hoffmeister*.

Distribution: Western Himalaya, to Pakistan, 4,500–4,800 m; Tibet.

Representative specimens: "Hab. Tibet. Occ. " [towards Parang Pass, Sassan], *Thomson* (*Hooker & Thomson*, Herb. Ind. Orient., *Hymenolaena 9*) (K). Jammu & Kashmir: W. side of Ladak, *Stoliczka* (CAL); Parang Pass, *Thomson* (K); Parang Valley, *Thomson* (CAL); Ladak, *Edgeworth 4038* (CAL). Himachal Pradesh: Lahul, Chatru, *Nair 16,959* (BSD).

Note. Clarke (1879, p. 705) says of this: "Except as to the obtuse (not mucronate) ultimate segments of the leaflets, there seems no difference between this and *P. stellatum* type." Nasir expresses the view (personal communication, Sept. 20, 1980) that *P. govanianum* and *P. lindleyanum* should probably constitute different varieties or subspecies under *P. stellatum.*

17. **Pleurospermum album** C. B. Clarke ex H. Wolff in Feddes Repert. 27: 113. 1929.

Plants caulescent, erect, the stem stout, fistulose, simple or few branched, scaberulous or puberulent on leaf sheaths and in inflorescence, 15–55 cm tall, from a branched taproot. Leaves oblong to ovate, 5–12 cm long, 2–10 cm broad, ternate-pinnately or pinnately decompound, the ultimate divisions lanceolate, mucronate, 1–2 mm long; petiole 3–8 cm long, broadly dilated into an oblong-oval, scarious and usually scaberulous sheath 2–5 cm long; cauline leaves 2–5, like the basal, mostly sessile with prominently inflated sheaths. Peduncles terminal, stout, 2–20 cm long, usually with one or more slender peduncles from upper leaf axils. Involucre of about 8 oblong-oval to obovate, broadly white-scarious-margined, scaberulous, denticulate bracts 1.5–4 cm long, with a broad green center and a pinnately dissected foliaceous apex. Rays 8–12, stout, spreading-ascending, scaberulous, 2–6 cm long. Umbellets 25–40-flowered, the mature pedicels flattened, scaberulous, spreading-ascending, 5–8 mm long. Involucel of ca. 8 oblong to obovate, broadly white-scarious and fimbriate-margined, scaberulous bracteoles 10–15 mm long, with a broad green center projected into a simple or pinnately dissected foliaceous apex, exceeding umbellets. Flowers greenish-brown; petals oblanceolate to obovate, short clawed, with a conspicuous white margin and a narrower inflexed apex; calyx teeth lanceolate, acute, to 1 mm long; stylopodium obtusely conical, about equal to the short styles. Fruit ellipsoid, ca. 5 mm long, ca. 2 mm broad, the mericarps compressed dorsally in transection, glabrous, the ribs unequally winged, the lateral a little broader than the dorsal. Vittae 2 or 3 in intervals, 4–6 on commissure. Commissural face plane. June–Oct.

Type locality: "Ost-Himalaya: Flora of Chumbi and Phari: Jo heu hu," 1879, *Dungboo s. n.* (CAL, isotype).

Distribution: Sikkim Himalaya, 4,000–4,500 m; Tibet.

Representative specimens: Sikkim: Ka-poop, S. of Jelop ha, *King's collector 73* (BM, K); Bhoktan, Jongri, *King's collector* (CAL); w/o definite locality, *Thomson* (CAL).

18. **Pleurospermum corydalifolium** Aitch. & Hemsley, Jour. Linn. Soc. Bot. 18: 62. 1880.

Hymenolaena corydalifolium (Aitch. & Hemsley) Rech. f. & Riedl, Biol. Skr. 13(4): 88. 1963.

Plants slender, simple or few branched, delicately striate. Leaves oblong to triangular-ovate, to 12 cm long, bipinnate, the leaflets subrhomboidal, deeply palmatisect, the ultimate divisions narrowly ovate to oval, acute or obtuse; petiole slender, to 20 cm long, dilated and scarious-sheathing. Peduncles slender. Involucre of 2, or occasionally 3, unequal, oblong, entire, broadly white-scarious-margined, greenish-veined bracts up to 2 cm long and 6 mm broad, equal to or exceeding rays. Rays 3–7, ca. 3 mm long, spreading-ascending. Umbellets many flowered, the pedicels short but slender. Involucel of several white or rose-tinted, green-veined, ovate-elliptic, entire or apically lobed bracteoles, longer than pedicels. Flowers white; petals orbicular, with an inflexed apex; calyx teeth prominent, acute; styles elongate, often reflexed. Fruit ovate-oblong, subterete, the ribs distinctly winged. Vittae 1 or 2 in intervals. Aug.

Type locality: "Common ... on shelving rocks and in dampish localities, Sikaram, Shendtai," Afghanistan, *Aitchison 953*.

Distribution: Afghanistan and Pakistan to Kashmir, 9,000–14,000 feet.

Representative specimen: Jammu & Kashmir: Baltistan, *Duthie 13,783* (CAL).

19. **Pleurospermum amabile** Craib & W. Smith, Trans. Bot. Soc. Edinb. 26: 154. 1913.

Plants low, short caulescent, the stem very stout, fistulose, usually simple, glabrous, 15–45 cm tall from a stout taproot. Leaves ovate to triangular-ovate, 3–8 cm in diameter, pinnately or ternate-pinnately decompound, the ultimate divisions linear to lanceolate, acute, 1–3 mm long; petiole slender, 2–5 cm long, broadly dilated into a scarious ovate sheath 1–3 cm long; cauline leaves rather numerous and usually overlapping, mostly sessile with greatly dilated, purple-veined sheaths. Peduncles terminal, very stout, 2–6 cm long. Involucre of 8–12 oblong to obovate, broadly white-scarious and denticulate-margined, purple-veined, sheathlike bracts 3–5 cm long, terminating in a dissected foliaceous apex, about equal to the rays. Rays 20–50, stout, spreading-ascending or ascending, 3–5 cm long. Umbellets 20–30-flowered, the mature pedicels stout, angled, spreading-ascending, 3–5 mm long. Involucel of 8–12 oblanceolate to obovate, broadly white-scarious and denticulate or erose-margined, purplish-veined bracteoles 6–12 mm long, terminating in a simple acuminate apex, exceeding umbellets. Flowers white or purplish (the whole plant appearing purplish!); petals lanceolate to oblanceolate, with a slightly inflexed acute apex; calyx teeth minute, ovate; stylopodium depressed, much shorter than styles. Immature fruit ovoid, subterete, the mericarps slightly compressed dorsally, glabrous, the ribs narrowly and subequally winged or subwinged. Vittae 2 or 3 in intervals, 4 on commissure. Commissural face concave. June–Aug.

Type locality: "At Chulong, Chumbi Valley, 15,000 ft.," Tibet, *Rohmoo 207* (E, holotype; CAL, K, isotypes).

Distribution: Eastern Himalaya of Bhutan and Sikkim, 3,900–4,500 m; Tibet.

Representative specimens: Sikkim: w/o definite locality, *Cave 207* (CAL). Bhutan: Thimpa-Naso Wasa La, *Bowes-Lyon 5129* (BM); Marlung, Tsampa, *Ludlow, Sherriff & Hicks 19,429* (BM); Shingbe (Me La), *Ludlow, Sherriff & Hicks 21,100* (BM).

20. **Pleurospermum wolffianum** H. Wolff, Feddes Repert. 27: 119. 1929.

Plants glabrous, the stems 1 to several, few branched and few leaved, usually 15 cm tall. Leaves slender, 1–2-ternate-bipinnate, the leaflets petiolulate, the ultimate divisions cuneate-obovate, 4–6 mm long, 1–3 mm broad, incisely serrate; cauline leaves like the basal, sessile on a broadly inflated sheath or shortly petiolate, ternate-bipinnate and exceeding the flowering umbels. Peduncles elongated. Involucre of several narrowly cuneate, broadly white-margined bracts with no or only small divided lamina, about equal to the rays. Rays about 10, ca. 1 cm long. Involucel of numerous lanceolate or oblanceolate, broadly white-margined, minutely lacerate, abruptly excurrent bracteoles, exceeding flowers. Umbellets barely 1 cm broad. Petals yellow, suborbicular, obtuse, scarcely emarginate. Fruit not known.

Type locality: "East Himalaya: Kalaeree, 16,000′," 1912, *Rhomoo 441* (K, holotype) (Wolff, 1929).

Distribution: Known only from the type collection.

55. CORTIA DC.

Prodr. 4: 186. 1830; C. B. Clarke in Hook. f.,
Fl. Brit. Ind. 2: 701. 1879.

Low, acaulescent to caulescent, caespitose or not, essentially glabrous or puberulent in foliage and scaberulous in inflorescence, or tomentose-velutinous throughout, perennial herbs arising from a fibrous caudex surmounting a taproot. Leaves rosulate or clustered toward base, petiolate, membranaceous, pinnately divided or decompound, the ultimate divisions small and narrow; petiole sheathing; cauline leaves few or none. Inflorescence of lax terminal compound umbel, the peduncle short to obsolete, and the umbellets often appearing to be simple umbels, or terminal umbel solitary and elongated. Involucre and involucel of several to numerous foliaceous, entire, trifid, or pinnately dissected bracts and bracteoles, or involucre lacking. Rays rather few, subequal (?) to very unequal, spreading-ascending. Umbellets several to many flowered, the flowers pedicellate. Flowers white or pinkish; petals obovate, with a narrower inflexed apex; calyx teeth prominent, linear or linearly lanceolate; stylopodium conical or low conical, the styles eventually reflexed. Pedicels rather slender, unequal or equal. Fruit ovoid or oblong-elliptic to orbicular, flattened dorsally, rounded at both ends, the mericarps strongly flattened dorsally, glabrous, the ribs prominent, or all equally or unequally winged, but wings narrower than body. Vittae 1 or 2 in intervals, 2 or 4 on commissure. Commissural face plane. Carpophore bipartite.

Type species: *Athamanta depressa* D. Don [*Cortia depressa* (D. Don) Norman].

Distribution: As interpreted here, a genus of 1–3 species of the higher Himalaya, from Sikkim to Pakistan.

Note. Leute (1969, 1970), in his masterly revision of *Ligusticum* and allied genera, construed the genus as comprising 7 Himalayan species, 5 of them usually referred to *Selinum*. He emphasized their mutual possession of prominent, slender calyx teeth, dorsally compressed fruit, and broadly winged lateral fruit ribs. However, he apparently overlooked the brief but penetrating analysis of *Cortia* and the publication of *Cortiella* by Norman (1937). Since the latter author had superior access to historical material in a very complicated taxonomic puzzle, we tentatively follow Nasir (1972) in deferring to Norman's judgment.

KEY TO SPECIES

1. Plants essentially stemless, cauline leaves lacking; involucre prominent, bracts pinnatisect. .... 1. *C. depressa*.
1. Plants with slender stems, a pair of cauline leaves present; involucre lacking. .... 2. *C. staintoniana*.

1. **Cortia depressa** (D. Don) Norman, Jour. Bot. 75: 96. 1937.

Athamanta depressa D. Don, Prodr. Fl. Nepal. 184. 1825.

Schulzia ? lindlei Wallich, Cat. 589. 1829, *nomen nudum*; *Cortia lindlei* DC., Prodr. 4: 187. 1830. Based on *Wallich 589*, "in Emodo Nepaliae ad Gossain-Than et Kamaon". *Cortia nepalensis* Norman, Jour. Bot. 67: 245. 1929; *Schultzia nepalensis* (Norman) Hiroe, Umbell. World, 765. 1979. Based on *Dhwoj 3*, "Goshai Kand, 17,000 ft.," Nepal. *Cortia oreomyrrhiformis* Farille & Malla in Farille, Cauwet-Marc & Malla, Candollea 40(2): 545. 1985. Based on *Cooper 3240*, "Bhutan, Phaiudiu Jimpu, 13,000′," 5 Aug. 1914 (E, holotype).

Plants low and essentially stemless from a fibrous perennial base, up to 30 cm tall. Leaves oblong, 2–10 cm long, 5–15 mm broad, bipinnatifid, the segments oblong or ovate, deeply incised, and the ultimate divisions linear, up to 3 mm long, 2 mm broad. Petiole elongate, 2–9 cm long. Peduncle stout, 6 cm long to obsolete. Involucre of 2–15 foliaceous, pinnatisect bracts up to 3 cm long. Rays 2–15, stout to slender, spreading-ascending, very unequal, 1.5–3.5 cm long, much longer in sessile than in pedunculate umbels. Umbellets 10–20-flowered, the mature pedicels subequal to unequal, 3–12 mm long. Involucel of 10–20 foliaceous, pinnatisect and laciniate bracteoles 8–13 mm long, longer than flowering pedicels. Fruit oblong-elliptic, 5–10 mm long, 3–6 mm broad. June–Oct.

Type locality: "Hab. in regione alpina Gosaingsthan dicta, in jugo Emodi," *Wallich 589.*

Distribution: Himalaya, from Bhutan and Sikkim to Pakistan, 3,000–5,600 m.

Representative specimens: Uttar Pradesh: Garhwal, near Bhowani, *Duthie* (CAL); Kumaon, *Wallich 589B* (CAL). Nepal: Gossanthain, *Wallich 589A* (CAL); Goshaikund, *Wigram 63* (CAL). Sikkim, Tosa, *W. W. Smith 4052* (CAL); Tankara, *Gammie 537* (CAL). Bhutan: Parshong Jimpu, *Cooper 1992* (E).

2. **Cortia staintoniana** Farille & Malla in Farille, Cauwet-Marc & Malla, Candollea 40(2): 545. 1985.

Plants rather slender, caulescent, ascending or erect, simple, tomentose-velutinous throughout, ca. 20 cm tall, from a fibrous base surmounting a woody taproot. Leaves oblong-oval, 3–4.5 cm long, 1–2.5 cm broad, pinnately lobed to pinnatifid, the lobes linearly lanceolate, acute, mucronulate; petiole slender, 4–8 cm long, narrowly long-sheathing, scarious; cauline leaves 1 pair, like the basal, but wholly long-sheathing. Peduncle solitary, terminal, ca. 8 cm long, tomentose. Involucre 0. Immature rays ca. 10, subequal (?), ca. 1 cm long, spreading-ascending, tomentose. Umbellets 15–25-flowered, the immature pedicels unequal, tomentose, 1–3 mm long. Involucel of 8–10 linearly lanceolate to oblanceolate, entire or trifid, ciliolate bracteoles to 1 cm long, exceeding the umbellet. Flowers "pale pink"; petals obovate, with a narrower inflexed apex, clawed, slightly radiant; calyx teeth prominent, green, linear, 1–1.5 mm long; stylopodium low conical, the immature styles very short, the ovary glabrous. Immature fruit ovoid, 1–1.5 mm long, ca. 1 mm broad, the mericarps apparently a little compressed dorsally, glabrous, the ribs prominent or shortly winged (?). Mature fruit and carpophore not seen. July.

Type locality: "E. Nepal: S. of Topke Gola, Arun-Tamur Watershed, 13,000 ft.," 9 Jul. 1956, *Stainton 894* (E, holotype; BM, isotype).

Distribution: Known only from the type collection.

Note. We find no evidence of the glandulosity mentioned in the original description. The flowers (petals, calyx) and very immature fruit resemble those of *Ligusticum hispidum* (Franchet) H. Wolff and *Selinum vaginatum* (Edgew.) C. B. Clarke. Placement of *C. staintoniana* here is strictly provisional, until additional and more mature specimens are available.

56. **CORTIELLA** Norman
Jour. Bot. 75: 94. 1937.

Low, acaulescent, caespitose, pubescent or puberulent perennial herbs from a stout taproot. Leaves rosulate, petiolate, membranaceous, pinnately decompound, the ultimate divisions linear; petiole slender, sheathing. Inflorescence of a congested, sessile, terminal compound umbel, usually with several small pedunculate lateral umbels. Involucre and involucel of several prominent, foliaceous, pinnately dissected bracts and bracteoles, or the latter merely lobed or entire. Rays of terminal umbel numerous, unequal, spreading-ascending, fewer and shorter in subsidiary umbels. Umbellets several flowered, not all maturing, the flowers pedicellate. Flowers white; petals ovate, slightly emarginate, with a narrower inflexed apex; calyx teeth prominent, linear; stylopodium low conical, the styles slender, diverging. Pedicels unequal. Fruit subquadrate-orbicular, compressed dorsally, truncate at apex, cordate to truncate at base, the mericarps flattened dorsally, glabrous, unequally winged, the lateral and 1 or 2 of the other ribs ± broadly winged. Vittae solitary in intervals, 2 on commissure (or the latter obscure). Commissural face plane. Carpophore obsolete.

Type species: *Cortia hookeri* C. B. Clarke [*Cortiella hookeri* (C. B. Clarke) Norman].

Distribution: A genus of 1 or more species endemic to the Eastern Himalaya of Sikkim, Bhutan, and Tibet.

1. **Cortiella hookeri** (C. B. Clarke) Norman,
Jour. Bot. 75: 94. 1937.

Cortia hookeri C. B. Clarke in Hook. f., Fl. Brit. Ind. 2: 702. 1879; *Schultzia hookeri* (C. B. Clarke) Hiroe, Umbell. World, 763. 1979.
Cortiella glacialis Bonner, Candollea 13: 320. 1951; *Pleurospermum glacialis* (Bonner) Hiroe, Umbell. World, 747. 1979. Based on *Wyss-Dunant 1174*, "Rive droit du glacier du Kangchendzönga," Sikkim, 1949 (BM, isotype).
Cortiella cauwetmarciana Farille & Malla in Farille, Cauwet-Marc & Malla, Candollea 40(2): 543. 1985. Based on *Cooper 4268*, "Bhutan: Kerigla [Kengla?], 15,000 feet". 1 Aug. 1915 (E, holotype).

Extremely dwarfed, stemless, caespitose, pubescent or puberulent perennial herb from a short crown surmounting a fibrous base. Leaves rosulate, oblong, 5–10 cm long (including petiole), 1–2-pinnate, the leaflets 5–10 mm long, the ultimate divisions linear, acute, 1–3 mm long, ca. 0.5 mm broad, the rachis puberulent. Petiole sheathing, puberulent. Peduncle of terminal umbel obsolete. Involucral bracts pinnatifid, scarcely distinguishable from foliage leaves. Rays of congested terminal umbel ca. 30, spreading-ascending, unequal, pubescent, 2–4.5 cm long, of lateral umbels 4–20, 3–7 mm long. Umbellets 15–20-flowered, not all maturing, the mature pedicels unequal, pubescent, 3–10 mm long. Involucel of several linear, entire or distally lobed bracteoles 3–10 mm long, often exceeding flowers. Fruit ca. 5 mm long, 4 mm broad, the lateral and usually 2 intermediate ribs winged in one mericarp, the lateral and one dorsal in the other. July–Sept.

Type locality: "SIKKIM: Jongri, 13,000 ft.," *Hooker* (*Hooker & Thomson*, Herb. Ind. Orient, *Cortia 1*) (K, holotype; GH) (Norman 1937).

Distribution: Sikkim Himalaya, 3,000–5,000 m; Tibet.

Representative specimens: Sikkim: w/o further locality, *Kurz* (CAL); Mathula, *Younghusband 175* (CAL); Jongri, *Hooker* (CAL).

57. **PRANGOS** Lindley

Quart. Jour. Sci. Arts 19: 7. 1825; C. B. Clarke in Hook. f., Fl. Brit. Ind. 2: 695. 1879.

Koelzella Hiroe, Umbell. Asia, 146. 1958.

Large, stout or mostly slender, erect, caulescent, branching, glabrous, scaberulous, or pubescent perennial herbs from a fibrous caudex surmounting a stout taproot. Leaves petiolate, membranaceous, pinnately or ternate-pinnately decompound, the ultimate divisions often elongate, filiform to lanceolate; petiole sheathing at base; cauline leaves reduced upward. Inflorescence of lax terminal and lateral compound umbels, the peduncles often elongate. Involucre and involucel of few to rather numerous narrow, entire, often caducous bracts and bracteoles. Rays few to rather numerous, subequal or unequal, spreading-ascending. Umbellets few to many flowered, not all maturing, the flowers pedicellate. Flowers yellow; petals oval to obovate, emarginate, with a narrower inflexed apex; calyx teeth evident, or calyx obsolete; stylopodium depressed to form broad, flattened disk, or obsolete, the styles moderately slender, reflexed. Pedicels slender, spreading-ascending. Fruit oblong to elliptic, subterete or slightly compressed dorsally, rounded at both ends, the mericarps compressed dorsally in transection, glabrous, the epicarp spongy, the ribs broadly and subequally thin winged, the wings often undulate or transversely folded. Vittae small, numerous, continuous around seed cavity. Commissural face concave to deeply sulcate. Carpophore bipartite.

Type species: *Prangos pabularia* Lindley.

Distribution: A Mediterranean to central Asian genus of perhaps two dozen species.

Note. This represents one facet of a closely interrelated group of Mediterranean to southwestern to central Asian plants that have been referred to several genera by various authors. Perhaps a dozen or so species possess the distinctly and often undulately winged ribs characteristic of Lindley's genus. In Flora Europaea, Tutin (1968) combined *Prangos* Lindley and *Hippomarathrum* Link under *Cachrys* Linn., as a single genus. Gruenberg-Fertig, Heyn & Herrnstadt (1973) and Herrnstadt & Heyn (1975 a, b) indicate that two quite different elements are involved and propose that they be recognized as *Cachrys* (including *Prangos*) and *Hippomarathrum* respectively. Herrnstadt & Heyn (1975 a, b) then show convincingly that in Israel, at least, the critical fruit characters supposedly distinguishing *Prangos* and *Cachrys* break down upon careful examination, and that these do not correlate with foliar distinctions. From a purely pragmatic viewpoint, however, it is easier to treat the single Indian species of this alliance under *Prangos*, of which it is the type species.

1. **Prangos pabularia** Lindley, Quart. Jour. Sci. Arts 19: 7. 1825; C. B. Clarke in Hook. f., Fl. Brit. Ind. 2: 695. 1879.

Koelzella pabularia (Lindley) Hiroe, Umbell. Asia 147. 1958; *Cachrys pabularia* (Lindley) Herrnst. & Heyn, Notes Roy. Bot. Gard. Edinburgh 33: 443. 1975.

Plants slender, erect, verticellately branched above, essentially glabrous, 0.5–2 m tall. Leaves rosulate but ascending, oblong, 20–70 cm long, 8–15 cm broad, pinnately decompound with linear to filiform, acute, ultimate divisions 5–20 mm long, 0.5–1 mm broad;

petiole stout, 5–20 cm long; cauline leaves similar to the basal, the upper often whorled. Involucre and involucel of 2–8 linear or linearly lanceolate, entire, often caducous bracts or bracteoles 2–8 mm long. Fertile rays 6–12, stout, spreading-ascending, unequal, 2–8 cm long. Umbellets many flowered, not all maturing, the mature pedicels slender, unequal, spreading-ascending, usually 7–15 mm long. Petals obovate; calyx teeth evident, triangular. Fruit oblong-elliptic, 10–15(–20) mm long, 5–9 mm broad (often with only one mericarp developing), the wings conspicuously corrugated-plicate, the vallecular surfaces tuberculate. Commissural face deeply sulcate, with a T-shaped furrow. Apr.–Oct.

Type locality: "in Indiae orientalis temperatis circa Imbal seu Draz," no collector indicated.

Distribution: Western Himalaya in Kashmir, 1,000–3,500 m; Tibet, Afghanistan, Central Asia.

Representative specimens: Jammu & Kashmir: Kishanganga Valley, *Keshavanand 1421* (DD); Bias, *Gammie* (DD); Lidder Valley, *Duthie 13,171* (DD); Upper Munda, *T. A. Rao 9179* (BSD); Karagbal, *Koelz 9263* (UC).

GROUP VI

1. Annuals; involucre and involucel both lacking; leaf divisions filiform; cultivated herb. .... 58. *Anethum.*

1. Perennials (or biennials); involucre present or lacking, involucel usually present; leaflets or leaf divisions (at least in our species) lanceolate or oblong to orbicular or obovate.

   2. Vittae visible neither on dorsal nor on commissural surfaces of fruit.

      3. Plants glabrous to variously pubescent, but not grayish-hirsute in our species and without a fibrous stem base; fruit wings neither hyaline nor with a tumid margin.

         4. Plants truly perennial; petioles and leaf divisions rarely decurrent and leaves not decompound; umbels terminal or lateral, none solely staminate.

            5. Flowers white; mericarps not closely appressed and separating readily; fruit wings broad, often broader than body. .... 59. *Angelica.*

            5. Flowers yellow; mericarps closely appressed, not separating readily; fruit wings much narrower than body. .... 60. *Peucedanum.*

         4. Plants mostly monocarpic; petioles and leaf divisions usually decurrent, or leaves decompound; proliferate staminate lateral umbels usually present. .... 61. *Ferula.*

      3. Plants grayish-hirtellous to tomentose, with a fibrous stem base; ovaries white-tomentose; fruit wings hyaline with a tumid, corky distal margin. .... 62. *Zosima.*

   2. Vittae clearly visible on dorsal surface of fruit, and usually also on commissural surface.

      6. Plants herbaceous; stems solitary to several, without persistent leaf bases.

         7. Plants from slender rootstocks; petals scarcely inflexed at apex; fruit only weakly compressed. .... 63. *Lalldhwojia.*

         7. Plants without rootstocks, from fibrous, thickened, or tuberous roots; petals with a conspicuously inflexed narrow apex; fruit strongly compressed.

            8. Roots fibrous or thickened, not tuberous; bracts and bracteoles, if present, not dimidiate.

               9. Bracts usually inconspicuous, smaller, caducous, or lacking; strengthening cells form a continuous layer about seed cavity and under ribs. .... 64. *Heracleum.*

               9. Bracts at least half as long as rays; strengthening cells beneath inner margin of wings only. .... 65. *Tordyliopsis.*

            8. Root tuberous; bracts, if present, and bracteoles dimidiate; calyx irregular.

               10. Exterior petal of marginal flowers only slightly if at all dilated; lateral fruit ribs thin winged; vittae extending only part way to base of fruit; strengthening cells apparently none. .... 66. *Polyzygus.*

10. Exterior petal of marginal flowers grossly dilated; lateral fruit ribs projecting but unwinged; vittae extending to base of fruit; strengthening cells prominent. .... 67. *Pinda.*

6. Plants shrubby; stems numerous from a caudex covered with persistent leaf bases. .... 68. *Semenovia.*

58. **ANETHUM** Linn.
Sp. Pl. 263. 1753.

Peucedanum sect. *Anethum* Benth. in Benth. & Hook. f., Gen. Pl. 1: 919. 1867.

Slender, erect, caulescent, branching, glabrous and glaucous annual herbs with strong anise odor, from subfusiform roots. Leaves petiolate, membranaceous, pinnately decompound, with filiform ultimate divisions; petiole slender, sheathing; cauline leaves reduced upward. Inflorescence of lax, terminal and lateral compound umbels, the peduncles elongate. Involucre and involucel lacking. Rays numerous, subequal, spreading-ascending. Umbellets numerous flowered, the flowers pedicellate. Flowers yellow; petals suborbicular with a narrower obtuse inflexed apex; calyx obsolete; stylopodium conical, the styles short, reflexed. Pedicels slender, unequal, spreading-ascending. Fruit ovoid to elliptic-oblong, compressed dorsally, the mericarps pentagonal in transection, glabrous, the lateral ribs narrowly winged, the dorsal ribs merely acute. Vittae solitary in intervals, 2–4 on commissure. Commissural face plane to slightly concave. Carpophore bipartite.

Type species: *Anethum graveolens* Linn.

Distribution: A monotypic genus of southern Asia, widely cultivated and adventive.

1. **Anethum graveolens** Linn. Sp. Pl. 263. 1753.

Peucedanum graveolens (Linn.) Hiern, Fl. Trop. Afr. 3: 19. 1877; C. B. Clarke in Hook. f., Fl. Brit. Ind. 2: 709. 1879, non S. Wats., 1871.
Anethum sowa Roxb. in Fleming, Asiat. Res. 11: 156. 1810; *Peucedanum sowa* (Roxb.) Kurz, Jour. Asiat. Soc. Bengal 46: 116. 1877. Based on *Roxburgh* (?), cultivated in Bengal.

Plants slender, alternately branched above, glabrous and glaucous, 30–170 cm tall. Leaves oblong to obovate, 10–40 cm long, 10–20 cm broad, pinnately decompound, the ultimate divisions filiform, 4–20 mm long, less than 0.5 mm broad; petiole 5–6 cm long; cauline leaves greatly reduced. Peduncles 4–16 cm long. Rays 5–45, slender, 3–10 cm long. Umbellets 15-50-flowered, the mature pedicels 6–10 mm long. Fruit ovoid to elliptic-oblong, glabrous, 4–5 mm long, 2–3 mm broad.

Type locality: "Inter Lusitaniae et Hispaniae segetes," collector unknown.

Distribution: Cultivated and sub-spontaneous throughout India; widely cultivated and occasionally escaped throughout Eurasia and the Americas.

Representative specimens: Punjab: w/o precise locality, *Clarke 134* (CAL). Uttar Pradesh: Dehradun, Rispana, *Babu 33,855* (BSD); Lucknow, *Das 33,916* (CAL). Nepal: W/o precise locality, *Wallich 595A* (CAL). West Bengal: W/o precise locality, *Kurz* (CAL). Assam: Khubri, *Chatterjee* (CAL). Tamil Nadu: "Peninsular India," *Wight 1179* (CAL).

59. ANGELICA Linn.

Sp. Pl. 250. 1753. C. B. Clarke in Hook. f.,
Fl. Brit. Ind. 2: 706. 1879.

Angelica subgen. *Angelicastrum* Drude in Engl. & Prantl, Pflanzenfam. *38*: 220. 1898.

Archangelica Hoffm. Gen. Umbell. 161. 1814; C. B. Clarke in Hook. f., Fl. Brit. Ind. 2: 707. 1879; *Angelica* subgen. *Archangelica* (Hoffm.) Maxim. ex Drude in Engler & Prantl, loc. cit.; *Angelica* sect. *Archangelica* (Hoffm.) Regel, Trudy Bot. Sada 5: 590. 1878, non *Archangelica* N. M. Wolff, 1776.

Very stout to slender, erect, branching, glabrous to tomentose perennial herbs from stout taproots. Leaves petiolate, membranaceous to somewhat fleshy or coriaceous, ternate-pinnately or pinnately compound, with broad or narrow, distinct or sometimes confluent, dentate, serrate, or lobed and usually rather large leaflets; petiole sheathing at base, the cauline sheaths often strongly inflated and the blades reduced or sometimes obsolete. Inflorescence of lax terminal and lateral compound umbels, the peduncles elongate. Involucre of a few to many narrow bracts, or these lacking. Rays usually numerous, spreading or spreading-ascending, equal or unequal, often webbed. Umbellets numerous to rather few flowered, the flowers pedicellate. Involucel of numerous narrow, usually entire bracteoles, or less commonly lacking. Flowers greenish to white, less commonly red or purple; petals oblong to obovate, with a narrower inflexed apex, glabrous or pubescent; calyx usually obsolete; stylopodium low conical to conical, the styles short to slender, spreading or recurved; ovaries glabrous to tomentose. Pedicels slender, spreading, often webbed. Fruit oblong to orbicular, rounded to emarginate at apex and base, strongly compressed dorsally, the mericarps flattened dorsally in transection, glabrous to tomentose, the dorsal ribs low, filiform or obtuse, to narrowly thin or corky winged, the lateral broadly winged, the wing narrower than to broader than body, the lateral wings of the two mericarps distinct and mericarps separating readily at maturity. Vittae 1–3 in intervals, 0–6 on commissure and adhering to pericarp, or numerous and adhering to seed, which becomes free in pericarp; vittae sometimes present also in ribs. Commissural face plane or slightly concave or convex. Carpophore bipartite.

Type species: *Angelica sylvestris* Linn.

Distribution: A large circumboreal genus which, depending upon how it is defined, may comprise 60–80 species; in India, it is confined to the Himalaya. Identity of some Indian material remains uncertain.

Note. Pimenov (1987) has recently described *A. harae*, based on a single incomplete but well illustrated specimen collected by Hara et al. in eastern Nepal. He likens it to *A. sikkimensis*, but cites foliar differences, especially in the shape of the terminal leaf divisions. From the description, the rays appear to be fewer and shorter. We have seen no material.

KEY TO SPECIES

1. Plants stout or robust, hispidulous or puberulent to pubescent; cauline leaves with conspicuously dilated sheaths; dorsal fruit ribs prominent, acute or obtuse.
    2. Leaflets ovate-lanceolate to oval, acute or acuminate; stylopodium low conical.
        3. Involucral bracts 4–12; dorsal fruit ribs obtuse; vittae solitary in intervals, 0–4 on commissure.
            4. Bracts and bracteoles reflexed; leaflets mucronulate-dentate; fruit oblong-oval, thick winged. .... 1. *A. glauca.*
            4. Bracts and bracteoles spreading, deciduous; leaflets serrate and lobed; fruit orbicular, thin winged. .... 2. *A. nubigena.*
        3. Involucral bracts few or 0; dorsal fruit ribs acute; vittae small, numerous, adhering to seed which becomes free in pericarp. .... 3. *A. roylei.*
    2. Leaflets lanceolate, longly-acuminate; stylopodium conical. . . 4. *A. cyclocarpa.*
1. Plants very slender, essentially glabrous; cauline leaves with scarcely dilated sheaths; dorsal fruit ribs inconspicuous, filiform. .... 5. *A. sikkimensis.*

1. **Angelica glauca** Edgew., Trans. Linn. Soc. 20: 53. 1851; C. B. Clarke in Hook. f., Fl. Brit. Ind. 2: 706. 1879.

Plants stout, erect, branched, aromatic, sparingly hispidulous on foliage and in inflorescence, 60–200 cm tall. Leaves triangular-ovate, 20–50 cm in diameter, 1-3-ternate or ternate-pinnate, the leaflets ovate-lanceolate to oval, 2–12 cm long, 1–6 cm broad, acuminate, coarsely mucronulate-dentate, paler and sometimes sparsely hispidulous beneath; petiole stout, to 20 cm long; upper cauline leaves with greatly reduced blades and broadly dilated sheaths. Peduncles stout, 10–30 cm long. Involucre and involucel of 5–10 linearly subulate, reflexed bracts or bracteoles 5–20 mm long. Rays 10–40, slender, subequal to unequal, 2–6 cm long, glabrous to hispidulous, spreading-ascending. Umbellets 20–40-flowered, the mature pedicels slender, unequal, spreading-ascending, 1–2 cm long. Flowers greenish-white; petals spatulate to obovate, glabrous or sparingly hispidulous; calyx obsolete; stylopodium low conical, about equal to the short, recurved styles; ovaries glabrous. Fruit oblong-oval, 7–14 mm long, 4–8 mm broad, emarginate at apex and base, glabrous, the dorsal ribs prominent but low and rounded, corky, the lateral broadly corky winged, the wing broader than body. Vittae solitary in intervals, 2–4 on the slightly concave commissure. June–Sept.

Type locality: "Himala, ad alt. ped. 8,000–10,000. Kamalori, Hattu, etc." *Edgeworth.* ["Kashmir," *Edgeworth 152* (K, lectotype)] (Nasir 1972).

Distribution: Western Himalaya, to Pakistan, Afghanistan, and Tibet, 2,500–4,000 m.

Representative specimens: Jammu & Kashmir: Tilail, *Duthie 13,897* (CAL). Himachal Pradesh: Wangtu Valley, *Brandis 3742* (CAL). Uttar Pradesh: Hemkund, *Bhattacharyya, 29,452* (BSD); Furkia, Pindari Moraine, *T. A. Rao 4579* (BSD).

2. **Angelica nubigena** (C. B. Clarke) P. K. Mukherjee, Bull. Bot. Surv. India 24(1–4): 42. 1983.

Heracleum nubigenum C. B. Clarke in Hook. f., Fl. Brit. Ind. 2: 713. 1879.

Plants stout, erect, branched, pubescent, 1–2 m tall. Basal leaves not seen, the lower cauline leaves to 35 cm long including petiole, 3-parted to ternate-pinnate or pinnate, the leaflets ovate-lanceolate or ovate, 5–10 cm long, 2–3 cm broad, serrate and often unequally lobed, decurrent; petiole dilated and sheathing at base. Peduncles stout, to 30 cm long. Involucre of 4–12 linear, deciduous, pubescent bracts to 1 cm long. Rays 30–50, stout, unequal, 3–16 cm long, pubescent, spreading-ascending. Umbellets 15–30-flowered, not all maturing, the mature pedicels slender, unequal, pubescent, spreading-ascending, 1–1.5 cm long. Involucel of 4–12 linear, pubescent bracteoles ca. 5 mm long. Flowers not known; calyx obsolete; stylopodium low conical, shorter than the slender styles. Fruit orbicular, 5–8 mm long, 4–6 mm broad, emarginate at apex and base, glabrous, the dorsal ribs prominent, obtuse, the lateral broadly thin winged, the wing slightly broader than body. Vittae solitary in intervals, lacking or fragmentary on the slightly convex commissure. Sept.(?).

Type locality: "SIKKIM, alt. 10,000–14,000 ft.; Chola Pass," *Hooker* (K); "Yakla Pass," *Clarke*.

Distribution: Sikkim Himalaya, 3,300–4,500 m..

Representative specimens: Sikkim: w/o precise locality, *Cave* (CAL), *Hooker* (CAL); Lachung, *Younghusband 1* (CAL).

3. **Angelica roylei** (Lindley) P. K. Mukherjee & Constance, Edinburgh Jour. Bot. 48(1): 44. 1991.

Archangelica roylei Lindley in Royle, Ill. Bot. Himal. Mount. 232. 1835.

Archangelica officinalis var. *himalaica* C. B. Clarke in Hook. f., Fl. Brit. Ind. 2: 707. 1879; *Angelica archangelica* var. *himalaica* (C. B. Clarke) Krishna & Badhwar, Jour. Sci. Industr. Res. 11: 251. 1952. Based on *Clarke*, *Atchison*, "KASHMIR; 8,000–13,000 ft."; *Thomson*, "Jogi; alt. 11,000 ft."; *Hooker*, "SIKKIM; Lachen, alt. 10,000–11,000 ft."

Archangelica oreadum Diels, Notizbl. Bot. Gart. Berlin 14: 353. 1939; *Angelica oreada* (Diels) Hiroe, Umbell. World, 1430. 1979. Based on *Troll 7976*, "Nanga-Parbat-Gebiet, Kalapani-Tal: zwischen Rattu und Kalipani. . . . 2,800–3,200 m."

Plants usually robust, erect, branched, aromatic, often a little scaberulous on foliage and puberulent in inflorescence, 1.5–3 m tall. Leaves ovate, 30–100 cm long, 2–3-pinnate, the leaflets ovate-lanceolate to ovate, 4–15 cm long, 2–10 cm broad, acute, crenate-dentate or serrate and often few lobed, the upper often decurrent on rachis; petiole stout, 8–30 cm long, broadly sheathing; cauline leaves reduced upward, with conspicuously dilated, scarious sheaths. Peduncles stout, 10–30 cm long. Involucre of a few linear bracts, or lacking. Rays 15–50, slender, subequal to unequal, 3–10 cm long, puberulent, spreading-ascending. Umbellets 20–40-flowered, the mature pedicels slender, spreading-ascending, 5–8 mm long. Involucel of numerous, linear, ciliolate bracteoles ca. 1 cm long. Flowers greenish-white or cream-colored; petals ovate to elliptic; calyx obsolete; stylopodium low conical, shorter than the slender, recurved styles; ovaries glabrous. Fruit oblong to subquadrate, 6–10 mm long, 3–6 mm broad, glabrous, the dorsal ribs prominent, acute, the lateral thick winged, the wing narrower than body (?). Vittae small, numerous, encircling and adhering to the seed, which becomes free in pericarp, vittae also in ribs. Commissural surface concave. June–Aug.

Type locality: "Hab. Urukta," *Royle* (?).

Distribution: Himalaya, from Sikkim to Kashmir; Pakistan; 2,500–4,000 m.

Representative specimens: "Himal. Bor. Occ.," *Thomson* (K). Jammu & Kashmir: Jozila, *Thomson* (K); Kashmir, *Clarke 28,702* (K); Tilail, *Clarke 30,675A, C, D* (K); Godai River Valley, Gilgit Agency, *Webster & Nasir 6514* (GH, UC); Karagbal, *Koelz 9258* (MICH, UC).

4. **Angelica cyclocarpa** (Norman) Cannon in Hara & Williams, Enum. Fl. Pl. Nepal 2: 184. 1979.

Archangelica cyclocarpa Norman, Jour. Bot. 67: 246. 1929.

Plants stout, aromatic, and apparently very large, at first densely fulvopubescent, later glabrescent. Basal and lower cauline leaves not known, the upper cauline leaves ternate-pinnate, with lanceolate leaflets 5.5–8 cm long, 1.5–3 cm broad, longly acuminate, cuneate at base, serrate and sometimes lobed, the terminal a little broader than the lateral and decurrent on rachis; petiole wholly and broadly dilated. Involucre lacking. Rays very numerous, unequal, the outer up to 12 cm long, the inner 4–7 cm long. Umbellets numerous flowered, the mature pedicels very slender, congested, up to 1.5 cm long. Involucel of linear bracteoles about equal to the pedicels. Flowers often polygamous, greenish-white; petals small, oblong; calyx obsolete; stylopodium conical, the slender styles reflexed. Fruit (barely mature) orbicular, 5 mm long, 4 mm broad, cordate at base, the dorsal ribs low but prominent, the lateral broadly winged. Vittae adhering to seed, those of commissure minute (4 ?), hidden by fleshy exocarp.

Type locality: "Barpak, 10,000–12,000 ft.," Nepal, *Dhwoj 28* (E, holotype).

Distribution: Himalaya of Nepal, 3,000–4,000 m.

Specimen examined: Nepal: Pongsing, *Dhwoj 254* (BM).

5. **Angelica sikkimensis** (C. B. Clarke) P. K. Mukherjee, Bull. Bot. Surv. India 24(1–4): 43. 1983.

Peucedanum sikkimensis C. B. Clarke in Hook. f., Fl. Brit. Ind. 2: 710. 1879.

Plants very slender, erect, freely branched, essentially glabrous, 30–90 cm tall. Leaves triangular-ovate, 8–30 cm in diameter, ternately or ternate-pinnately compound, the leaflets ovate-lanceolate, 1–5 cm long, 1–1.5 cm broad, acuminate, sharply serrate and lobed; petiole slender, 5–15 cm long, sheathing at base; cauline leaves reduced upward, with narrow leaflets and scarcely dilated sheaths. Peduncles slender, terminal and lateral, 5–30 cm long. Involucre of 1 or more linear bracts to 10 mm long, or usually lacking. Rays 8–12, slender, unequal, 3–5 cm long, spreading-ascending. Umbellets about 12-flowered, few fertile, the mature pedicels slender, unequal, 5–15 mm long, spreading. Involucel of 4–6 linear bracteoles 1–6 mm long, or sometimes lacking. Petals oval; calyx obsolete; stylopodium conical; ovaries glabrous. Fruit oblong-oval, 5–7 mm long, 4–6 mm broad, emarginate at apex and base, glabrous, the dorsal ribs filiform, inconspicuous, the lateral broadly thin winged, the wing much broader than body. Vittae 2 or 3 in intervals, 0–2 on commissure. Commissural face slightly convex. Oct.

Type locality: "SIKKIM; alt. 3,000–7,000 ft., frequent," *Hooker, Clarke*.

Distribution: Sikkim Himalaya, 1,000–2,500 m.

Representative specimens: Sikkim: w/o specific locality, *Hooker* (K), *King* (CAL); Lingdam, *Clarke 13,065H* (CAL), *13,065J* (BM); Riskee, *Clarke 4,987B* (CAL), *24,987C* (BM); Hee, *Clarke 13,074A* (BM); Kulhait, *Clarke 25,530B* (K).

60. PEUCEDANUM Linn.

Sp. Pl. 245. 1753; C. B. Clarke in Hook. f., Fl. Brit. Ind. 2: 708. 1879. *Palimbia* DC. Prodr. 4: 176. 1830.

Very stout to slender, erect, branching, glabrous to variously pubescent perennial (sometimes woody, rarely biennial or annual) herbs from stout taproots. Leaves petiolate, membranaceous, pinnate or ternate or ternately, ternate-pinnately, or pinnately compound, with broad to narrow, entire, serrate, dentate, incised, or lobed, usually rather large leaflets; petiole sheathing at base; cauline leaves reduced upward, the cauline sheaths somewhat inflated. Inflorescence of lax terminal, or terminal and lateral compound umbels, the peduncles elongate. Involucre commonly lacking, or of 1 to numerous narrow bracts. Rays few to numerous, equal or unequal, spreading-ascending, often webbed. Umbellets numerous flowered, the flowers pedicellate. Involucel of few to numerous, narrow, entire bracteoles, or lacking. Flowers sometimes polygamous, white, cream-colored, or yellow, often reddish; petals oval to obovate, with a narrower inflexed apex, emarginate or bilobed, glabrous to pubescent; calyx teeth minute (rarely prominent) to obsolete; stylopodium conical to low conical, the styles slender to short, spreading or recurved; ovaries glabrous. Pedicels slender, spreading-ascending, often webbed. Fruit oblong or pyriform to orbicular, rounded at apex and base, very strongly compressed dorsally, the mericarps flattened dorsally in transection, glabrous to pubescent, the dorsal ribs filiform to obsolete, the lateral usually thick winged, narrower than body. Vittae 1 to few in intervals, 2 to several on commissure. Commissural face plane to slightly concave. Carpophore bipartite.

Type species: *Peucedanum officinale* Linn.

Distribution: A large, poorly defined, and doubtless not wholly natural genus of the Old World (Eurasia, Africa), with perhaps 100–150 species.

Note. Of the 9 species attributed to this genus by Clarke (1879, pp. 708–711), one is referred to *Anethum*, a second to *Angelica*, and a third to *Heracleum*. In general, taxa with broad, thin, freely separating lateral wings are referred to *Angelica*, those with thick, previously conjoined wings narrower than or equal to the body to *Peucedanum*, and those with radiant flowers, hairy fruit, and abbreviated vittae to *Heracleum*. However, the last genus also includes some species (sect. *Tetrataenium*) with non-radiant flowers, glabrate fruit, and vittae essentially as long as the fruit. For a concise discussion of the boundaries of *Peucedanum*, see Flora Iranica 162: p. 442. 1987.

Peucedanum josephianum Wadhwa & Chowdhery, a new species, ostensibly of this genus, was recently described from Kerala (Indian Jour. Forestry 13(1): 78–79. 1990). The taxon, characterized as yellow-flowered, glabrous, and possessing white-margined leaflets, is contrasted with the very dissimilar white-flowered, densely puberulent to hirtellous maritime *P. japonicum* Thunb. Any close similarity appears to be confined to superficial resemblance in leaf form; the fruit, in our opinion, is too immature to permit satisfactory generic assignment. Because of numerous similarities in root and in floral and inflorescence characters, we think it may represent our *Karnataka benthamii* (p. 74) despite the discrepancy in flower color.

Norman (in BM) tentatively referred the specimen, *Kingdon-Ward 20,103*, from Di Chu Gorge, Assam, to *Peucedanum praeruptorum* Dunn. The specimen is immature (and nearly glabrous) and its assignment here, as well as inclusion of this taxon in the Indian flora, are both doubtful.

KEY TO SPECIES

1. Leaflets or divisions of basal and lower cauline leaves all entire.
   2. Plants slender; leaves 3-parted to simply pinnate; involucre usually present; flowers yellow. .... 1. *P. dhana.*
   2. Plants robust; leaves biternate; involucre usually lacking; flowers white. .... 2. *P. dehradunense.*
1. Leaflets or divisions of basal and lower cauline leaves serrate, incised, lobed, or pinnatifid.
   3. Umbellets 5–10-flowered, the pedicels divaricate.
      4. Leaves triangular-ovate, biternate or ternate-pinnate, the leaflets serrate to lobed; fruit oval to elliptic; commissural vittae 2 or 0. .... 3. *P. ramosissimum.*
      4. Leaves oblong-ovate, 2–4-pinnate, the leaflets serrulate; fruit orbicular; commissural vittae 4–6. .... 4. *P. anamallayense.*
   3. Umbellets many (usually more than 10-) flowered, the pedicels spreading-ascending to erect.
      5. Leaflets linear to oblong-lanceolate; involucel 0, or deciduous; lateral fruit wings thick. .... 5. *P. nepalense.*
      5. Leaflets broadly lanceolate to ovate; involucel of several persistent, ultimately reflexed bracteoles; lateral fruit wings thin. .... 6. *P. nagpurense.*

1. **Peucedanum dhana** C. B. Clarke in Hook. f., Fl. Brit. Ind. 2: 709. 1879.

Peucedanum dhana Buchanan-Hamilton in Wallich, Cat. 244. no. 7216. 1832, *nomen nudum.*

Plants low to slender, erect, few branched, glabrous, 15–60 cm tall, from a stout taproot. Leaves clustered at base, oblong to oval, pinnately or ternately parted to simply pinnate or subbipinnate, the 3–5 leaflets linearly oblong or lanceolate-elliptic to orbicular, acute at apex, cuneate at base, entire; petiole short to slender, 1–15 cm long, narrowly sheathing at base; cauline leaves few, greatly reduced upward, the sheaths slightly dilated. Peduncles slender, terminal and lateral, 3–10 cm long. Involucre of 3–6, linear to lanceolate, acute or acuminate bracts 2–4 mm long. Rays 5–10, slender, unequal, glabrous, 2–12 cm long, spreading-ascending. Involucel of 3–8 linear to lanceolate bracteoles 2–3 mm long. Umbellets 15–20-flowered, the pedicels slender, subequal to unequal, spreading-ascending, 2–6 mm long. Flowers yellow, regular; petals ovate with a narrower inflexed apex; calyx teeth mostly obsolete; stylopodium low conical, the styles rather short, erect; ovaries glabrous. Fruit subquadrate to elliptic, 7–8 mm long, ca. 5 mm broad, glabrous, the dorsal ribs filiform to obscure, the lateral thick, narrower than to equal the body, the mericarps not separating readily. Vittae 1 to several in intervals, 2–6 on commissure. Commissural face plane to slightly concave. Carpophore bipartite.

KEY TO VARIETIES

1. Leaflets linearly oblong to lanceolate-elliptic, narrow and elongate, remote, membranaceous. ....a. var. *dhana.*
1. Leaflets oblong to suborbicular, shorter and broader, crowded, somewhat coriaceous. ....b. var. *dalzellii.*

1a. var. **dhana**

Leaves pinnatifid to pinnate, membranaceous. Fruit truncate at base. Vittae solitary in dorsal intervals, 2 in lateral intervals, 4–6 on commissure. Feb.–May.

Type locality: "Goruckpore," Uttar Pradesh, *Buchanan-Hamilton* (CAL, isotype).

Distribution: Occurring disjunctly in the Western Himalaya and Nepal, and in the hills of the Eastern Ghats, peninsular India, from the plains to 3,000 m.

Representative specimens: Uttar Pradesh: Mala, Pilibhit, *Malhotra 37,142* (CAL); Garhwal, Dun, *Inayat Khan 25,896* (DD). Nepal: Bhotekosi, *Stainton 4463* (BM). Andhra Pradesh. Araku Valley, *Balakrishnan 515* (CAL). Bangladesh. Dinajpur, *Clarke 26,408C* (BM).

lb. var. **dalzellii** C. B. Clarke in Hook. f., Fl. Brit. Ind. 2: 709. 1879.

Pastinaca glauca (DC.) Dalzell, Hooker's Jour. Bot. Kew Gard. Misc. 4: 293. 1859, non *Peucedanum glaucum* DC. 1830, vel Link 1821.

Leaves 3-parted to pinnate, coriaceous. Fruit often emarginate at both apex and base. Vittae 1 or 2 in all intervals, 2–6 on commissure. Mar.–May, July–Aug.

Type locality: "Bombay," Bombay Presidency, *Dalzell*.

Distribution: Hills of central and western India, 250–950 m.

Representative specimens: Orissa: Delaisara, *Mooney 3784* (DD); Jashpur, *Mooney 1333* (DD). Andhra Pradesh: Araku Valley, *Suneraj 21,325* (MH). Madhya Pradesh: Jubbalpur, *Beddome* (CAL). Maharashtra: Bombay, *Dalzell* (CAL, DD); Belgaum, *Talbot 3811* (CAL).

2. **Peucedanum dehradunense** Babu, Indian Forester 96: 535. 1970.

Plants robust, erect, simple or sparsely branched above, glabrous, 50–105 cm tall, from an elongated fusiform taproot bearing nonfibrous leaf sheaths at summit. Leaves mostly rosulate, triangular-ovate, 1–3 cm long, biternate, the leaflets oblong-lanceolate, 1–4 cm long, 5–7 mm broad, obtuse, entire, paler beneath; petiole slender, 5–10 cm long, the sheath white-margined; cauline leaves 1 or 2, or usually lacking. Peduncles slender, terminal and lateral, 30–40 cm long. Involucre usually lacking. Rays 5–10, slender, unequal, 3.5–5 cm long. Involucel of 5 linearly subulate, deciduous bracteoles 3–3.5 mm long. Umbellets 15–40-flowered, the mature pedicels ca. 5 mm long. Flowers white; petals oblong-ovate, with a narrower inflexed apex; calyx obsolete; stylopodium conical, about equal to the recurved styles; ovaries glabrous. Fruit ellipsoid, 9–10 mm long, 4 mm broad, glabrous, the dorsal ribs prominent, the marginal narrowly thick winged, the wings about equal to the body. Vittae 1 or 2 in intervals, 4–6 on commissure. Commissural face plane. Carpophore bipartite, the halves filiform. Mar.–June.

Type locality: "UTTAR PRADESH. Dehra Dun, Bindal Ravines, *C. R. Babu 35058* (CAL, holotype; BSD, isotypes)."

Distribution: Himalaya of Uttar Pradesh; Tibet (?).

Representative specimens: Uttar Pradesh: Dehra Dun, *Babu 33,058* (BSD, CAL, K).

3. **Peucedanum ramosissimum** (DC.) C. B. Clarke in Hook. f., Fl. Brit. Ind. 2: 709. 1879.

Selinum ramosissimum Wallich, Cat. 578. 1829, *nomen nudum*; *Palimbia ramosissima* DC., Prodr. 4: 176. 1830.

Plants very slender, erect or sprawling, sparsely leafy and freely branching, glabrous or nearly so, *Daucus*-scented perennial herbs, from a slender taproot. Leaves triangular-ovate, 8–15 cm in diameter, biternate or ternate-pinnate, the leaflets narrowly lanceolate to ovate, 1–4 cm long, 5–15 mm broad, acute or the terminal often acuminate, sharply serrate to lobed; petiole very slender, 2.5–10 cm long, narrowly sheathing at base; upper cauline leaves greatly reduced and with narrower divisions. Inflorescence open paniculate. Peduncles very slender, terminal and lateral, 3–8 cm long. Involucre lacking, or of an occasional minute bract. Rays 3–8, very slender, unequal, divaricate, 5–25 mm long, glabrous. Involucel of 1 to several linearly lanceolate bracteoles to 2 mm long, or lacking. Umbellets 5–10-flowered, many not maturing, the mature pedicels filiform, unequal, divaricate, 3–7 mm long. Flowers greenish-yellow to white; petals oblong-elliptic, with a narrower inflexed apex; calyx teeth minute; stylopodium low conical, about equal to the reflexed styles; ovaries glabrous. Fruit oval to elliptic, 5–7 mm long, 4–5 mm broad, usually emarginate at apex and base, the dorsal ribs filiform, the lateral broadly thin winged, the wing narrower than the black body. Vittae solitary in intervals, 2 on commissure, or the latter lacking. Carpophore bipartite. Aug.–Sept.

Type locality: "in Indiae montibus Pandua prov. Sillet confinibus," *Wallich 578*.

Distribution: Khasia Hills of northeastern India, 1,300–2,600 m; Bangladesh; China.

Representative specimens: Megalaya: Khasia, *Griffith 2616* (CAL); Khasia, *Hooker & Thomson* (Herb. Ind. Orient) (CAL, UC, W); Mawphlang, *Chand 2188* (MICH, UC), *2319* (MICH, UC); Mawryngkneng, *Koelz 28,329* (MICH, UC), *Chand 4771* (MICH, UC); Shillong, *Clarke 38,891* (CAL).

4. **Peucedanum anamallayense** C. B. Clarke in Hook. f., Fl. Brit. Ind. 2: 710. 1879.

Plants very slender, erect, branching, glabrous to puberulent, 30–60 cm tall, from a taproot. Leaves oblong-ovate, 10–20 cm long, including the slender petiole, 2–4-pinnate, the leaflets linear to narrowly oblong, 2.5–6 cm long, 3–5 mm broad, acute, serrulate; cauline leaves reduced upward, the sheaths slightly dilated. Peduncles slender, terminal and lateral, 6–12 cm long. Involucre lacking, or of few minute, linearly oblong bracts. Rays 4–8, very slender, subequal, divaricate, 1–3.5 cm long. Involucel of 3–5 ovate-lanceolate bracteoles ca. 2.5 mm long. Umbellets 8–10-flowered, the mature pedicels slender, subequal, divaricate, 3–6 mm long. Petals with a narrower inflexed apex; calyx teeth minute, lanceolate; stylopodium conical (?), the styles slender; ovaries glabrous. Fruit orbicular, ca. 7 mm long, 6 mm broad, emarginate at apex and base, the dorsal ribs filiform, the lateral broadly thin winged, the wings narrower than to equal to the brownish-yellow body. Vittae 1 or 2 in intervals, 4–6 on commissure. Commissural face plane. Carpophore bipartite. Nov.

Type locality: "ANAMALLAY MTS.," Tamil Nadu, *Wight, Beddome.*

Distribution: Apparently confined to area of type locality.

Representative specimens: Tamil Nadu: "India Orientalis," [Anamalays], *Wight 1180* (CAL, K); Anamalays, *Beddome 3375* (BM); Pulney Hills, *Beddome 3379* (BM).

5. **Peucedanum nepalense** P. K. Mukherjee,
Indian Forester 95: 567. 1969.

Selinum glaucum Wallich, Cat. 580. 1829, *nomen nudum*; *Peucedanum glaucum* DC., Prodr. 4: 179. 1830; C. B. Clarke in Hook. f., Fl. Brit. Ind. 2: 710. 1879, non Link, 1821.

Plants slender, erect, glabrous, 50–120 cm tall, from a fusiform taproot. Leaves 15–40 cm long, the leaflets lanceolate-oblong to linear, 3–5 cm long, 5–7 mm broad, acute or acuminate, serrate or the uppermost entire (?); petiole slender, 5–15 cm long, broadly sheathing at base; lower cauline leaves like the basal, the upper 1–3-foliate, or reduced to inflated sheaths. Peduncles slender, terminal or lateral 10–40 cm long. Involucre lacking, or occasionally of 1 or 2 minute bracts. Rays 6–16, unequal, 2–5 cm long, spreading-ascending. Involucel lacking, or of 1–5 deciduous linear bracteoles 1.5–3.5 mm long. Umbellets numerous flowered, the mature pedicels slender, unequal, spreading-ascending, 5–12 mm long. Petals oblong-ovate to obovate, with a narrower inflexed apex, often slightly unequal; calyx obsolete; stylopodium conical, the styles rather slender; ovaries glabrous. Fruit subquadrate or oblong-elliptic, 9–11 mm long, 4–7 mm broad, emarginate at apex and base, the dorsal ribs filiform to obscure, the lateral broadly thick winged, the wings about equal to the body. Vittae 1 (or 2?) in intervals, 2 or more on commissure. Commissural surface plane. Carpophore bipartite. Mar.–July.

Type locality: "Nipal," *Wallich 580* (CAL, holotype).

Distribution: Central Himalaya in Nepal and Uttar Pradesh.

Representative specimens: Uttar Pradesh: Dehra Dun, *Wallich 580* (CAL). Nepal: w/o specific locality, *Wallich 580* (BM, CAL, K), *Thapi 7483* (CAL); Dudhkosi, Wapsa Khani, *Stainton 6606* (BM).

6. **Peucedanum nagpurense** (C. B. Clarke) Prain,
Bengal Fl. 1: 540. 1903.

Peucedanum glaucum var. *nagpurense* C. B. Clarke in Hook. f., Fl. Brit. Ind. 2: 710. 1879.

Plants stout, erect, glabrous, 1–2 m tall, from a stout taproot. Leaves 20–35 cm long, biternate, the leaflets ovate, 7–15 cm long, 5–8 cm broad, acuminate, rounded or truncate at base, petiolulate, coarsely serrate; petiole 10–25 cm long; cauline leaves with linearly lanceolate or occasionally narrower divisions. Peduncles stout, 5–15 cm long. Involucre lacking, or of 1 or few oblong acuminate bracts to 1 cm long. Rays 8–12 (–16), 1–4 cm long, unequal, spreading-ascending. Involucel of 3–7 linearly setaceous bracteoles to 8 mm long, reflexed in fruit. Umbellets 14–25-flowered, the mature pedicels spreading-ascending. Flowers green or brownish; petals oblong-lanceolate, with a narrower inflexed apex; calyx teeth minute, triangular; stylopodium conical (?), the styles slender; ovaries glabrous. Fruit elliptic to obovate, 8–10 mm long, emarginate at apex, emarginate or truncate and narrowed at base, the dorsal ribs filiform, the lateral thin winged, the wings equal to or narrower than the body. Vittae solitary in intervals, 2 on commissure or obsolete. Commissural face slightly concave. Carpophore bipartite. Oct.-Dec.

Type locality: "Chota Nagpore, alt. 1,500-2,000 ft.; Ramghur and Bagoda," *Clarke*; "Behar, at Monghyr," *Lockwood*.

Distribution: Mountains and hills of Central India.

Representative specimens: Nepal: Siwalik Hills, near Kuepani, *Polunin, Sykes & Williams 5923* (BM). Bihar: Monghyr, *Lockwood* (K); Bagoda, *Clarke 24,936A* (K); Ramnagar, *Clarke 24,982B* (K); Tundi Hills, *Campbell* (CAL, CUH); Ranchi, *Haines 32* (CAL). Madhya Pradesh, Jashpurnagar, Bilaspur Distr., *Arora 7231* (CAL).

61. **FERULA** Linn.

Sp. Pl. 246. 1753; C. B. Clarke in Hook. f., Fl. Brit. Ind. 2: 707. 1879; Korovin, Gen. *Ferula*, 91 pp. 48 pls. 1947.

Narthex Falc., Trans. Linn. Soc. 20: 285. 1851; *Ferula* subgen. *Narthex* (Falc.) Drude in Engl. & Prantl, Pflanzenfam. 38. 229. 1898.

Usually robust, erect, branching, glabrous or pubescent perennial or biennial and often ± monocarpic herbs from taproots. Leaves mostly rosulate, membranaceous to coriaceous, petiolate, ternately, quinately, or pinnately divided to decompound, the leaflets or divisions various, linearly filiform to obovate, entire to lobed or pinnatifid, petiolulate to conspicuously decurrent on rachis; petiole stout, the basal sheath usually large and inflated; cauline leaves with usually dilated sheaths. Inflorescence paniculate or corymbose, often irregular and proliferating, with terminal pedunculate to sessile compound perfect umbels accompanied by lateral simple to compound staminate or polygamous umbels. Involucre and involucel lacking or inconspicuous. Rays few to numerous, subequal or unequal, spreading or spreading-ascending, sometimes proliferating. Umbellets few to numerous flowered, the flowers pedicellate. Flowers usually yellow or yellowish, rarely white, polygamous; petals lanceolate to oval or obovate, with a narrower inflexed apex; calyx usually obsolete; stylopodium low conical to depressed, the styles short to elongated. Pedicels short to slender. Fruit oblong to elliptic, rounded to narrowed at apex and base, strongly compressed dorsally, the mericarps flattened dorsally in transection, glabrous, the dorsal ribs filiform, the lateral narrowly somewhat thick winged, the wings much narrower than the body. Vittae various, few and large to small and numerous, sometimes branching, often obliterated in fruit. Commissural face plane. Carpophore bipartite.

Type species: *Ferula communis* Linn.

Distribution: A very large genus of 100–130 species extending from central Asia westward throughout the Mediterranean region and into northern Africa. Western Himalaya of India lies at the extreme eastern end of its range. (See map, fig. 12, in Bernardi 1979.)

KEY TO SPECIES

1. Leaves pinnately or ternate-pinnately compound with large narrow leaflets or ultimate divisions several centimeters long.
    2. Leaf divisions sessile, obtuse, longly decurrent on rachis.
        3. Inflorescence of regular compound umbels; leaflets finely crenate-serrate. .. .. 1. *F. jaeschkeana.*
        3. Inflorescence irregular, of terminal and compound and axillary simple or imperfectly compound umbels; leaflets entire, sinuate, or incised. .... 2. *F. narthex.*
    2. Leaf divisions mostly petiolulate, subacute, not longly decurrent.
        4. Leaves glabrous on both sides, tripinnate, the margins subentire; vittae 2 or 3 in intervals. .... 3. *F. thomsonii.*

4. Leaves puberulent beneath, ternate-bipinnatisect, the margins serrate; vittae solitary in intervals. .... 4. *F. wolfii.*

1. Leaves decompound with small, broad ultimate divisions mostly less than 1 cm in length, lobed to pinnatifid .... 5. *F. ovina.*

1. **Ferula jaeschkeana** Vatke, Ind. Sem. Hort. Berol., append 2. 1876; C. B. Clarke in Hook. f., Fl. Brit. Ind. 2: 708. 1879.

Peucedanum jaeschkeanum (Vatke) Baillon, Hist. Pl. 7: 185. 1880.
Ferula jaeschkeana var. *parkeriana* O. Schulz, Notizbl. Bot. Gart. Berlin 11: 877. 1933; *F. parkeriana* H. Wolff ex O. Schulz, Notizbl. Bot. Gart. Berlin 11: 877. 1933, *nomen in synon.* Based on *R. R. Stewart 9786*, "Kashmir, Sonamarg ... alt. 9,000–11,000 ft.," and *R. R. Stewart 10,755*, "Hazara, Abbatabad, alt. 4,500 ft."

Plants very stout and robust, erect, much branched above, pilose in foliage to glabrescent, 1–3 m tall, monocarpic, from a massive taproot crowned with persistent fibrous leaf bases. Leaves triangular-ovate, 20–40 cm in diameter, ternate-pinnately or pinnately compound, the leaflets narrowly oblong to oblong-elliptic, 5–15 cm long, 1.5–5 cm broad, obtuse, finely crenate-serrate, glabrous above, pilose beneath, longly decurrent on rachis; petiole stout, 10–25 cm long, dilated into an oblong sheath; cauline leaves reduced upward, with prominent inflated sheaths. Inflorescence large, paniculate or corymbose, the peduncles terminal and lateral, 15 cm long to obsolete. Involucre and involucel lacking. Rays 8–25, stout, subequal, 2.5–5 cm long, spreading. Umbellets 10–20-flowered, the mature pedicels stout, spreading-ascending, 2–3 mm long. Flowers polygamous, yellow; petals lanceolate; calyx obsolete; stylopodium depressed, flat, much shorter than the slender, reflexed styles. Fruit oblong to oval, 15–20 (–40 ?) mm long, 6–10(–20 ?) mm broad, rounded at apex and base, the dorsal ribs filiform, the lateral winged, the wings much narrower than the body. Vittae solitary in intervals, 2–6 on commissure. Mar.–Aug.

Type locality: "Himal. Bor. Occ.," *Jaeschke 66.*

Distribution: Western Himalaya to Tibet, Pakistan, Afghanistan, and Turkestan, 1,800–4,500 m.

Representative specimens: Jammu & Kashmir: Kistawar, *Thomson* (CAL); Gulmarg, *Duthie 13,022* (CAL, DD). Himachal Pradesh: Lahul, *Jaeschke 111* (CAL); Kunawar, *Royle 93/31* (DD). Uttar Pradesh: Ganges Valley, *Duthie 1139* (CAL, DD).

2. **Ferula narthex** Boiss., Fl. Orient. 2: 994. 1872; C. B. Clarke in Hook. f., Fl. Brit. Ind. 2: 707. 1879.

Peucedanum narthex (Boiss.) Baillon, Bot. Med. 2: 1010. 1884.
Narthex asafoetida Falc., Trans. Linn. Soc. 20: 285. 1851, non *Ferula assafoetida* Linn. 1753.

Plants stout, erect, branched, pubescent in foliage, 1.5–2.5 m tall, monocarpic, from a stout fusiform root. Leaves ovate, 30–60 cm in diameter, ternate-pinnately compound, the leaflets linearly oblong, obtuse, oblique, entire, sinuate or incisely lobed, ca. 10 cm long, 2.5 cm broad, decurrent on the rachis; cauline leaves broadly sheathing. Inflorescence large, paniculate or corymbose, the terminal peduncles short and stout; slender axillary peduncles bearing simple or imperfectly compound, few-rayed umbels, arising in cauline leaf sheaths. Involucre and involucel lacking. Rays of terminal umbels (10–)25–30. Flowers polygamous, yellow; petals oblong-ovate; calyx obsolete. Fruit elliptic to obovate, 8–15 mm long, 4–8(–12) mm broad, rounded at apex and base, the dorsal ribs filiform, the lateral narrowly winged, the wings much narrower than the body. Vittae 1 (rarely 2) in intervals, 4–6 on commissure. June–Sept.

Type locality: "Astori Valley & Mussorah, behind Kashmir," *Falconer*.

Distribution: Western Himalaya, Kashmir to Baltistan and Tibet.

Representative specimen: Jammu & Kashmir: Godai, near Astor, *Aitchison* (CAL).

3. **Ferula thomsonii** C. B. Clarke in Hook. f.,
Fl. Brit. Ind. 2: 708. 1879.

Plants stout, erect, branched, glabrous. Leaves 2-3-pinnate, the leaflets remote, ca. 1.5–5 cm long, 1–1.5 cm broad, lanceolate, subacute, subentire, mostly petiolulate. Inflorescence corymbose, the terminal peduncles stout, "the umbels and flowers as in the terminal umbel of *P. narthex*". Pedicels 6–8 mm long, clavate at apex. Fruit elliptic, 6–8 mm long, narrowed at apex and base, the dorsal ribs filiform, the lateral narrowly winged, narrower than body. Vittae small, 2 or 3 in intervals.

Type locality: "KASHMIR; Banahal, alt. 5,000–6,000 ft.," *Thomson* (*Hooker & Thomson*, Herb. Ind. Orient., *Dorema 5*) (CAL).

Distribution: Kashmir. Only the type collection examined.

Note. Clarke's description, combining Thomson's flowering specimens with an infrutescence "of this, or an allied *Ferula* from Kashmir at Newgunga," may comprise disparate elements, leaving the identity and validity of the taxon in doubt.

4. **Ferula wolffii** Korovin, Gen. *Ferula*, 35. 1947.

Plants erect, branched, puberulent in foliage, monocarpic. Leaves ternate-bipinnatisect, the leaflets oblong, ca. 10 cm long, 4 cm broad, rounded at apex, serrate, narrowly decurrent at base; cauline leaves subtended by ovate-triangular sheaths. Inflorescence laxly paniculate, the branches alternate, the terminal peduncle short to lacking, the two lateral peduncles elongate. Involucre and involucel lacking. Rays of terminal umbels up to 13. Umbellets up to 14-flowered. Flowers yellow; petals ovate, plane, the acute incurved apex 2 mm long; calyx obsolete. Fruit (immature) elliptic, thick, striate at the ribs. Vittae 1 in the intervals, 0 on the commissure.

Type locality: "India: montes himalayenses occidentales, in prov. Chamba, Pendshab. Lahol in val. fl. Baga . . . Chamba," ca. 800 m. *Lace*.

Distribution: Northwestern Himalaya in Himachal Pradesh; known only from the type collection. No specimens examined.

5. **Ferula ovina** (Boiss.) Boiss., Fl. Orient. 2: 986. 1872.

Peucedanum ovinum Boiss., Diagn. Pl. Nov. 2: 61. 1845.
Peucedanum thomsonii C. B. Clarke in Hook. f., Fl. Brit. Ind. 2: 711. 1879, non *Ferula thomsonii* C. B. Clarke 1879. Based on *Thomson*, "KASHMIR; alt. 5,000–9,000 ft., KISHTWAR and BANAHAL" (*Hooker & Thomson*, Herb. Ind. Orient., *Peucedanum 5*) (CAL, K).

Ferula stewartiana var. *affghanica* O. Schulz, Notizbl. Bot. Gart. Berlin 11: 877. 1933. Based on *Griffith 2617*, "Afghanistan."

Plants slender to stout, erect, rigid, branched, glabrous to hispidulous or scabrous-prickly throughout, perennial, 0.5–2 m tall. Leaves coriaceous when dry, broadly ovate, 15–20 cm in diameter, pinnately or ternate-pinnately decompound, the ultimate divisions oblong to ovate 1–5(–10) mm long, obtuse, lobed to pinnatifid, hispidulous; petiole short, inflated; cauline leaves few, with inflated, oval-lanceolate sheaths. Inflorescence paniculate, the lower branches alternate, the upper verticillate, the umbels often proliferating, the peduncles variable, from 13 cm long to obsolete. Involucre and involucel lacking or inconspicuous. Rays (3–)6–16, subequal, 2–3 cm long. Umbellets 7–14-flowered, the mature pedicels 5–9 mm long. Flowers yellowish; petals elliptical; calyx obsolete; stylopodium low conical, the styles elongate, divergent. Fruit oval, 7–10 mm long, the dorsal ribs filiform, the lateral narrowly winged. Vittae solitary (–3?) in intervals, 4–8 on commissure. May–July.

Type locality: "Hab. in rupestribus prope ruinas Persepolitanas," Iran, *Kotschy 283*.

Distribution: Kashmir to Pakistan, Afghanistan, and central Asia. Only the type collection of *P. thomsonii* has been examined.

Note. Hamal and Koul (1986) announced the rediscovery and collection of *Peucedanum thomsonii* at Khalen, 5 km E. of Kishtwar Plateau (*Hamal & Koul 176 & 183*, CAL).

62. ZOSIMA* Hoffm.

Gen. Umbell. 1: 145. 1814; C. B. Clarke in Hook. f., Fl. Brit. Ind. 2: 717. 1879.

Pastinaca sect. *Zosimia* (Hoffm.) Calest., Webbia 1: 247. 1905.

Stout, erect, short caulescent and subscapose to elongate, branching, grayish-white-hirsute or hirtellous herbs, from a fibrous collar surmounting a massive branched rhizome or taproot. Leaves petiolate, thickish, simple to 3-pinnate or pinnately decompound, with small oval to linearly oblong, simple, lobed, or trifid ultimate divisions; petiole stout, sheathing; cauline leaves similar to the basal, reduced upward. Inflorescence of lax terminal and lateral compound umbels, the peduncles stout, elongate. Involucre and involucel of usually numerous, linear to lanceolate, scarious, pubescent bracts or bracteoles. Rays few to numerous, stout, unequal, spreading-ascending. Umbellets few to many flowered, the flowers pedicellate. Flowers whitish (or yellowish, or greenish); petals obovate, rounded or emarginate with a narrower inflexed apex, the marginal sometimes radiant; calyx teeth prominent and linear, but caducous, to obsolete; stylopodium low conical with undulant margin, the styles short to slender, reflexed. Pedicels stout, unequal, spreading-ascending or ascending. Fruit oval to obovate, rounded and emarginate at apex, rounded at base, flattened dorsally and complanate, the mericarps flattened dorsally in transection, hirtellous or puberulent, the dorsal and intermediate ribs filiform or obsolete, the lateral broadly winged and the fruit thus surrounded by a continuous entire wing as broad as the body, that is hyaline proximally and tumid and corky distally. Vittae large, solitary and completely occupying intervals, 2 on commissure. Commissural face plane. Carpophore bipartite.

Type species: *Heracleum absinthifolium* Vent. [*Zosima absinthifolia* (Vent.) Link].

Distribution: A small genus of southwestern and central Asia, only one species of which is reported from India.

1. **Zosima absinthifolia** (Vent.) Link, Enum. Hort. Berol. 1: 274. 1821.

Heracleum absinthifolium Vent., Choix de Pl. pl. 7. 1803; *Tordylium absinthifolium* (Vent.) Pers., Syn. Pl. 1: 314. 1805; *Pastinaca absinthifolia* (Vent.) Calest., Webbia 1: 247. 1905.

Zosimia orientalis Hoffm., Gen. Umbell. 1: 148. 1814. Based on the preceding, "Habitat in Oriente, Persia. Occurrit in regionibus caucasicis, circa montem Beschtau et in Iberia, circa Tiflin," no collector indicated.

Plants stout, erect, simple or branched, grayish-hirtellous to tomentose throughout, 10–100 cm tall. Leaves oblong to ovate, 5–35 cm long, 1–15 cm broad, 2- or 3-pinnate or pinnately decompound with small, linearly oblong or lanceolate ultimate divisions; petiole 2–20 mm long; cauline leaves few, reduced upward. Peduncles terminal and lateral, stout, 20–40 cm long. Involucre and involucel of 5–10 linearly lanceolate, scarious, hirsutulous to tomentose bracts or bracteoles, 2–5 mm long. Rays usually 10–35, slender to stout, unequal 3–8(–15) cm long, hirsutulous or hispidulous. Umbellets 8–40-flowered, the mature pedicels stout, unequal, hirsutulous to tomentose, to 10 mm long. Flowers white;

* Also written as *Zosimia* and *Zozimia*.

petals obovate; calyx teeth evident, linear, caducous; stylopodium low conical, with a conspicuously crenulated margin, the styles very slender, reflexed and spreading; ovary white-tomentose. Fruit elliptic to obovate, 8–12 mm long, 5–8 mm broad, hirsutulous to tomentose, pubescent also on commissure. Apr.–July.

Type locality: Described from cultivated plants raised from seed collected between Baghdad and Kermanshah (Shiskin 1951, p. 191).

Distribution: Iraq and Turkey to Iran, Afghanistan, and Pakistan, only reaching extreme western India.

Representative specimens: Jammu & Kashmir: Lachalung, Kiangchu, Rupshu, *Koelz 6651* (MICH, UC). Maharashtra: Dooband, Bombay Presidency, *Dalzell 978* (CAL).

63. **LALLDHWOJIA** Farille
Rev. Gen. Bot. 91: 27. 1984.

Slender, erect, usually sparsely branching, lightly appressed-pubescent perennial herbs from slender rootstocks. Leaves mostly basal, petiolate, membranaceous, 3-foliate or 3-partite with broad, dentate, trilobed leaflets; petiole slender, narrowly sheathing; cauline leaves few, reduced upward. Inflorescence of lax terminal and lateral compound umbels, the peduncles elongate. Involucre and involucel usually lacking, or rarely of 1–3 caducous narrow bracts or bracteoles. Rays few, unequal to subequal, spreading to spreading-ascending. Umbellets few flowered, the flowers pedicellate. Flowers dark red; petals ovate to obovate, scarcely inflexed at apex; calyx teeth minute, pubescent; stylopodium broadly to more narrowly conical, the styles short, spreading. Pedicels slender, spreading to spreading-ascending. Fruit oval to oblong-oval, slightly narrowed or rounded at apex, rounded at base, lightly compressed dorsally and not constricted at commissure, the mericarps compressed dorsally in transection, glabrous, the dorsal and intermediate ribs obsolete, the lateral narrowly to more broadly winged. Vittae 1–4 in intervals, 2–6 on commissure, unequal in length and some not reaching apex and/or base of mericarp. Commissural face plane, sclerenchyma discontinuous or continuous (?) about seed cavity and in wings. Carpophore bipartite.

Type species: *Lalldhwojia staintonii* Farille.

Distribution: A genus of two species, recently described from the high Himalaya of Nepal, Sikkim, and probably Bhutan.

KEY TO SPECIES

1. Plants 20–30 cm tall; lateral leaflets sessile; rays unequal; fruit oval, the lateral wings broad and thin. .... 1. *L. staintonii.*
1. Plants 10–20 cm tall; lateral leaflets petiolulate; rays subequal; fruit oblong-oval, the lateral wings narrow and thick. .... 2. *L. cooperi.*

1. **Lalldhwojia staintonii** Farille, Rev. Gen. Bot. 91: 28. 1984.

Plants slender, flexuous, erect, simple or few branched, purplish, sparingly appressed white-pubescent, 20–30 cm tall. Leaves ovate, 3–5 cm long, 2–4 cm broad, 3-(–5) parted or -foliate, the leaflets ovate to orbicular, 1.5–2 cm long, 2–3 cm broad, rounded at apex and base, dentate and lobed, the margin ciliolate; petiole slender, 10–20 cm long, with a narrow sheath; cauline leaves few, reduced upward, the sheaths scarcely dilated. Peduncles slender, terminal and lateral, 8–15 cm long. Rays 3–6, slender, unequal, 1–5 cm long. Umbellets 2–5-flowered, the mature pedicels slender, unequal, 5–20 mm long. Flowers dark red. Calyx teeth ca. 0.3 mm long. Fruit oval, ca. 5 mm long, 2–3 mm broad, rounded at both apex and base. Aug.–Sept.

Type locality: "Versant nord de Lamjung (massif de l'Annapurna), Nepal," 3,800–4,000 m, 24 Sept. 1981, *M. A. Farille 81-565*.

Distribution: Known only from the region of Annapurna, Nepal.

Representative specimen: Nepal: Annapurna Himalaya, Siti Khola, *Stainton, Sykes & Williams 6581* (BM).

2. Lalldhwojia cooperi Farille, Rev. Gen. Bot. 91: 31. 1984.

Plants slender, flexuous, erect, simple or few branched, sparingly appressed white-pubescent, 10–20 cm tall. Leaves ovate to orbicular, 2–3 cm long, 3–4 cm broad, 3-foliate, the leaflets orbicular, dentate and trilobed, petiolulate, the margin ciliolate; petiole slender, 4–10 cm long, with a narrow sheath; cauline leaves few and reduced upward, the sheaths scarcely dilated. Peduncles slender, terminal and lateral, 8–15 cm long. Rays 2–5, slender, subequal, 1–3 cm long. Umbellets 1–3-flowered, the mature pedicels unequal, 5–20 mm long. Flowers unknown. Calyx teeth pubescent on veins. Fruit oblong-oval, ca. 5 mm long, 1.5 mm broad, narrowed at apex, rounded at base.

Type locality: "'Bharang,' Sikkim (on border of Bhutan)," 1 Sept. 1913, *R. E. Cooper 772*.

Distribution: Known only from the type collection. No specimen seen.

64. HERACLEUM Linn.

Sp. Pl. 249. 1753; C. B. Clarke in Hook. f., Fl. Brit. Ind. 2: 711. 1879.

Pastinaca sect. *Heracleum* Calest., Webbia 1: 246. 1905. *Sphondylium* Miller, Gard. Dict. Abr. ed. 4. 1754.

Heracleum sect. *Tetrataenium* DC., Prodr. 4: 191. 1830; *Tetrataenium* (DC.) Mandenova, Trudy Tbilissi Bot. Inst. 20: 16. 1959; Notulae Syst. Geogr. Inst. Bot. Tbilissi 41: 43–49. 1986.

Stout or slender, erect, branching, mostly pubescent perennial herbs from taproots or fascicled fibrous roots. Leaves petiolate, usually membranaceous, ternately or pinnately compound or merely lobed to pinnatifid, the leaflets broad, crenate or serrate (to variously cleft or lobed; petiole sheathing, the sheath often conspicuously inflated; cauline leaves reduced upward. Inflorescence of lax terminal and lateral compound umbels, the peduncles elongate. Involucre of narrow, usually caducous bracts, or lacking. Rays few to numerous, unequal, spreading-ascending. Umbellets rather numerous flowered, the flowers pedicellate. Involucel usually of several narrow, persistent bracteoles, or these less commonly caducous or lacking. Flowers white or less frequently yellow, sometimes reddish-tinged; petals lanceolate to obcordate, with a narrower inflexed apex, the outer radiant, enlarged, emarginate to bilobed, or all petals subequal and none radiant; calyx teeth evident, linear to lanceolate, or calyx obsolete; stylopodium conical to low conical, the styles short to slender and elongated; ovary usually pubescent. Pedicels rather slender, usually unequal. Fruit oblong-oval to obovate, usually rounded or emarginate at apex, rounded or narrowed at base, strongly compressed dorsally, the mericarps flattened dorsally in transection, pubescent to glabrate, the dorsal ribs filiform, the lateral thin winged and nerved near outer margin. Vittae prominent, visible from dorsal surface, linear to subclavate, commonly extending only part way from stylopodium to base of mericarp, 1 or 2(–3) in intervals, usually 2 or 4 (6–8?) on commissure. Commissural face plane, strengthening cells continuous about seed cavity and under ribs. Carpophore bipartite.

Type species: *Heracleum sphondylium* Linn.

Distribution: A large predominantly Eurasian genus (1 species in North America, 2 or more in East Africa), of perhaps 60–75 species, as here defined. Indian species, several of which are poorly distinguished or problematical, are confined to one of three areas: (1) Himalaya, (2) northeastern India (Assam region), or (3) peninsular India.

Note. We follow Mandenova (1959) in excluding *Tordyliopsis* DC. and *Semenovia* Regel & Herder (*Platytaenia* Nevski & Vived.) from *Heracleum*, and she concurs with our removal of *H. pedatum* Wight and *H. concanense* Dalzell (personal communication).

She goes further, however, in concluding that the remaining taxa we have here included in *Heracleum* should probably all be referred to *Tetrataenium* (DC.) Mandenova because of their possession of well-developed calyx teeth, better developed and narrowly carinate dorsal and intermediate fruit ribs, and elongate, truly claviform and septate dorsal vittae (Mandenova *et al.*, 1977). "En prenant compte des données obtenues je crois pouvoir dire que le genre *Heracleum* (proper) n'est pas présenté á Himalaya et dans les monts d'Indoustan, de Ceylan et de l'Indo-Chine, mais agrees s'il vous plait, mes assurances sinceres que je n'auvrai absolument nien contre si dans 'Conspectus' le genre *Heracleum* sera pris sensu lato" (personal correspondence, Dec. 8, 1982).

She may very well be correct in her conclusions; palynological and biochemical evidence appear to lend them some support (Mandenova *et al.*, 1977). Because we have difficulty in applying some of the criteria mentioned to the imperfect material available to us, and because Dr. Mandenova's work is continuing—we think it unwise to attempt to anticipate the results of her studies.

We therefore fall back on the disclaimer that: "... the genus *Tetrataenium* is not very clearly defined and investigation of much more numerous populations is needed; the existence of subgroups revealed by this pluridisciplinary research and apparently not correlated with the area disjunction of the genus (the Himalayas and southern India, with a gap in North India), must be understood before any satisfactory classification is put forward" (Mandenova *et al.*, 1977, p. 675).

KEY TO SPECIES

1. Plants of dry areas with rough, hispidulous pubescence; leaves oblong, simply pinnate with 5 or 7 oblong-oval to orbicular, dentate leaflets 0.5–5 cm in diameter. .... 1. *H. pinnatum.*

1. Plants mostly mesophytic with soft hirsute, hirtellous villous, or tomentose pubescence to glabrate; leaves oblong to broadly ovate or orbicular, but if oblong, the leaflets larger or of different shape.
    2. Fruit wings definitely broader than body; E. Himalaya and NE. India.
        3. Plants slender; stylopodium depressed. .... 6. *H. wallichii.*
        3. Plants stout; stylopodium conical or low conical.
            4. Cauline leaf sheaths inconspicuous; bracteoles linear or 0; vittae 1 in intervals, 3 or 4 on commissure; Nepal. .... 7. *H. nepalense.*
            4. Cauline leaf sheaths moderately dilated; bracteoles ovate-lanceolate; vittae 1–3 in intervals, 2 on commissure; Assam region. .. 11. *H. burmanicum.*
    2. Fruit wings narrower than to equal width of body.
        5. Leaflets linearly lanceolate to narrowly oblong, less than 20 mm broad.
            6. Leaf margins spinulose-serrate, not cartilaginous; calyx teeth prominent, ca. 1 mm long; stylopodium conical; fruit oval to orbicular. .... 4. *H. sublineare.*
            6. Leaf margins subentire, cartilaginous; calyx obsolete; stylopodium low conical; fruit oblong-oval. .... 5. *H. cachemiricum.*
        5. Leaflets oblong to orbicular, 20 mm or more broad (if narrower, an involucre lacking and the few flowers all fertile).
            7. Plants weak and reclining; calyx teeth prominent, to 1 mm long. .... 3. *H. canescens.*
            7. Plants stout or slender, low to elongated, but neither weak nor reclining; calyx teeth merely evident to obsolete.
                8. Leaves merely lobed, ternately or pinnately parted, to pinnatifid, without distinct leaflets, or occasionally pinnate (see no. 9).
                    9. Plants elongated, 30–100 cm tall; basal leaves neither crowded nor with conspicuously overlapping sheaths; Himalaya.

10. Plants slender; leaves and leaflets ovate; rays 7–12; pedicels 10–15 mm long; flowers white; fruit small, orbicular. .... 2. *H. jacquemontii.*

10. Plants stout; leaves and leaflets orbicular; rays 10–16; pedicels 5–12 mm long; flowers yellowish-white; fruit large, obovate. .... 9. *H. obtusifolium.*

9. Plants low, 25–50 cm tall; basal leaves congested, with conspicuously overlapping sheaths; Nilgiri Hills, peninsular India. .... 13. *H. hookerianum.*

8. Leaves pinnately, ternately, or ternate-pinnately compound, with distinct leaflets at least in radical leaves.

11. Flowers white, the outer normally conspicuously radiant; Himalaya and NE India.

12. Cauline leaf sheaths inconspicuous; fruit small, ca 5 mm long. .... 2. *H. jacquemontii.*

12. Cauline leaf sheaths at least moderately dilated; fruit 7–10 mm long.

13. Plants stout; leaves membranaceous; commissural vittae 0–4.

14. Involucre inconspicuous, caducous, or lacking; fruit wings narrower than body.

15. Leaves narrowly oblong, ca. 17 cm long; fruit (immature) oblong, ca. 7 mm long; commissural vittae 4, subclavate. .... 10. *H. lallii.*

15. Leaves oblong to ovate, 10–40 cm long; fruit oval to orbicular, 8–10 mm long; commissural vittae 2, linear, or 0. .... 8. *H. candicans.*

14. Involucre of 5–8 ovate-lanceolate bracts 5–15 mm long; fruit wings equal to body. .... 11. *H. burmanicum.*

13. Plants slender; leaves lightly coriaceous; commissural vittae 6; Assam region. .... 12. *H. biternatum.*

11. Flowers yellow, yellowish-white, or greenish-white, only sporadically or not at all radiant; peninsular India.

16. Basal leaves simple, 3–5-lobed or pinnatifid, the petioles crowded and strongly overlapping. .... 13. *H. hookerianum.*

16. Basal leaves pinnately, ternately, or ternate-pinnately compound with distinct leaflets, the petioles scarcely overlapping.

17. Plants glabrous or nearly so, the leaves often shiny.

18. Leaves 1–2-ternate, -ternate-pinnate, or simply pinnate; rays mostly 6–10; fruit oval to orbicular, broadly winged; vittae 2 or 3 in intervals, 2–4 on commissure. .... 16. *H. aquilegifolium.*

18. Leaves bipinnate; rays 10–20; fruit obovate, narrowly winged; vittae 1 or 2 in intervals, usually 6–8 on commissure. .... 17. *H. grande.*

17. Plants grayish-hirsute or hirtellous to tomentose at least on lower leaf surface, rarely glabrate, not shiny.

19. Involucre evident, of several linearly-lanceolate to obovate bracts.

20. Leaflets rounded at apex; rays 10–30. . . . 14. *H. rigens.*

20. Leaflets acute or acuminate; rays 4–12.

21. Ovaries and fruit softly pubescent; calyx teeth evident. . . . . 18. *H. dalgadianum.*

21. Ovaries and fruit glabrous; calyx teeth obsolete. . . . . 19. *H. candolleanum.*

19. Involucre inconspicuous or lacking.

22. Cauline leaves numerous; umbellets 15–20-flowered; vittae subclavate. . . . . 15. *H. sprengelianum.*

22. Cauline leaves few; umbellets 10–15-flowered; vittae linear. . . . . 20. *H. ceylanicum.*

1. **Heracleum pinnatum** C. B. Clarke in Hook. f., Fl. Brit. Ind. 2: 712. 1879.

Tetraenium pinnatum (C. B. Clarke) Mandenova, Notulae Syst. Geogr. Inst. Bot. Tbilissi, 42: 12. 1991.

Plants rather slender, erect, few branched, hispidulous throughout, 30–100 cm tall from a stout root. Leaves oblong, 9–25 cm long, 4.5–15 cm broad, pinnate, the 5–7 leaflets oblong-oval to orbicular, 15–50 cm long, 0.5–4 cm broad, mostly sessile, mucronulate-dentate and the terminal and larger leaflets frequently lobed, hispidulous; petiole slender, 3–14 cm long, rather narrowly sheathing; cauline leaves few, greatly reduced upward, with moderately dilated oval sheaths. Peduncles slender, terminal and lateral, 10–30 cm long. Involucre of 1 to several linear or oblong pubescent bracts to 5 mm long, or lacking. Rays 10–30, slender, unequal, 3–10 cm long, hispidulous. Umbellets 10–30-flowered, the mature pedicels slender, unequal, hispidulous, spreading-ascending, 3–10 mm long. Involucel of 5–8 filiform to linearly oblong, hispidulous bracteoles 2–4 mm long. Flowers white; petals oval to obovate, hispidulous, the outer radiant; calyx teeth minute; stylopodium conical, a little shorter than the recurved styles; ovary hispidulous. Fruit obovate, 7–10 mm long, 5–7 mm broad, sparsely hispidulous to glabrate, the dorsal ribs evident, filiform, excurrent, the lateral broadly winged, the wings 1/2–3/4 width of body. Vittae large, solitary in intervals, extending 2/3–3/4 length of body, 2 of ± equal length on commissure. July–Sept.

Type locality: "KASHMIR northwards to the Karakorum, alt. 10,000–16,000 ft.," *Thomson*, *Clarke* [Hab. Tibet Occident., *Thomson* (*Hooker & Thomson*, Herb. Ind. Orient., *Heracleum 14*) (K, holotype; CAL, NY, UC, W, isotype)] (Nasir 1972).

Distribution: Western Himalaya to Baltistan and Tibet, 3,000–5,300 m.

Representative specimens: Jammu & Kashmir: Nubra, *Thomson* (K); Ladak, *Thomson* (K); Pangong, *Thomson* (K); Karakorum, *Clarke 30,124A, B, 31,039A, 30,371, 30,388a* (K); Gudhai Valley, Astor Distr., *Duthie* (CAL, DD): Indus Valley near Leh, *Stoliczka* (CAL).

2. **Heracleum jacquemontii** C. B. Clarke in Hook. f., Fl. Brit. Ind. 2: 712. 1879.

Plants slender, erect, few branched, villous with lax white hairs, 30–70 cm tall. Leaves ovate, 5–8 cm long, tripartite or pinnate, the three leaflets ovate, 1–3 cm long, ca. 2 cm broad, acute or subobtuse, sessile, serrate or lobed; petiole slender, 3–15 cm long, narrowly sheathing; cauline leaves few, reduced upward, the sheaths inconspicuous. Peduncles slender, terminal and lateral. Involucre lacking or inconspicuous. Rays 7–12, slender, unequal, pubescent, 1.5–5 cm long. Umbellets 10–12-flowered, the mature pedicels slender, 10–15 mm long. Involucel of 4–8 linear bracteoles, 5–8 mm long. Flowers white, the marginal petals radiant, emarginate; calyx obsolete; stylopodium low conic, shorter than styles; ovary pubescent. Fruit orbicular, 2.5–3 mm in diameter, minutely pubescent, the dorsal ribs filiform, not excurrent, the lateral winged with a narrow, continuous wing much narrower than body. Vittae 1 or 2 in intervals, extending 2/3 length of body, 4 on commissure.

Type locality: "NORTH-WEST HIMALAYA," *Jacquemont 421* (K, P).

Distribution: Western Himalaya.

Representative specimen: Only the type collection examined.

3. **Heracleum canescens** Lindley in Royle, Ill. Bot. Himal. Mount. 232. 1835; C. B. Clarke in Hook. f., Fl. Brit. Ind. 2: 713. 1879.

Heracleum hirsutum Edgew., Trans. Linn. Soc. 20: 57. 1851. Based on *Edgeworth 163*, "Hab. Himala, ad alt. Ped. 6000–7000 . . . Simla, etc."

Plants slender, weak, erect or ascending, branched, 30–100 cm tall, thinly (densely on juvenile parts) white-hirsute or hirtellous. Leaves oblong to oval, 5–20 cm long, 5–10 cm broad, pinnate, the 5–7 leaflets oblong to ovate, 1.5–8 cm in diameter, usually shortly petiolulate, acute or obtuse, crenate-serrate, the larger trilobed, hirsute beneath, hirsutulous above; petiole slender, 4–12(-20) cm long, shortly and rather narrowly sheathing; cauline leaves reduced upward, with inconspicuous sheaths. Peduncles slender, terminal and lateral, 10–35 cm long. Involucre of 1–5 minute, caducous, hirsute bracts up to 3 mm long, or lacking. Rays 6–15, slender, unequal, 1.5–7 cm long, hirsute. Umbellets 10–15-flowered, the mature pedicels slender, unequal, spreading-ascending, 4–8 mm long, hirsute. Involucel of 4–8 linear or lanceolate, hirsute bracteoles 2.5–3 mm long. Flowers white, the petals oval to obovate, the outer strongly radiant, unequally emarginate; calyx prominent, the teeth linear, ca. 1 mm long; stylopodium low conical, shorter than the recurved styles; ovary hirsutulous. Fruit oval to obovate, 5–8(-12) mm long, 4–7 mm broad, sparsely hirsute, the dorsal ribs filiform, the lateral broadly thin winged, the wing equal to or slightly narrower than body. Vittae large, linear, solitary in intervals, extending 2/3 length of body, 2 of ± equal length on commissure. July –Oct.

Type locality: "Hab. Mussooree," "N. W. India," *Royle* (K). (Nasir 1972).

Distribution: Western Himalaya, to Pakistan and Tibet (?), 2,000–3,000 m.

Representative specimens: Himachal Pradesh: Choor, *Royle 93/24* (DD); Simla, *Raizada 7213* (DD), *19,524* (DD), *Schlich* (CAL). Uttar Pradesh: Chakrata, *Mukherjee* (CUH); Mussoorie, *Raizada 197* (DD), *5531* (DD).

4. Heracleum sublineare C. B. Clarke in Hook. f., Fl. Brit. Ind. 2: 713. 1879.

Tetrataenium sublineare (C. B. Clarke) Mandenova ex Farille, Cauwet-Marc & Malla, Candollea 40(2): 562. 1985; Mandenova, Notulae Syst. Geogr. Inst. Bot. Tbilissi, 41: 47. 1986.

Plants slender, erect, branched, hirsute to hispid or hirsutulous, 30–105 cm tall. Leaves ovate, ca. 15 cm in diameter, bipinnate, the leaflets lanceolate, 2.5–10 cm long, 5–15 mm broad, subsessile, acute, spinulose-serrate, hispid and hirsute to glabrate beneath; petiole slender, to 5 cm long, narrowly sheathing; cauline leaves simply pinnate with narrower leaflets and slightly dilated, lanceolate sheaths. Peduncles slender, mostly terminal, 50–100 cm long. Involucre of one to several filiform, hirsute bracts 5–8 mm long, or lacking. Rays 4–8, slender, unequal, 1–3.5 mm long, hirsute. Umbellets 6–16-flowered, the mature pedicels slender, spreading-ascending, ca. 5 mm long. Involucel of 2–6 linear, hirsute bracteoles 3–5 mm long. Flowers white (or reddish-tinged), the petals obovate, unequally emarginate, the marginal (often ?) strongly radiant and almost bifid; calyx prominent, the teeth linear, ca. 1 mm long; stylopodium conical, shorter than the slender, spreading styles; ovary hirtellous. Fruit oval to orbicular, 5–6 mm long, ca. 4 mm broad, glabrous or glabrescent, the dorsal ribs filiform, the lateral broadly thin winged, the wing narrower than body. Vittae linear, solitary in intervals, extending nearly to base, the 2 commissural about 3/4 as long as body and close to center, sometimes accompanied by abbreviated additional ones. July–Oct.

Type locality: "?Sikkim, alt. 10,000–13,000 ft.," *Hooker* (*Hooker & Thomson*, Herb. Ind. Orient., *Pastinaca 4*).

Distribution: Eastern Himalaya, from Bhutan to Nepal and Tibet, 3,500–4,800 m.

Representative specimens: Sikkim: Lachen, *Hooker* (K); Lachung, *Hooker* (K); w/o precise locality, *Hooker* (CAL); Yeumthong, *Gammie* (CAL); Gnatong, *Smith 4358* (CAL). Bhutan: Tarola, above Ha, *Gould 1181* (DD).

5. Heracleum cachemiricum C. B. Clarke in Hook. f., Fl. Brit. Ind. 2: 712. 1879.

Plants slender, erect, few branched, hirtellous or hispidulous throughout, 30–100 cm tall, from a slender taproot. Leaves oblong to triangular-ovate, 1–3 cm long, 4–10 cm broad, bipinnate or ternate-pinnate, the leaflets linearly lanceolate to oblong, 1.5–9.5 cm long, 3–12 mm broad, sessile, entire or remotely dentate, cartilagineous-margined; petiole slender, 8–35 cm long, very narrowly sheathing; cauline leaves few, greatly reduced upward, the sheaths narrow and inconspicuous. Peduncles slender, mostly terminal, 5–15 cm long. Involucre of 1-several narrow, scarious, dark-veined, hirsutulous bracts, or lacking. Rays 5–12, slender, unequal, 2–4 cm long, hirsutulous. Umbellets 10–25-flowered, the mature pedicels slender, unequal, spreading-ascending, 3–8 mm long. Involucel of 5–8 linear to lanceolate bracteoles like the bracts, 2–6 mm long. Flowers white, the petals with purple veins; petals obovate, unequally emarginate, the marginal radiant; calyx obsolete; stylopodium low conical, shorter than the spreading styles; ovary hirsutulous. Fruit oblong-oval, 5–10 mm long, 4–6 mm broad, hirsutulous to glabrescent, the dorsal ribs filiform, the lateral narrowly thin winged, the wing narrower than body. Vittae large, linear, solitary in intervals, extending nearly to base (dorsals longer than intermediates), the 2 commissural subequal, central. Apr.–July.

Type locality: "MURREE," *Falconer*; "JAMU, alt. 5,000–6,000 ft.," *Thomson*; "KASHMIR; Nowgunge, alt. 8,000 ft.," *Clarke*. ["Murree," *Falconer 295/H* (K, lectotype) (Nasir 1972)].

Distribution: Western Himalaya, Kashmir to Pakistan, 1,600–2,600 m.

Representative specimens: Jammu & Kashmir: Jammu Hills, *Thomson* (K); Murree, *Thomson* (K); Nowgunge, *Clarke 29,169* (K); near Pajidar, *Duthie 14,571* (CAL). Himachal Pradesh: Bre Forest, Chamba, *Lace 1531* (DD); Khajar, *Rau 2587* (BSD). Uttar Pradesh: Mussoorie, *Royle 93/78* (DD); Lambatach, *Gamble 24,931* (CAL, DD).

Note. We discovered tardily that Alava (Flora Iranica 161: 491, 162. pl. 453) transferred this species to *Tricholaser* Gilli, as *T. cachemiricum* (C. B. Clarke) Alava, on the basis of the occurrence of both perfect and staminate flowers in the same umbellet, the slightly tumescent marginal fruit wings, and subequal dorsal vittae extending nearly to the fruit base. He makes a persuasive case.

6. **Heracleum wallichii** DC. Prodr. 4: 195. 1830; C. B. Clarke in Hook. f., Fl. Brit. Ind. 2: 712. 1879.

Heracleum diversifolium Wallich, Cat. 574. 1829, *nomen nudum*.
Tetrataenium wallichii (DC.) Mandenova, Notulae Syst. Geogr. Inst. Bot. Tbilissi 41: 47. 1986.
Heracleum wallichii var. *elatior* C. B. Clarke in Hook. f., Fl. Brit. Ind. 2: 713. 1879. Based on *Wallich* (?), "Nipal."

Plants slender, erect, branched, hirsute to hirtellous, 50–150 cm tall. Leaves ovate-triangular, 30–50 cm in diameter, biternate, the leaflets ovate, 5–8 cm long, 1.5–2 cm broad, the first divisions petiolulate, the second mostly sessile, serrate and the larger trilobed, hirsutulous on both surfaces; petiole slender, up to 50 cm long, narrowly sheathing; cauline leaves tripartite or pinnate, with narrower divisions and narrower sheaths. Peduncles slender, mostly terminal, 50–80 cm long. Involucre of 1 or 2 long, narrow bracts, or lacking. Rays 6–16, slender, unequal to subequal, hirsutulous to glabrescent, 2.5–7 cm long. Umbellets 8–20-flowered, the mature pedicels slender, hirsutulous, spreading-ascending, 8–12 mm long. Involucel of several linear bracteoles 5–10 mm long, or lacking. Flowers white; petals emarginate, the outer strongly radiant; calyx teeth prominent, linear; stylopodium depressed, the styles diverging; ovary hairy. Fruit obovate, 8–12 mm long, 7–11 mm broad, glabrous or glabrescent, the dorsal ribs filiform, the lateral broadly winged, the wing about twice width of the body. Vittae subclavate, solitary in intervals, about half as long as body, the commissure with 2 ± abbreviated ones, or lacking. Aug.–Sept.

Type locality: "Mont. alto Nepalia," *Wallich 574*.

Distribution: Eastern Himalaya, Nepal to Sikkim, 3,000–5,300 m.

Representative specimens: Nepal: w/o precise locality, *Wallich 574A* (CAL); Tamur Valley, Mewa Khola Topke Gola, *Stainton 1319* (AAH, BM). West Bengal: Tonglo, *Thomson* (CAL). Sikkim: Jongri, *Anderson 629* (CAL), *Mukherjee* (CUH).

Note. I have seen no specimens of var. *elatior*, which Clarke distinguishes by the plants being more hairy, having lobed rather than divided upper cauline leaves, more numerous bracts and rays, and more prominent bracteoles, but I suspect that it may belong here. (P.K. M.)

7. **Heracleum nepalense** D. Don, Prodr. Fl. Nepal, 185. 1825; C. B. Clarke in Hook. f., Fl. Brit. Ind. 2: 714. 1879.

Tetrataenium nepalense (D. Don) Mandenova, Trudy Tbilissi Bot. Inst. 20: 17. 1959. *Heracleum nepalense* var. *bivittatum* C. B. Clarke in Hook. f., Fl. Brit. Ind. 2: 714. 1879, non *H. bivittatum* Boiss. 1903. Based on "Sikkim and Bhotan, common," no collector indicated.

Plants stout, shrubby, erect, branched, sparsely hirsute and hirtellous, 0.7–2.3 m tall. Leaves oblong-ovate, the lower to 80 cm long, 5 cm broad, pinnate or ternate-pinnate, the leaflets ovate, 4–15 cm long, 2–10 cm broad, obtuse or somewhat acute, coarsely serrate, the larger lobed to pinnatifid (?), sparsely hirsute beneath; petiole 10–15 cm long; cauline leaves reduced upward, the upper acutely trilobed to -divided, the sheaths narrow and inconspicuous. Peduncles stout, terminal and lateral (lateral umbels often staminate, or largely so), 10–15 cm long. Involucre of 3–5 linear, hirsutulous, caducous bracts up to 12 mm long, or lacking. Rays 8–20, slender, unequal, 1.5–7.5 cm long, hirsutulous. Umbellets 8–20-flowered (many flowers sterile), the mature pedicels filiform, unequal, hirsutulous, 8–20(–25) mm long. Involucel of several linear bracteoles like the bracts, 5–12 mm long, or lacking. Flowers white; petals obovate, the outer radiant (or not ?); calyx evident, the teeth linear, ca 0.5 mm long; stylopodium low conical, a little shorter than styles; ovary hirsutulous. Fruit orbicular to obovate, 6–11 mm in diameter, emarginate, rounded at base, glabrous, the dorsal ribs evident, filiform, the lateral broadly thin winged, the wing broader than body. Vittae large, subclavate (?), solitary in intervals, extending 1/2–2/3 length of body (dorsals the longer), 2–4 on commissure, the outer much abbreviated or lacking, the dorsal about 2/3 length of body. July–Sept.

Type locality: "Nepalia," *Wallich 576* (CAL, K).

Distribution: Eastern Himalaya, West Bengal to Nepal, 2,400–4,000 m.

Representative specimens: Nepal: Lahare Powa, Dhaibong, *Mukherjee 231* (CUH). West Bengal: Bhootan, *Griffith 262/1* (CAL); Darjeeling, *Gamble 3916/83* (CAL), *Mukherjee 3709* (CUH, UC); Gairibas, *Mukherjee 290* (CUH); Sinchal, *Anderson 626* (CAL), *Cave 886* (CAL). Sikkim: w/o precise locality, *Hooker* (CAL, GH, K), *Kurz* (CAL); Lachung, *Hooker* (K), *Ribu & Rhomoo 4741* (CAL).

8. **Heracleum candicans** DC., Prodr. 4: 192. 1830; C. B. Clarke in Hook. f., Fl. Brit. Ind. 2: 714. 1879.

Heracleum candicans Wallich, Cat. 573. 1829, *nomen nudum*; *Tetrataenium candicans* (DC.) Mandenova, Notulae Syst. Geogr. Inst. Bot. Tbilissi 41: 44. 1986.

Plants stout, erect, branched, sparsely hirsute to densely hirtellous or white-tomentose, 80–200 cm tall, from a massive taproot. Leaves oblong to ovate, 10–40 cm long, 10–25 cm broad, pinnate or ternate-pinnate, the 5–9 leaflets oblong to ovate, 5–10(–20?) cm long, 2–10 cm broad, the terminal usually confluent and decurrent, mostly obtuse, truncate or cuneate at base, doubly serrate or dentate, the larger often lobed to subpinnatifid, sparsely hirsutulous to glabrate above, densely gray-hirsutulous to white-tomentose beneath; petiole stout, 5–35 cm long, the sheath short and broadly dilated; cauline leaves reduced upward, with fewer and narrower divisions, and a dilated oval to orbicular sheath. Peduncles stout, terminal and lateral (lateral umbels frequently staminate, or largely so), hirsute above, 8–25 cm long. Involucre of 1–3 linear bracts 2–10 mm long, or lacking. Rays 10–40, slender, unequal, 2–10 cm long, hirsute or hirsutulous. Umbellets 10–35-flowered, many central flowers sterile, the mature pedicels filiform, unequal, spreading-ascending, hirsutulous, 5–10 mm long. Involucel of 3–8 linear to lanceolate, hirsutulous bracteoles 2–10 mm long. Flowers white; petals oblanceolate to obovate, unequally emarginate, the outer radiant; calyx evident, the teeth linear, to 0.5 mm long; stylopodium conical, shorter than the spreading styles; ovary hirsutulous. Fruit oval to orbicular (pyriform), 8–10 mm long, 5–7 mm broad, emarginate, somewhat tapering at base, glabrous, the dorsal ribs filiform, the lateral broadly thin winged, the wing narrower than body. Vittae linear, solitary in intervals, extending 2/3 length of body, 2 on commissure, extending 2/3 length of body (rarely lacking ?). June–Sept.

Type locality: "in India septentr. ad Deyradoon et Kumaon," Uttar Pradesh, *Wallich 573* (CAL, K).

Distribution: Himalaya, from Bhutan to Kashmir, Tibet, and Pakistan, 2,000–4,000 m.

Representative specimens: Jammu & Kashmir: Banihal Mountains, *T.A. Rao 7599* (BSD); Kashmir, *Duthie 13,263* (CAL); Sonamarg, *Duthie 13,604* (DD). Himachal Pradesh: Choor, *Royle* (CAL); Dalhousie, *Clarke 23,233c* (CAL); Kardang, *Mukherjee 330* (CUH, UC); Lahul, *Bor 14,679* (DD). Uttar Pradesh: Alli, Chamoli, *N. P. Singh 14,846* (BSD); Bashahr, *Lace 1160* (CAL, UC); Mussoorie, *Pant 39,786* (BSD).

9. **Heracleum obtusifolium** DC., Prodr. 4: 191. 1830; C. B. Clarke in Hook. f., Fl. Brit. Ind. 2: 714. 1879.

Tetraenium obtusifolium (DC.) Mandenova, Notulae Syst. Geogr. Inst. Bot. Tbilissi 42: 13. 1991

Plants stout, erect, branched, hirsute or hirtellous to white-tomentose, 30–100 cm tall. Leaves oblong, cordate, 60–90 cm long, 20–30 cm broad, pinnately parted or pinnatifid to simply pinnate, the 3 or 5 leaflets ovate-orbicular, 5–8 cm long, 4–6 cm broad, crenate-dentate, the larger often lobed, sparsely hirsutulous above, white-tomentose beneath; petiole 10–15 cm long, shortly but broadly sheathing; cauline leaves reduced upward, trilobed or -parted, the sheaths relatively inconspicuous. Peduncles rather stout, terminal and lateral, 10–20 cm long. Involucre of 2–5 narrow, leafy, caducous bracts 10–25 mm long. Rays 10–20, slender, unequal, 2.5–7.5 cm long. Umbellets 10–20-flowered, the mature pedicels 5–12 mm long. Involucel of 4–8 narrow bracteoles 5–12 mm long. Flowers yellowish-white; petals obovate, the outer radiant or not; calyx teeth evident, linear; stylopodium low conic, the styles short; ovary pubescent. Fruit obovate, 8–12 mm long, 4–6 mm broad, the dorsal ribs slender, the lateral winged, the wing 1/3 the body. Vittae linear, solitary in intervals, extending 3/4 length of body, 4 on commissure.

Type locality: "in Nepalis," *Wallich 577* (CAL, K).

Distribution: Eastern Himalaya, Bhutan to Nepal, ca. 2,400 m.

Representative specimens: Nepal: Larjung, Kali Gandaki Valley, *Stainton, Sykes & Williams 5553* (AAH, BM). Sikkim: w/o precise locality, *Cave 270* (CAL); Lachen, *Hooker* (K); Teesta Valley, *Younghusband* (CAL); Thangu, *Lepcha collector 2846* (CAL); Lachoong, *Mukherjee 623* (CUH). Bhutan: w/o precise locality, *Griffith 262/1* (CAL).

Note. Cannon (1979) comments: "Probably only a minor variant of *H. candicans*, but further investigation is needed to establish its status."

10. **Heracleum lallii** Norman, Jour. Bot. 67: 247. 1929.

Plants stout, erect, sparsely branched, pubescent throughout. Leaves narrowly oblong, to 17 cm long, pinnate, the 7 leaflets ovate, up to 6 cm long, 4 cm broad, acute to obtuse, rounded or cordate, to sometimes cuneate at base, irregularly crenate-serrate, often incised or shallowly lobed; petiole ca. 33 cm long; cauline leaves similar to the basal, sessile on the sheath. Involucre of a few (caducous ?) narrow bracts. Rays ca. 13, slender, unequal, 4–8 cm long. Mature pedicels 3–6 mm long. Involucel of inconspicuous linear bracteoles. Petals narrowly oblong, the outer radiant and broadly obcordate; calyx teeth minute, triangular; stylopodium conical, shorter than the slender styles. Fruit (immature) oblong, 7 mm long, 4 mm broad, the dorsal ribs filiform, the lateral narrowly winged, the wing narrower than body. Dorsal vittae extending to base, the intermediate paired, shorter, 4 on commissure, subclavate and not extending to base.

Type locality: "Hab. Booshki, 13,000–14,500 ft.," Nepal. *Dhwoj 30* (E, holotype).

Distribution: Himalaya of Nepal, 3,000–4,200 m.

Representative specimen: Nepal: Booshki, *Dhwoj 132* (BM).

Note. Cannon (1979) remarks, "apparently confined to Nepal and possibly only a small variant of *H. candicans*." The holotype, however, is at E, not at BM.

11. **Heracleum burmanicum*** Kurz, Jour. Asiat. Soc. Bengal 2: 309. 1872; C. B. Clarke in Hook. f., Fl. Brit. Ind. 2: 714. 1879.

*Tetrataenium burmanicum** (Kurz) Mandenova, Notulae Syst. Geogr. Inst. Bot. Tbilissi 41: 46. 1986.

Plants rather stout, erect, branched, sparsely hirsute and hirtellous to tomentose, 35–120 cm tall. Leaves triangular, 10–20 cm in diameter, ternate-pinnate or 1–2-pinnate, the leaflets ovate-lanceolate to ovate, 3–8 cm long, 2–6 cm broad, mostly petiolulate, acute or obtuse, cordate to cuneate at base, serrate and the larger lobed, hirsute to tomentose beneath, hirsutulous above. Petiole short, 1–2 cm long, sheathing; cauline leaves reduced upward, pinnate or trifid to 3-parted, the sheaths short and moderately dilated. Peduncles terminal and lateral, 3–10 cm long. Involucre of 5–8 ovate-lanceolate bracts 5–15 mm long. Rays 6–20, slender, unequal, 4–9 cm long, hirsutulous. Umbellets 10–15-flowered, the mature pedicels to 5 mm long. Involucel of 5–10 bracteoles like the bracts, 5–8 mm long. Flowers white; petals obovate, emarginate, the outer radiant; calyx teeth linear, 1–2 mm long; stylopodium conical, the styles very short; ovary hirsutulous. Fruit oval to obovate, 8–10 mm long, 4–7 mm broad, the dorsal ribs evident, acute, the lateral broadly winged, the wing as broad as or broader than body. Vittae 1–3 in intervals, extending 1/2–3/4 length of body, 2 on commissure 1/2 as long as body, often with additional fragmentary vittae. July–Aug.

Type locality: Pegu, Yomah, Burma, 830–1,000 m, *Kurz 1940*.

Distribution: Khasia and Naga Hills of northeastern India, 830–1,500 m; Burma.

* Also written "*barmanicum*" and "*birmanicum*."

Representative specimens: Nagaland: Kohima, Naga Hills, *Koelz 26,219* (MICH, UC); Kohima, *Clarke 40,932A* (CAL). Meghalaya: Shillong, *Clarke 38,809* (CAL); Khasia, *Hooker & Thomson* (Herb. Ind. Orient., *Heracleum 5*) (GH, NY).

12. **Heracleum biternatum** W. Smith, Rec. Bot. Surv. India 6(2): 37. 1913.

Plants slender, erect, sparsely branched, scantily hirsute and hirtellous throughout, 60–120 cm tall. Leaves lightly coriaceous, oval to triangular-ovate, 1.5–2.5 cm long, 1–2 cm broad, 1–2-ternate or ternate-pinnate, the leaflets oblong to ovate-lanceolate or broadly ovate, 3–12 cm long, 2–8 cm broad, mostly petiolulate, acute or obtuse, truncate to cuneate at base, finely serrate to coarsely crenate-dentate and the larger lobed, hirsute beneath on veins, hirsutulous above; petiole slender, 10–15 cm long, shortly and broadly sheathing; cauline leaves reduced upward, the upper ternate or trilobed, the sheaths shortly and moderately dilated. Peduncles slender, terminal and lateral, 8–35 cm long, hirsute at summit. Involucre of 2–8 linear- to ovate-lanceolate, hirsutulous, caducous bracts 5–8 mm long, or lacking. Rays 7–12, slender, unequal, 2–8 cm long, hirsutulous. Umbellets 15–25-flowered, the mature pedicels slender, unequal, spreading-ascending, 5–10(–30??) mm long. Involucel of 3–8 bracteoles like the bracts, 2–6(–20 ?) mm long. Flowers white; petals oval to obovate, the marginal radiant; calyx obsolete; stylopodium conical, longer than or about equal to the very short, divergent styles; ovary hirsutulous. Fruit oval to obovate, 7–10 mm long, 4–8 mm broad, glabrous or nearly so, the dorsal ribs filiform, the lateral thin winged, the wing narrower than body. Vittae linear, 1–3 in intervals, extending halfway or nearly to base, 6 on commissure, extending nearly to base (the outer apparently branching). Sept.

Type locality: "Burma: At Maymyo Hill, Upper Burma," *Khan 333*; "in the Southern Shan States," *MacGregor*.

Distribution: Khasia and Naga Hills of Assam and Burma, 1,500–2,000 m.

Representative specimens: Meghalaya: Mawphlang, Khasi Hills, *Chand 7240* (MICH, UC), *Koelz 23,816* (MICH, UC).

13. **Heracleum hookerianum** Wight & Arn., Prodr. Fl. Ind. Orient. 373. 1834; C. B. Clarke in Hook. f., Fl. Brit. Ind. 2: 715. 1879.

Pastinaca hookeriana Wight, Ic. Pl. Ind. Orient. 3: 12, pl. 1010. 1843; *Tetrataenium hookerianum* (Wight. & Arn.) Mandenova, Trudy Tbilissi Bot. Inst. Bot. 18. 1959.

Plants low, erect, few branched, hirsute to tomentose, often glabrate above, 25–50 cm tall. Leaves clustered at base with overlapping sheaths, ovate to orbicular, 3–12 cm long, 2–10 cm broad, deeply 3–5-lobed or pinnatifid, obtuse, cordate at base (rarely slightly peltate), the lobes oval to orbicular, 2.5–7 cm long, 2–6.5 cm broad, crenate-dentate, the larger again lobed, with narrow sinuses, hirsute above, tomentose beneath; petiole short and stout, hirsute, 3–12 cm long; cauline leaves reduced upward, the upper mostly dilated bladeless sheaths. Peduncles rather stout, terminal and lateral, 5–25 cm long. Involucre of ca. 5 linearly lanceolate, caducous bracts 4–8 mm long. Rays 4–10, slender, unequal, hirsute, 1–3.5 cm long. Umbellets 10–15-flowered, the mature pedicels rather slender, ca. 5 mm long. Involucel of 5–8 (rarely 0!) bracteoles like the bracts, 3–5 mm long. Flowers yellowish-white, red-tinged; petals lanceolate-oblong, the outer radiant, obovate, or not radiant; calyx teeth minute, linear; stylopodium conical, longer than the short styles; ovary pubescent. Fruit broadly elliptic, 6–8 mm long, 4–6 mm broad, puberulent, rounded at apex and base, the dorsal ribs low, filiform, the lateral thin winged, the wing about as broad as body. Vittae solitary in intervals, extending nearly to base, 2 on commissure, extending nearly to base and with or without 2 shorter additional ones.

Type locality: "Mountains in South of the Peninsula" [Peninsula Indiae Orientalis], *Wight 1184* (CAL, GH, K, NY).

Distribution: Apparently endemic to the Nilgiri Hills of peninsular India, 2,300–2,700 m.

Representative specimens: Tamil Nadu: Pykarah, *Meebold 11,623* (CAL); Avalanche, Nilgiri Distr., *Vivekananthan 42,930* (MH); Dodabetta, *Gamble 12,987* (CAL).

14. **Heracleum rigens** DC. Prodr. 4: 191. 1830; C. B. Clarke in Hook. f., Fl. Brit. Ind. 2: 715. 1879.

Conium rigens Wallich, Cat. 575. 1829, *nomen nudum*; *Pastinaca rigens* (DC.) Wight, Ic. Pl. Ind. Orient. 3: 12, pl. 1009. 1843; *Tetrataenium rigens* (DC.) Mandenova, Trudy Tbilissi Bot. Inst. 20: 17. 1959.

Heracleum rigens var. *multiradiatum* Gamble, Fl. Madras 1: 565. 1919. Based on *Gamble* (?), "W. Ghats, in the Pulney Hills, at lower levels."

Heracleum rigens var. *elongatum* Gamble, Fl. Madras 1: 565. 1919. Based on *Bourne*, "W. Ghats, in the Pulney Hills."

Plants rather low, erect, few branched, grayish-hirsute and hirtellous to tomentose, 30–120 cm tall. Leaves oblong-oval, 7.5–30 cm long, 5–15 cm broad, pinnate, the 3-5(-7) leaflets oval to obovate, 2–7 cm in diameter, rounded at apex, rounded to cordate at base, crenate-serrulate, the larger often lobed, the terminal leaflet trilobed, hirsutulous above, tomentose beneath; petiole short, 3–15 cm long, broadly sheathing; cauline leaves rather numerous, like the basal but reduced upward. Peduncles stout, mostly terminal, 10–50 cm long. Involucre of several ovate-lanceolate hirsutulous bracts 10–18 mm long. Rays 10–30, slender, unequal, 5–12 cm long, hirsutulous. Umbellets ca. 15-flowered, the mature pedicels short, 2–5 mm long, spreading-ascending, hirsutulous. Involucel of several bracteoles like the bracts, to 7 mm long. Flowers yellow; petals oval, the outer radiant or not; calyx teeth evident, ovate; stylopodium low conical, about equal to the short styles; ovary hirsutulous. Fruit elliptic to obovate, 8–10 mm long 5–7 mm broad, hirsutulous to glabrate, the dorsal ribs filiform, the lateral winged, the wing narrower than body. Vittae 1 or 2 in intervals, extending 3/4 to base, 2 or 4 of equal length on commissure, but 2 often with additional fragments. Aug.–Sept.

Type locality: "Penninsular India," *Heyne* (*Wallich 575*).

Distribution: Nilgiri and Pulney Hills, peninsular India, 2,000–2,500 m.

Representative specimens: "Peninsular India," *Heyne* (*Wallich 575*) (CAL). Tamil Nadu: Naterikal, Tinnevally Distr., *Barber 10,954* (MH); Pulney, *Wight 1188* (CAL); Dodabetta, *Gamble 13,136* (CAL). Kerala: Munnar, *Mukherjee 162* (CUH). Karnataka: Mysore, *Talbot 2342* (CAL); Bababudan Hills, *Mukherjee 607* CUH, UC).

15. **Heracleum sprengeleanum** Wight & Arn., Prodr. Fl. Ind. Orient. 372. 1834; C. B. Clarke in Hook. f., Fl. Brit. Ind. 2: 716. 1879.

Pastinaca sprengeliana (Wight. & Arn.) Wight, Ic. Pl. Ind. Orient. 3: 12, pl. 1008. 1843; *Tetrataenium sprengelianum* (Wight. & Arn.) Mandenova, Trudy Tbilissi Bot. Inst. 20: 17. 1959.

Plants stout, erect, much branched, sparsely hirsute and hirtellous 45–200 cm tall. Leaves ovate, 15–40 cm long, 10–30 cm broad, pinnate or ternate-pinnate, the leaflets ovate to orbicular, acute, rounded or truncate at base, 8–13 cm in diameter, crenate-serrate and lobed to pinnatifid, hirsute or hirsutulous or glabrate; petiole slender, 10–30 cm long, abruptly and broadly sheathing; cauline leaves numerous, like the basal but reduced upward, with broadly inflated sheaths. Peduncles short, terminal and lateral, 3–8 cm long. Involucre of 1 or 2 small caducous bracts, or lacking. Rays 15–20, slender, unequal, 3–7 cm long, hirsutulous to glabrate. Umbellets 15–20-flowered, many flowers sterile, the mature pedicels slender, subequal, spreading-ascending, 5–8 mm long. Involucel of 3–5 linearly lanceolate to lanceolate bracteoles ca. 5 mm long. Flowers white; petals lanceolate, the outer radiant, suborbicular and bilobed, or not radiant; calyx obsolete; stylopodium conical, about equaling the short styles; ovary ± hirsutulous. Fruit elliptic, 9–12 mm long, 5–8 mm broad, a little narrowed at apex and base, sparsely hirsutulous to glabrate, the dorsal ribs filiform, the lateral thin winged, the wing narrower than body. Vittae large, subclavate, solitary in intervals, extending 3/4 to base, 2 or 4 of equal length on commissure. July–Oct.

Type locality: "Nilghiri mountains," *Wight 1204*.

Distribution: Hills of peninsular India, 900–1,800 m.

Representative specimens: Tamil Nadu: peninsular India, *Wight 1193* (CAL); Kodaikanal, *Bourne 484* (CAL); Coonoor, *Sebastine 4016* (MH). Kerala: Munnar, *Deb 30,789* (MH).

16. **Heracleum aquilegifolium** C. B. Clarke in Hook. f., Fl. Brit. Ind. 2: 715. 1879.

Plants rather stout, erect, branched, glabrous to scaberulous, 60–100 cm tall, from a fusiform root. Leaves broadly ovate, 1-2-ternate, -ternate-pinnate, or simply pinnate, the leaflets ovate, 5–10 cm in diameter, trilobed to entire and irregularly serrate-denticulate, glabrous; petiole elongate, 10–20 cm long, dilated and sheathing at base; cauline leaves few, ternate or the upper often undivided, the uppermost reduced to sheaths. Peduncles terminal and lateral, 3–8 cm long, forming a nearly naked raceme. Involucre of 3–6 lanceolate bracts, acute, 6–10 mm long. Rays 6–10(–15), 2–6 cm long. Umbellets 6–10-flowered, the mature pedicels unequal, 3–7 mm long. Involucel of 3–5 linear to lanceolate, acute to obtuse bracteoles ca. 5 mm long. Flowers yellow; petals subequal, not radiant; calyx minute, the teeth ovate; stylopodium low conical, the styles short; ovary glabrous. Fruit broadly oval to orbicular, 6–10 mm long, 6–8 mm broad, glabrous, the dorsal ribs distinct, shortly excurrent, the lateral broadly winged. Vittae linear, 2 or 3 in intervals, extending nearly to base, 2–4 on commissure. Sept.–Oct.

Type locality: "CONCAN; tropical region," *Law*, *Stocks* (*Hooker & Thomson*, Herb. Ind. Orient, *Heracleum 11* (CAL, GH).

Distribution: Western hills (ghats) of peninsular India.

Representative specimens: Peninsular India, *Wight 1194* (CAL, GH). Tamil Nadu: Anamalai, *Wight 1186* (CAL, GH).

17. **Heracleum grande** (Dalzell & Gibson) Mukhopadhya [P. K. Mukherjee], Science & Culture 39: 542. 1973.

Pastinaca grande Dalzell & Gibson, Bombay Fl. 107. 1861; *Peucedanum grande* (Dalzell & Gibson) C. B. Clarke in Hook. f., Fl. Brit. Ind. 2: 710. 1879; *H. grandis* Almeida, Indian Forester 111(3): 160. 1985; *Tetrataenium grande* (Dalzell & Gibson) Mandenova, Notulae Syst. Geogr. Inst. Bot. Tbilissi 42: 14. 1991.

Plants stout, erect, sparsely branched, glabrous, to 1 m tall, from a large woody root. Leaves mostly basal, ovate, 6–8 cm long, bipinnate, the leaflets oval, 2.5–7.5 cm in diameter, trilobate and the lobes triangular to simple, crenate-serrate, shiny on both surfaces; petiole elongate, 3–5 cm long, sheathing at base; cauline leaves 1 or 2, 1–2-ternate, reduced upward, the sheaths strongly dilated. Peduncles terminal and lateral. Involucre prominent, of 5 oblong to obovate, usually obtuse bracts ca. 1 cm long. Rays 10–20, stout, unequal, 2.5–7.5 cm long. Umbellets 10–20-flowered, the mature pedicels 8–13 mm long. Involucel of 5 oblong, acute or obtuse bracteoles ca. 5 mm long. Flowers yellow; petals oblong-ovate, subequal, non-radiant; calyx teeth short, triangular; stylopodium low conical, the styles short; ovary glabrous or nearly so. Fruit obovate, 10–15 mm in diameter, obtuse at apex, narrowed toward base, glabrous or nearly so, the dorsal ribs prominent but obtuse, the lateral narrowly winged. Vittae 1 or 2 in intervals, extending nearly to base of the lateral abbreviated, 4 on commissure, abbreviated. July.

Type locality: "BOMBAY GHAUTS," *Dalzell*.

Distribution: Western hills (ghats) of peninsular India.

Representative specimens: Maharashtra: Bombay, *Dalzell* (CAL); Matheran, *Mukherjee 612* (CUH).

18. **Heracleum dalgadianum** Almeida, Indian Forester 111(3): 158. 1985.

Plants erect, softly pubescent, to 1.5 m tall, from a stout, somewhat swollen tuberous root. Leaves ovate, 20–25 cm long, 12–16 cm broad, pinnate or ternate-pinnate, the leaflets ovate-lanceolate, 8–10 cm long, 4–5 cm broad, shortly petiolulate or sessile, longly acuminate, oblique at base, spinulose-serrate, the larger frequently lobed, softly pubescent especially on veins beneath; petiole 30–35 cm long, narrowly sheathing at base; cauline leaves like basal, reduced upward with inconspicuous sheaths. Peduncles terminal, slender. Involucre of several (5 or 6?) linearly lanceolate, pubescent bracts 1–1.5 cm long. Rays 4–10, unequal, 4–7 cm long, pubescent. Umbellets 6–8-flowered, the mature pedicels slender, subequal, spreading-ascending. Involucel bracteoles like the bracts, shorter than pedicels. Flowers yellow; petals oblong-lanceolate, longly acuminate, subequal; calyx teeth small, linear; stylopodium conical, shorter than the persistent erect styles; ovary softly pubescent. Fruit orbicular-obovate, 8–10 mm long, 6–8 mm broad, softly pubescent, the dorsal ribs evident, the lateral broadly winged, the wing slightly thickened at outer margin. Vittae 1–3 in intervals, extending length of body, 4–8 on commissure, abbreviated. Sept.–Oct.

Type locality: MAHARASHTRA: "apud Amboli Savantwadi," 11 Sept. 1980, *S. M. Almeida 2815* (BLAT, holotype).

Distribution: Savantwadi Region, Sindhudurg Distr., Maharashtra, on hill slopes.

Other specimen examined: Maharashtra: Amboli, *Almeida 1627* (BLAT).

19. **Heracleum candolleanum** (Wight & Arn.) Gamble, Fl. Madras 1: 564. 1919.

Pastinaca candolleana Wight & Arn., Prodr. Fl. Ind. Orient. 372. 1834; *Heracleum rigens* var. *candolleana* (Wight. & Arn.) C. B. Clarke in Hook. f., Fl. Brit. Ind. 2: 715. 1879. *Pastinaca ligusticifolia* Wight, Ill. Ind. Bot. 2: pl. 116. 1831, non Calest. 1905; *Heracleum rigens* var. *ligusticifolium* (Wight) C. B. Clarke in Hook. f., Fl. Brit. Ind. 2: 716. 1879; *H. sprengelianum* var. *ligustifolium* [sic] Fyson, Fl. Nilgiri & Pulney Hill-Tops 1: 176. 1915, non *H. ligusticifolium* M. Bieb. 1808; *H. courtallense* Gamble, Fl. Madras 1: 565. 1919. Based on "*Wight*! cat. n. 1203," w/o locality. ["W. Ghats, at Courtallum in Tinnevally," fide Gamble, 1919].

Tetraenium candolleanum (Wight, & Arn.) Mandenova, Notulae Syst. Geogr. Inst. Bot. Tbilissi 42: 11. 1991.

T. ligusticifolium (Wight, & Arn.) Mandenova, Notulae Syst. Geogr. Inst. Bot. Tbilissi 42: 11. 1991.

Plants rather stout, erect, branched, hirsute or hirtellous to glabrate, 30–100 cm tall. Leaves ovate, 25–35 cm long, 15–25 cm broad, ternate or ternate-pinnate, the leaflets oblong to ovate, 3–10 cm long, 2–5 cm broad, the larger petiolulate, acute or acuminate, shallowly or deeply serrate and frequently pinnately lobed, hirsute or hirsutulous on both surfaces; petiole 5–15 cm long; cauline leaves reduced upward, the blade ternate-pinnate, the sheath moderately dilated. Peduncles 2–12 cm long. Involucre of ca. 5 linear-lanceolate, inconspicuous, caducous or persistent bracts. Rays 6–12(–20), slender, unequal, 1.5–5 cm long, hirsute or hirsutulous. Umbellets 15–20-flowered, the mature pedicels rather slender, subequal, hirsutulous, spreading-ascending, 4–7 mm long. Involucel of ca. 8 linearly lanceolate bracteoles like the bracts, 3–5 mm long. Flowers apparently either yellow or white; petals lanceolate, nonradiant (?); calyx teeth minute, ovate, or obsolete; stylopodium low conical, the styles ca. 1 mm long; ovary glabrous. Fruit elliptic to obovate, 5–11 mm long, 4–8 mm broad, glabrous, the dorsal ribs acute, filiform, the lateral winged, the wing narrower than to equal to the body. Vittae 1 or 2 in intervals, extending 2/3–3/4 to base, 2 or 4 on commissure, ± abbreviated. July–Sept.

Type locality: Peninsular India, "*Wight*! cat. n. 1202" (K). ["W. Ghats, from Nilgiris to the Anamalai and Pulney Hills, above 6,000 ft. common," fide Gamble, 1919].

Distribution: Hills and mountains of peninsular India, 1,500–2,300 m.

Representative specimens: Tamil Nadu: w/o specific locality, *Wight 1191* (CAL, GH), *1192* (CAL, GH), *1202* (K); Courtallum, *Wight 1181* (GH, K); Nilgiri, *Miquel* (*Hohenacker, Pl. Ind. Orient. 1104*) (BM); Pannaikadu, Kodaikanal, *Sebastine 25,010* (MH); Ootacamund, *Cleghorn* (MH); Vannathiparai, Madurai Distr., *Jacob 77,206* (MH). Kerala: Lockhert Gap, Kootayam Distr., *Sebastine 16,510* (MH). Karnataka: Bababudan Hills, *Mukherjee 608* (CUH).

20. **Heracleum ceylanicum** (Gardner) C. B. Clarke in Hook. f., Fl. Brit. Ind. 2: 716. 1879.

Peucedanum ceylanicum Gardner, Calcutta Jour. Nat. Hist. 8: 14. 1847; *Palimbia cey-*

lanica (Gardner) Hiroe, Umbell. World, 619. 1979; *Tetrataenium ceylanicum* (Gardner) Mandenova, Notulae Syst. Geogr. Inst. Bot. Tbilissi 41: 46. 1986.

Plants slender, erect, branched, sparsely and softly hirtellous to glabrate (?), 80–150 cm tall. Leaves oblong-ovate, 10–30 cm long, 10–15 cm broad, pinnate, the 5–9 leaflets oblong-lanceolate to ovate, 3–12 cm long, 1–6 cm broad, acute or obtuse, the lower petiolulate, crenate-serrate and pinnately lobed to pinnatifid, usually softly hirsutulous on both surfaces; petiole slender, 5–25 cm long, abruptly dilated and broadly sheathing; cauline leaves few, reduced upward, the uppermost consisting of moderately dilated, bladeless sheaths. Peduncles rather stout, terminal and lateral, 8–40 cm long. Involucre lacking. Rays 8–10, slender, unequal, 3.5–9 cm long, hirsutulous to glabrate. Umbellets 10–15-flowered, few fertile, the mature pedicels slender, unequal, 5–15 mm long, spreading-ascending. Involucel of 1–6 linear to linearly lanceolate bracteoles 2–6 mm long, or lacking, usually reflexed. Flowers white or greenish-white; petals oblong-oval, subequal and nonradiant, or the outer strongly radiant (?); calyx teeth evident, lanceolate; stylopodium conical, about equal to the short styles; ovary hirsutulous. Fruit oval-obovate, 8–11 mm long, 4–6 mm broad, glabrous, the dorsal ribs filiform, the lateral broadly thin winged, the wing narrower than body. Vittae large, solitary in intervals, extending 2/3–3/4 to base, 4 (arcing and joining at apex and base, plus additional fragments) on commissure. (Feb.–) May–Sept. (–Dec.).

Type locality: "HAB. — In open rocky, grassy places on the descent from the Horton Plains to Galagama, in the district of Saffragam," Sri Lanka, *Gardner*.

Distribution: Pulney Hills of peninsular India; Sri Lanka, 2,000–2,500 m. No Indian material has been examined. Gardner described the species as "annual," but in cultivation at Berkeley it behaved as a perennial and persisted for several years. (L. C.)

Doubtful species

Tetrataenium himalayense Mandenova, Notulae Syst. Geogr. Inst. Bot. Tbilissi 42:13.1991. — We have seen no specimen of this species and hence its generic placement could not be ascertained.

65. TORDYLIOPSIS DC.

Prodr. 4: 199. 1830; Mandenova, Trudy Tbilissi Bot. Inst. 20: 18. 1959.

Tordylioides Wallich ex DC., Prodr. 4: 199. 1830. *nomen nudum in synon.*

Moderately stout, erect, simple or sparsely branching, softly hirsute perennial herbs from a base beset with withered leaf sheaths surmounting a taproot. Leaves petiolate, membranaceous, simply pinnate, with large oval to obovate, coarsely dentate or trilobed leaflets; petiole slender, sheathing at base; cauline leaves few, like the basal but reduced upward. Inflorescence of terminal and often 1 or more lateral compound umbels, the peduncles stout. Involucre of several narrow, tomentose, mostly subentire bracts about half as long as rays, or longer. Rays few, stout, unequal, tomentose, spreading-ascending or ascending. Involucel of bracteoles like the bracts. Umbellets rather numerous flowered, the flowers pedicellate. Flowers greenish-white or purplish; petals dimorphic, sparsely hirsute dorsally, the outer radiant, obcordate, emarginate, the others much smaller, ovate-lanceolate, with a narrower inflexed apex; calyx teeth evident, unequal; stylopodium conical, the styles slender, erect or recurved; ovaries hirsute. Pedicels slender, hirsute, unequal, spreading-ascending. Fruit elliptic, rounded at apex and base, flattened dorsally and complanate, the mericarps flattened dorsally in transection, sparsely hirsute to glabrate, the dorsal ribs filiform, equidistant, the lateral broadly winged and the fruit thus surrounded by a continuous entire thin wing narrower than the body. Dorsal vittae thin, linear, unequal, extending more than half way to base of fruit, solitary in intervals, the commissural 0–4 and often much abbreviated, to obsolete. Strengthening cells beneath inner margin of lateral wings. Commissural face plane. Carpophore bipartite.

Type species: *Tordyliopsis brunonis* DC.

Distribution: A monotypic genus of the Himalaya.

1. **Tordyliopsis brunonis** DC., Prodr. 4: 199. 1830.

Tordylioides ? brunonis Wallich, Cat. 590. 1829, *nomen nudum*; *Heracleum brunonis* (DC.) C. B. Clarke in Hook. f., Fl. Brit. Ind. 2: 713. 1879.

Plants herbaceous, 30–100 cm tall, softly hirsute throughout, the stem usually solitary, moderately leafy. Basal and lower cauline leaves oblong-oval, 8–15 cm long, 5–10 cm broad, pinnate with 1–3 pairs of opposite, oval to obovate, subsessile, coarsely dentate or lobed leaflets 2–6 cm long, 1.5–4 cm broad; petiole 8–12 cm long. Peduncles 5–12 cm long. Involucre of 5–8 lanceolate, acuminate, hirsute bracts 1–2 cm long. Rays 5–8, hirsute, 1.5–2.5 cm long. Umbellets 12–18-flowered, the mature pedicels hirsute, unequal, 5–10 mm long. Involucel of 5–8 lanceolate, hirsute bracteoles 5–20 mm long. Flowers greenish-white or purplish; petals apically inflexed, the outer conspicuously radiant; calyx teeth linear, unequal; styles slender, 2–3 mm long. Fruit 8–10 mm long, 5–7 mm broad, the lateral wings very thin. Vittae often markedly unequal, or the commissural even obsolete. July–Sept.

Type locality: "in Kamaon Nepaliae versus Emodum," *Wallich 590* (CAL, K).

Distribution: Himalaya, from Sikkim to Kumaon, 3,000–5,000 m.

Representative specimens: Himachal Pradesh: Manali, *Bor 15,549* (DD); Laka, *Edgeworth* (CAL). Uttar Pradesh: Ralam Glacier, *Duthie 2962* (DD); Kedarnath, *Mukherjee* (CUH, UC); Milam, *Strachey & Winterbottom* (GH). Nepal: Tamur Valley, Mewa Khola

Topke Gola, *Stainton 935* (BM, GH); E of Chelike Paher, *Stainton, Sykes & Williams 3725* (BM, GH). Sikkim: w/o specific locality, *Hooker* (CAL, GH); Tangu, *Smith & Cave 2286* (CAL).

66. **POLYZYGUS Dalzell**

Hooker's Jour. Bot. Kew Misc. 2: 260. 1850; C. B. Clarke in Hook. f.
Fl. Brit. Ind. 2: 697. 1879.

Slender, erect, branching, hirtellous to hispidulous perennial herbs from an ovoid or oblong tuber. Leaves petiolate, membranaceous, 1-2-ternately or ternate-pinnately compound, with broad, mucronulate-serrate and often lobed leaflets; petiole slender, narrowly sheathing; cauline leaves reduced upward. Inflorescence of lax terminal and lateral compound umbels, the peduncles elongate. Involucre usually lacking. Rays few, unequal, spreading-ascending. Umbellets few flowered, the flowers pedicellate. Involucel dimidiate, of several narrow bracteoles. Flowers whitish (?), homomorphic; petals oblong to oval, with a narrower inflexed apex; calyx teeth minute, unequal; stylopodium low conical, the styles short to slender, spreading. Pedicels slender, unequal, spreading-ascending. Fruit ellipsoid, rounded at apex and base, strongly compressed dorsally, the mericarps flattened dorsally in transection, glabrous, the dorsal ribs low and filiform, the lateral narrowly thin winged. Vittae prominent, 1 or 2 in intervals, extending nearly to base, 2–4 subequal or abbreviated vittae on commissure. Strengthening cells apparently lacking. Commissural face planoconcave. Carpophore bipartite.

Type species: *Polyzygus tuberosus* Dalzell ex Walpers.

Distribution: A monotypic genus endemic to the Western Ghats of peninsular India.

1. **Polyzygus tuberosus** Dalzell ex Walpers, Ann. Bot. Syst. 2: 715. 1852; C. B. Clarke in Hook. f., Fl. Brit. Ind. 2: 698. 1879; P. K. Mukherjee & Constance, Kew Bull. 41(1): 228. 1985.

Heracleum concanense var. ? *stocksii* C. B. Clarke in Hook. f., Fl. Brit. Ind. 2: 716. 1879. Based on *Stocks* (*Hooker & Thomson*, Herb. Ind. Orient., *Heracleum 10*), "Concan" (K, lectotype; K, P, isotypes).

Plants slender, erect, few branched, 20–50 cm tall from a short pseudoscape surmounting a tuber. Leaves triangular-ovate, 5–20 cm in diameter, 1-2-ternate or ternate-pinnate, the primary divisions longly petiolulate, the petiolulate or sessile leaflets lanceolate to ovate, 1–4 cm long, 0.5–3 cm broad, acute, hispidulous to hirtellous, especially on veins on both surfaces; petiole 3–10 cm long, with a very narrow sheath. Cauline leaves several, the sheaths scarcely dilated. Peduncles 3–20 cm long, hispidulous at summit. Rays of fertile umbels 2–10, slender, unequal, 1–4 cm long, scaberulous to glabrate. Fertile umbellets ca. 15-flowered, few flowers fertile, the mature pedicels 5–12 mm long, hispidulous. Involucel of 2–5 linearly lanceolate hispidulous bracteoles 1–3 mm long. Petals subequal or the marginal slightly radiant; styles variable, 1–2 mm long. Fruit elliptic, 4–5 mm long, 3–4 mm broad, glabrous, the dorsal ribs low, the lateral narrowly thin winged. Sept.–Nov.

Type locality: "In prov. Malwan," "Bombay," *Dalzell* (K, lectotype; CAL, K).

Distribution: Western Ghats of peninsular India.

Representative specimens: Maharashtra: Canara & Mysore, *Law* (K); Deccan, *Bell 4165* (K); Panorama Point, Matheran, *Mukherjee 611* (CUH, UC).

67. **PINDA** P. K. Mukherjee & Constance
Kew Bull. 41(1): 224. 1986.

Slender, erect, simple or sparsely branching, pubescent perennial herbs from a swollen tuberous base. Leaves petiolate, membranaceous, ternate-pinnate or bipinnate, with rather large serrate and often lobed leaflets; petiole sheathing; cauline leaves few, reduced upward. Inflorescence of lax terminal and lateral compound umbels, the peduncles usually elongate, or sometimes abortive. Involucre of 1 or 2 broad sheathlike bracts, or lacking. Rays few, unequal, spreading-ascending. Involucel dimidiate, of several oblong or ovate bracteoles. Umbellets few flowered, the flowers pedicellate. Flowers white with colored veins, dimorphic, the marginal radiant and perfect, the central regular and usually staminate; petals oval with a narrower inflexed apex, the external marginal petal obcordate and greatly enlarged; calyx with 2 persistent outer teeth, the others obsolete; stylopodium low conical, exceeding the short spreading styles; ovaries glabrous. Pedicels rather short, unequal, spreading-ascending. Fruit oval to elliptic, rounded at apex and base, strongly compressed dorsally, the mericarps flattened dorsally in transection, glabrous, the ribs low and rounded, unwinged, the lateral a little broader than the dorsal. Vittae prominent, 1–3 in intervals and extending to base of fruit, 2 or more abbreviated vittae on commissure and small vittae also in ribs. Strengthening cells on the slightly concave commissural face and in ribs thus discontinuous about seed cavity. Carpophore bipartite.

Type species: *Heracleum concanense* Dalzell [*Pinda concanensis* (Dalzell) P. K. Mukherjee & Constance].

Distribution: A monotypic genus endemic to the Western Ghats of peninsular India.

1. **Pinda concanensis** (Dalzell) P. K. Mukherjee & Constance,
Kew Bull. 41(1): 226. 1985.

Heracleum concanense Dalzell, Hooker's Jour. Bot. Kew Gard. Misc. 2: 260. 1850; C. B. Clarke in Hook. f., Fl. Brit. Ind. 2: 716. 1879.
Heracleum pinda Dalzell & Gibson, Bombay Fl. 107. 1861. Based on *Dalzell*, "On Hursur and Hurrychunder hill-forts."
Heracleum grandiflorum Dalzell & Gibson, Bombay Fl. 108. 1861. Based on *Dalzell*, "On Hursur and Hurrychunder hill-forts."

Plants slender, erect, sparsely branched above, hirsute to hispidulous, 40–120 tall, from a conspicuously swollen tuberous base. Leaves ovate to triangular-ovate, 5–8 cm long, 5–6 cm broad, ternate-pinnate or bipinnate, the leaflets ovate-lanceolate to ovate, petiolulate, mostly acute, the larger frequently 1–3-lobed, mucronulate-serrate, sparsely hirsute, 1–2 cm long, 0.5–1 cm broad; petiole 3–7 cm long, abruptly dilated and broadly sheathing; cauline leaves reduced upward, the uppermost often reduced to oblong-oval, bladeless sheaths. Peduncles slender, 2–10 cm long, or sometimes abortive. Rays usually 4–10, slender, unequal, 1–5 cm long, hispidulous. Umbellets 6–12-flowered, about half fertile, the mature pedicels 2–5 mm long. Involucel bracteoles 3–5, oblong to ovate, venose, 3–6 mm long. Strongly radiant marginal flowers 3–8, perfect, the regular central and usually staminate flowers 3–10; petals 1–1.5 mm long, the radiant marginal petal 6–10 mm in diameter; calyx teeth small, ovate; functional styles ca. 1 mm long, slightly exceeding the low conical, massive stylopodium. Fruit 4.5–6 mm long, 2.5–3 mm broad, glabrous, the ribs prominent, unwinged. July–Aug.

Type locality: "In collibus Concanensibus," Bombay Presidency, *Dalzell* (K, holotype; CAL, isotype).

Distribution: Western Ghats of peninsular India, 600–1,200 m.

Representative specimens: Maharashtra: Malabar, Concan, *Stocks & Law* (CAL, K, P); Khandala, *Meebold 8810* (CAL), *Santapau 28,061* (BLAT); Mahabaleswar, *Ansari 116,596* (BSI), *116,556* (BSI); Matheran, *Wadhawa 109,737* (BSI); Dhak Forest, Junnan, *Hemadri 107,478* (BSI).

68. SEMENOVIA Regel & Herder

Bull. Soc. Nat. Mosc. 39(2): 78. 1866; Mandenova, Trudy Tbilissi Bot. Inst. 20: 19. 1959.

Platytaenia Nevski & Vived., Trudy Bot. Inst. Akad. Nauk SSR, ser. 1, Fl. Sist. Vyss. Rast. 4: 270. 1937, non Kuhn 1882.

Stout, shrubby, erect, branching, pubescent to glabrous perennial herbs, the stems numerous, from a woody base beset with numerous persistent leaf bases surmounting a stout taproot. Leaves mostly basal, petiolate, membranaceous, pinnate to bipinnate, with linear to oval, dentate, lobed, or divided leaflets; petiole slender, sheathing; cauline leaves greatly reduced upward. Inflorescence of lax terminal and usually also lateral compound umbels, the peduncles elongate. Involucre of several narrow bracts, or lacking. Rays few, slender, subequal or unequal, spreading-ascending. Umbellets rather few flowered, the flowers pedicellate. Involucel of several narrow bracteoles, or lacking. Flowers white or yellowish-white (rarely yellow or purplish); petals obovate with a narrower inflexed apex, deeply but unequally emarginate or bilobed, usually pubescent dorsally, the outer petals commonly radiant; calyx teeth minute, or calyx obsolete; stylopodium depressed conical, the styles short or slender; ovary pubescent or glabrous. Pedicels few, unequal, spreading-ascending. Fruit oblong or elliptic, rounded at apex and base, flattened and complanate, the mericarps flattened dorsally in transection, pubescent or rarely glabrous, the dorsal ribs prominent, filiform, but usually masked by hairs, equidistant, the lateral broadly winged and the fruit thus surrounded by a continuous entire thick wing narrower than body. Vittae thin, linear, extending nearly to base of fruit, solitary in intervals, 2 on commissure. Strengthening cells continuous about seed cavity, under ribs, and extending into wings. Commissural face plane. Carpophore bipartite.

Type species: *Semenovia transiliensis* Regel & Herder.

Distribution: A central Asian genus of perhaps two dozen species, one of which extends to northwestern India.

1. **Semenovia thomsonii** (C. B. Clarke) Mandenova,
Trudy Tbilissi. Bot. Inst. 20: 23. 1959.

Heracleum thomsonii C. B. Clarke in Hook. f., Fl. Brit. Ind. 2: 711. 1879; *Platytaenia lasiocarpa* (Boiss.) Mandenova ssp. *thomsonii* (C. B. Clarke) Rech. f. & Riedl, Biol. Skr. 13(4): 104. 1963; *Semenovia lasiocarpa* (Boiss.) Mandenova ssp. *thomsonii* (C. B. Clarke) Alava in Rechinger Fl. Iranica 162: 488. 1987.
Heracleum thomsonii var. *glabrior* C. B. Clarke in Hook. f., Fl. Brit. Ind. 2: 712. 1879; *Platytaenia lasiocarpa* ssp. *thomsonii* var. *glabrior* (C. B. Clarke) Nasir, Fl. W. Pakistan 20: 135. 1972. Based on *Thomson* (*Hooker & Thomson, Herb Ind. Orient., Pastinaca 2*), "KASHMIR" [Ladak] (K). (Nasir 1972).

Plants woody at base, 10–80 cm tall, glabrous to villous or tomentulose, the stems few to numerous, slender, branching, nearly naked. Leaves clustered at base, oblong to oval, 5–15 cm long, 2–4 cm broad, pinnate to bipinnate, with 3–6 pairs of opposite, oblong to orbicular, shortly petiolulate to sessile, mucronate-serrate or lobed to pinnatisect leaflets 0.5–3 mm long, 0.5–2 mm broad; petiole slender, 3–10 cm long, scarious-sheathing at base; cauline leaves few, greatly reduced upward. Peduncles slender, terminal and lateral, 5–35 cm long. Involucre of 3–8 linear pubescent bracts 2–5 mm long. Rays 3–12, slender, unequal, 2–4 cm long, tomentulous to glabrate. Umbellets 10–12-flowered, the mature pedicels unequal, spreading-ascending, 3–10 mm long. Involucel of 4–8 linear, pubescent bracteoles like the bracts, 2–2.5 mm long. Flowers white to yellowish-white; petals unequally emarginate, the outer radiant; calyx teeth minute, triangular; styles slender, 1.5–2 mm long, reflexed; ovaries tomentose. Fruit elliptic, 4–6 mm long, 3–4 mm broad, ± white-villous, the ribs filiform. July–Sept.

Type locality: "LADAK and PITI; alt. 11,000–15,000 ft.," *Thomson* (*Hooker & Thomson*, Herb. Ind. Orient., *Pastinaca 1*) (K, lectotype, CAL) (Rechinger f. & Riedl 1963).

Distribution: Northwestern Himalaya, to Afghanistan and Pakistan, 2,500–4,000 m.

Representative specimens: Jammu & Kashmir: Cashmere, *Jacquemont* (CAL); Tilail, *Clarke 30,727a* (CAL). Himachal Pradesh: Ladak, *Thomson* (*Hooker & Thomson*, Pl. Ind. Orient., *Pastinaca 1, 2* (CAL, K); Lahul, *Jaeschke 99* (CAL); Tandi, Lahul, *Mukherjee 334* (CUH, UC).

REFERENCES AND PERTINENT LITERATURE

Almeida, R. M.

1985. A note on *Heracleum* from Maharashtra; a new species and a new combination. *Indian Forester* 111(3): 158–160.

Babu, C. R.

1970. *Peucedanum dehradunensis* Babu (Apiaceae)—a new species from N. W. Himalaya. *Indian Forester* 96: 535.

Bentham, G.

1867. Umbelliferae. In: *Genera plantarum* 1: 859–931 (G. Bentham and J. D. Hooker, eds.).

Bernardi, L.

1979. Tentamen revisionis generis *Ferulago*. *Boissieria* 30: 1–182.

Blasco, F., *et al.*

1978. Étude multidisciplinaire du genre *Vanasushava* (Wight) Mukherjee & Constance. *Actes 2e Sympos. Internat. Ombellifères, Perpignan*, pp. 664–674.

Boissier, E.

1872. Umbelliferae. *Flora orientalis* 2: 819–1090.

Boissieu, H. de

1906. Note sur quelques Ombellifères de la Chine, d'après les collections du Muséum d'Histoire naturelle de Paris. *Bull. Soc. Bot. France* 53: 418–439.

Burkill, I. H.

1965. Chapters on the history of botany in India. Calcutta. 245 pp. (Reprinted from *Jour. Bombay Nat. Hist. Soc.* 1957–1962.)

Buwalda, P.

1936. The Umbelliferae of the Netherlands Indies. *Blumea* 2(3): 119–220. (Reprint, pp. 1–102.)

Buwalda, P. and P. Groningen

1949. Umbelliferae. *Flora Malesiana*, ser. 1, 4(2): 113–140.

Calder, C. C.

1937. *An Outline of the Vegetation of India*, pp. 71–89.

Candolle, A. P. de

1829. Collection de mémoire sur la famille des Ombellifères. *Coll. Mém.* 5: 1–184, pls. 1–19.

1830. *Umbelliferae*: *Prodromus systematis naturalis regni vegetabilia* 4: 1–683.

Candolle, R. de and A. Radcliffe-Smith

1981. Nathaniel Wallich, MD, PhD, FRS, FLS, FRGS, (1786–1854) and the Herbarium of the Honourable East India Company, and their relation to the de Candolles of Geneva and the *Great Prodromus. Bot. Jour. Linn. Soc.* 83: 325–348 (portrait).

Cannon, J. F. M.

1978. Umbelliferae. *Flora Zambesiaca* 4: 555–621.

1978. Umbelliferae. In: *An Enumeration of the Flowering Plants of Nepal* 2: 184–190 (H. Hara and L. H. J. Williams, eds.).

Carbonnier, J. and D. Molho

1977. Contribution phytochimique à l'étude de la position systématique de *Vanasushava pedata* (Wight) Mukherjee et Constance (*Heracleum pedatum* Wight). *Bull. Mus. Hist. Nat.*, ser. 3, 484: 1–17.

Cauwet-Marc, A. M.

1977. Anatomie comparée du fruit de *Vanasushava pedata* (Wight) Mukherjee et de quelques genres voisins. *Newsletter Umbell.* 2: 30–34.

1979. Contribution de la caryologie à la connaissance et de la phylogenie du genre *Bupleurum* L. *Candollea* 34/1: 49–86.

Cauwet-Marc, A. M., J. Carbonnier, and M. Farille

1980. Contribution à l'étude des ombellifères du Népal. I. *Candollea* 35/2: 497–510.

Cerceau-Larrival, M.-T., *et al.*

1977. Contribution palynologiques à l'étude de *Vanasushava pedata.* Ombellifère archaïque du sud de l'Inde. *Pollen et Spore* 19: 285–297.

Cerceau-Larrival, M.-T. and M. C. Carbonnier-Jarreau

1980. Characteristiques tectales de certaines Ombellifères indiennes et himalayennes. Relations avec le climat et la biogeographie. *Mém. Mus. Nat. Hist. Nat.*, ser. B, Bot. 27: 39–43.

Chakrabarty, C. and P. K. Mukherjee

1986. Studies on *Bupleurum* in India. I. Secretory structure in leaves; II. SEM observations of leaf surface. *Feddes Repert.* 97: 479–487, 489–496.

Chang, H.-t.

1985. *Selinum* Linn. *Flora Reipublicae Popularis Sinicae* 55(2): 225–228.

Chatterjee, D.

1939. Studies on the endemic flora of India and Burma. *Jour. Asiat. Soc. Bengal* 2(5): 19–67.

Clarke, C. B.

1879. Umbelliferae. In: *Fl. Brit. Ind.* 2: 665–720. (J. D. Hooker, ed.).

1884. Notes on the flora of Parasnath. *Jour. Linn. Soc.* 21: 252–255.

1886. Botanical observations made in a journey to the Naga Hills. *Jour. Linn. Soc.* 21: 384–391.

1889. On the plants of Kohima and Muneypur. *Jour. Linn. Soc.* 25: 1–107.

Collett, H.

1921. Umbelliferae. *Flora Simlensis*, 2nd ed., pp. 204–216.

Constance, L. and R. H. Shan

1948. The genus *Osmorhiza* (Umbelliferae), a study in geographic affinities. *Univ. Calif. Publ. Bot.* 23(3): 111–156.

Cooke, T.

1901. *The Flora of the Presidency of Bombay* 1: 560–573.

Dalzell, N. A.
1850. Contributions to the botany of western India. *Hooker's Jour. Bot.* 2: 260.

Dalzell, N. A. and A. Gibson
1861. *The Bombay Flora*, pp. 105–108.

Deb, D. B. and R. M. Dutta
1969. An addition to Umbelliferae from North East India. J. Sen Memorial vol.; *Bull. Bot. Soc. Bengal*, 421–423.

De Caisne, J.
1844. Umbelliferae. In: *Voyage dans l'Inde botanique* 4: 69–71 (V. Jacquemont, ed.).

Dharke, J. S.
1964. Morphological studies in the Order Umbellales: Umbelliferae. I. *Psammogeton biternatum* Edgew. *Agra Univ. Jour. Res. Sci.* 13: 125–134.
1968. Morphological studies in the order Umbellales: Umbelliferae. II. *Anethum graveolens* Linn. *Jour. Ind. Bot. Soc.* 47: 13–21.

Don, D.
1825. Prodromus florae nepalensis sive enumeratio vegetabilium, que in itinere per nepalium prope dictam et regiones conterminas. . . . *Umbelliferae*, pp. 182–185.

Drude, O.
1898. Umbelliferae. In: *Die natürlichen Pflanzenfamilien* 38: 63–250, 271 (A. Engler and H. Prantl, eds.).

Duthie, J. F.
1903. *Flora of the Upper Gangetic Plain* 1: 388–398.

Edgeworth, M. P.
1846. Descriptions of some unpublished species of plants from North-Western India. *Proc. Linn. Soc.* 1: 252–253.
1851. On some unpublished species of plants from North-Western India. *Trans. Linn. Soc.* 20: 51–58.

Eichler, H.
1987. Nomenclatural and bibliographical survey of *Hydrocotyle* L. (Apiaceae). *Feddes Repert.* 98: 1–51, 145–196, 273–351.

Farille, M. A.
1984. Apiaceae himalayenses. 2. *Rev. Gen. Bot.* 91: 27–34.

Farille, M. A., A.-M. Cauwet-Marc, and S. B. Malla
1984. Apiaceae himalayenses. I. *Bull. Soc. Bot. France* 131. *Lettres Bot.* 1984(1): 63–73.
1985. Apiaceae himalayenses. III. *Candollea* 40(2): 509–562.

Fleming, J.
1810. *Catalogue of Indian Medicinal Plants and Drugs*. 72 pp.

Franchet, A.
1894. Notes sur quelques Ombellifères du Yunnan. *Bull. Soc. Philom. Paris* VIII. 6: 104–146.

Froebe, H. A.
1979. *Die Infloreszenzen der Hydrocotyloideen (Apiaceae). Tropische und subtropische Pflanzenwelt.* Akad. Wissens. & Lit. Mainz 29, 679 pp.
1980. Randmusterbildung und Synorganisation bei strahlenden Apiaceendolden. *Plant. Syst. & Evol.* 133: 223–237.

Fu, K.-t. and Y.-c. Ho

1979. *Pleurospermum* Hoffm. *Flora Reipublicae Popularis Sinicae* 55(1): 133–184.

Fyson, P. F.

1915. *The Flora of the Nilgiri and Pulney Hill-tops* 1: 167–177.

1932. *The Flora of the South Indian Hill Stations* 1: 245–157.

Gamble, J. S.

1919. *Flora of the Presidency of Madras* 1(3): 554–566.

Gandhi, K. N.

1976. Apiaceae. In: *Flora of Hassan District, Karnataka, India*, pp. 414–418 (C. J. Saldanha and D. H. Nicolson, eds.).

Goel, A. K. and U. C. Bhattacharyya

1985. *Pimpinella diversifolia* DC. var. *sarmentifera* Goel & Bhattacharya (Apiaceae) — a new variety from Garhwal Himalaya. *Jour. Econ. & Taxon. Bot.* 6(1): 213–215.

Gruenberg-Fertig, I., C. C. Heyn, and I. Herrnstadt

1973. Typification of *Cachrys libanotis* L. (Umbelliferae). *Taxon* 22(4): 425–434.

Guyot, M.

1978. Contribution phytodermique (types stomatiques, morphologie des poils) à l'étude du *Vanasushava pedata* (Wight) Mukherjee et Constance, ombellifère du Sud de l'Inde. *Rev. Gen. Bot.* 85: 11–19.

Haines, H. H.

1920. Some new species of plants from Bihar and Orissa. *Jour. Asiat. Soc. Bengal* II. 15: 309.

1922. *The Botany of Bihar and Orissa* 1(3): 404–413.

Hamal, I. A. and A. K. Koul

1983. Studies on Kashmir Himalayan Umbelliferae — II: Origin and affinities. *Jammu Univ. Rev.* 1: 33–48.

1986. *Peucedanum thomsonii* Cl. — a rare Indian umbellifer rediscovered. *Bull. Bot. Surv. India* 26(3 & 4): 236.

1988. Cytotaxonomic analysis of the Himalayan species of the genus *Torilis* (Apiaceae). *Pl. Syst. Evol.* 159: 185–192.

Handel-Mazzetti, H.

1933. Umbelliferae. *Symbolae Sinicae*, Botanische Ergebnisse der Expedition der Akademie der Wissenschaften in Wien nach Südwest-China 1914/1918, 7: 706–730.

Hara, H.

1971. Umbelliferae. In: *Fl. East. Himalaya* 2: 91–92 (H. Hara, ed.).

Hedge, I. C. and J. M. Lamond

1980. Notes on Umbelliferae: some Asiatic Scandiceae. *Notes Roy. Bot. Gard. Edinburgh* 38: 251–257.

Hedge, I. C., J. M. Lamond, *et al.*

1972. Umbelliferae. In: *Flora of Turkey*, 4: 265–538 (P.H. Davis, ed.).

Hedge, I. C., J. M. Lamond, K. H. Rechinger, *et al.*

1987. Umbelliferae. In: *Flora Iranica* 162: 555 pp. (K. H. Rechinger, ed.).

Herrnstadt, I. and C. C. Heyn

1975a. New combinations in *Cachrys* L. as a genus distinct from *Hippomathrum* Link. *Notes Roy. Bot. Gard. Edinburgh* 33(3): 441–443.

1975b. A study of *Cachrys* populations in Israel and its application to generic delimitation. *Bot. Notiser* 128: 227–234.

Hiroe, M.

1958. *Umbelliferae of Asia* (excluding Japan). I, 219 pp.

1979. *Umbelliferae of World*, 2128 pp. Ariake Book Co., Tokyo, Japan.

Hiroe, M. and L. Constance

1958. Umbelliferae of Japan. *Univ. Calif. Publ. Bot.* 31: 144 pp., 75 figs.

Hooker, J. D. and T. Thomson

1855. *Introductory Essay to the Flora Indica*, 260 pp.

Kanai, H. and H. Hara

1966. Umbelliferae. In: *Fl. East. Himalaya*, pp. 228–232 (H. Hara, ed.).

Kanjilal, U.N., P. C. Kanjilal, and A. Das

1938. *Flora of Assam* 2: 338–343.

Kaul, M. K.

1975. Contribution to the Umbelliferae of Kashmir. *Jour. Bombay Nat. Hist. Soc.*, 72(3): 692–715.

Kitamura, S.

1960. *Flora of Afghanistan*. Results of the Kyoto University Scientific Expedition to the Karakoram and Hindukush, 1955, 2: 281–291.

1964. *Flora of West Pakistan and Afghanistan*, pp. 100–113.

Kljuykov, E. V.

1986. A taxonomic survey of the genus *Sinodielsia* Wolff including a description of two new species. *Feddes Repert.* 97: 11–12, 753–757.

Kljuykov, E. V., M. V. Pimenov and V. N. Tikhomirov

1976a. A revision of the genus *Aulacospermum* Ledeb. (Umbelliferae-Apioideae). 1. Limits and system of the genus. *Bjull. Moskovsk. Obsc. Isp. Prir., Otd. Biol.* 81(4): 75–89.

1976b. Ibid. 2. Review of species. *Op. cit.* 81(5): 61–68.

1976c. *Elaeosticta* Fenz., a self-contained genus of the family Umbelliferae, distinct from *Scaligeria* DC. *Op. cit.* 81(6): 83–94.

Klotzsch, F.

1862. Umbelliferae. In: *Die botanischen Ergebnisse der Reise seiner Königl. Hoheit des Prinzen Waldemar* von *Preussen in den Jahren* 1845 *und* 1846, pp. 146–150 (F. Klotzsch and A. Yarcke, eds.).

Koch, W. E. J.

1824. Genera tribuumque plantarum Umbelliferarum nova dispositio. *Nova Acta Acad. Leop. Carol.* 12: 55–156.

Korovin, E. P.

1947. Generis *Ferula* (Tourn.) L. monographia illustrata. Taschkent, 91 pp., 48 pls. Monogr. Acad. Sci. USSR, Tashkent.

1950 [1973]. *Scaligeria*. In: *Flora of the USSR*, 16: 204–218 [146–156] (B. K. Shishkin, ed.).

Koul, A. K., P. Koul and I. A. Hamal
1986. Insects in relation to pollination of some Umbellifers. *Bull. Bot. Surv. India* 28: 39–42.

Kozo-Poljansky, B.
1915a. Observations sur le genre *Sium* L. sensu DC. *Bull. Soc. Imper. Naturalistes Moscou* II. 28: 170–186.
1915b. Epitome Bupleurorum Rossiae. *Acta Hort. Petrop.* 30: 135–345.
1916. Sciadophytorum systematis lineamenta. *Bull. Soc. Nat. Mosc.* II. 29: 23–222.

Krishna, S. and R. L. Badhwar
1952. Aromatic plants of India. Umbelliferae. *Jour. Sci. & Indian Res.* 11(12): 245–266, 267–290.

Leute, G.-H.
1969, 1970. Untersuchungen über den Verwandtschaftskreis der Gattung *Ligusticum* L. (Umbelliferae). I. *Ann. Naturhistor. Mus. Wien* 73: 55–98; II. *Op. cit.* 74: 457–519.

Lindley, J.
1835. Notes upon some of the Himalayan Umbelliferae. In: *Ill. Fl. Himal. Mount.*, pp. 232–233, pls. 51–52 (J. F. Royle, ed.).

Liou, S.-I. and R.-h. Shan
1980. A preliminary study on the Chinese *Acronema*. *Acta Phytotax. Sinica* 18: 194–204.

Liu, T. S., C. Y. Chao, and T. I. Chuang
1962. Umbelliferae of Taiwan. *Quart. Jour. Taiwan Museum* 14(1 & 2): 15–47.

Lowry, P. P. II & A. G. Jones
1985. Systematics of *Osmorhiza* Raf. (Apiaceae: Apioideae). *Ann. Mo. Bot. Gard.* 71(4): 1128–1171.

Ludlow, F.
1976. Reliquiae botanicae himalaicae. *Bull. Brit. Mus. [Nat. Hist.], Bot.* 5(5): 269–289, pls. 30–36.

Mandenova, I. P.
1959. Materiali po sistematike Tribi Pastinaceae K.-Pol. emend. Manden. (Umbelliferae-Apioideae). *Akad. Nauk Gruzinskoi SSR. Bot. Inst. Trudy* 20: 3–57.
1986. De genere *Tetrataenium* (DC.) Manden. (Umbelliferae) notulae systematicae. I. *Notulae Syst. Geogr. Inst. Bot. Tbilis.* 41: 43–49.
1991. De genere *Tetrataenium* (DC.) Manden. (Umbelliferae) notulae systematicae. II. *Notulae Syst. Geogr. Inst. Bot. Tbilis.* 42:11–14.

Mandenova, I. P., *et al.*
1977. Contribution à 1'etude de genre *Tetrataenium* (DC.) Manden. *Actes 2e Sympos. Internat. Ombelliferes, Perpignan*, 675–726.

Mittall, S. P.
1955. A contribution to the morphology of *Centella asiatica* (Linn.) Urban, and some other related species. *Jour. Ind. Bot. Soc.* 34: 248–261.

Mukherjee, P. K.
1968. Notes on *Hydrocotyle* (1). *Hydrocotyle javanica* Thunb. vs. *H. nepalensis* Hook. *Proc. 56th Ind. Sci. Congr.* III, p. 336.
1969a. *Hydrocotyle himalaica* P. K. Mukherjee, a new *Hydrocotyle* from India. *Indian Forester* 95(7): 470.

1969b. Some name changes in Indian Umbelliferae. *Indian Forester* 95(8): 95.
1970. On the identification and typification of *Pimpinella urceolata* Watt ex Banerji (Apiaceae). *Bull. Bot. Surv. India* 12(1–4): 77–79.
1971. A note on the correct identity of *Pimpinella radiata* W. W. Sm. *Indian Forester* 94(1): 55.
1972. Taxonomic Studies on Indian Umbelliferae. Ph. D. Thesis, University of Calcutta, 458 pp. +28 pls.
1978. A resume of Indian Umbellifers. *Actes 2ᵉ Sympos. Internat. Ombelliferes, Perpignan*, pp. 47–70.
1983a. Nomenclatural transfers in Indian Umbelliferae. *Bull. Bot. Surv. India* 24(1–4): 42–45.
1983b. Distribution and collection of rare umbellifers in India. In: *An Assessment of Threatened Plants of India*, pp. 202–206. [Bot. Surv. India] (S. K. Jain and R. R. Rao eds.).
1987, 1988. Apiaceae. In: *Red Data Book of Indian Plants*, 1: 33–40, 2: 19–30. [Bot. Surv. India] (M. P. Nayar and A. R. K. Sastry, eds.).

Mukherjee, P. K. & R. N. Banerjee
1979. A new distributional record of *Pternopetalum tanakae* (Franch. & Sav.) Hand.-Mazzt. from Bhutan. *Bull. Bot. Surv. India* 18(1–4): 227–228.

Mukherjee, P. K. and L. Constance
1974. *Vanasushava*, an old South Indian umbellifer renamed. *Kew Bull* 29(3): 593–596.
1986a. A new disposition of the South Indian *Heracleum concanense* Dalzell (Umbelliferae). *Kew Bull.* 41(1): 223–229.
1986b. Two new genera of Indian Umbelliferae/Apiaceae. *Brittonia* 38(2): 145–149.
1991. New taxa and transfers in Indian Umbelliferae. *Edinburgh Jour. Bot.* 48(1): 41–44.

Mulligan, G. A.
1980. The genus *Cicuta* in North America. *Canad. Jour. Bot.* 58: 1755–1767.

Nasir, E.
1955. The Bupleura (Umbelliferae) of North-West Himalaya. *Univ. Calif. Publ. Bot.* 27: 417–445.
1957. A key to the genera and the species of Umbelliferae of N. W. Himalaya and West Pakistan. *Biologia* 3(1): 56–72.
1972. Umbelliferae. In: *Fl. West Pakistan* 20: 1–169 (E. Nasir and S. I. Ali, eds.).

Nayar, M. P. & R. N. Banerjee
1972. A new species of *Bupleurum* Linn. (Umbelliferae) from Andhra Pradesh, India. *Bull. Bot. Surv. India* 12(1–4): 255–256.

Norman, C.
1929. Umbelliferae from Nepal. *Jour. Bot.* 67: 245–247.
1937. *Cortia Hookeri* C. B. Clarke. *Jour. Bot.* 75: 93–96.
1938a. New genera in Umbelliferae. *Jour. Bot.* 76: 198–202.
1938b. The genus *Trachydium*. *Jour. Bot.* 76: 229–233.
1940. Notes on Sino-Himalayan Umbelliferae. *Jour. Bot.* 78: 229–232.

Pimenov, M. G.
1977. *Vicatia* DC.—a new genus of the family Umbelliferae in the flora of the U.S.S.R. *Bot. Zhurn.* 62: 1321–1326.

1987. A new species of the genus *Angelica* L. from Nepal, Himalaya. *Feddes Repert.* 98(7–8): 379–382.

Pimenov, M. G. and L. Constance
1985. Nomenclature of suprageneric taxa in Umbelliferae/Apiaceae. *Taxon* 34(3): 493–501.

Pimenov, M. G. and E. V. Kljuykov
1987. *Neoconopodium*—a new genus of the Umbelliferae from the Himalaya. *Feddes Repert.* 98(7–8): 373–378.

Pimenov, M. G. and O. V. Posorelova
1974. Systematic position of the genera *Ammi* L. and *Visnaga* Mill. *Rastit. Resursy* 10: 216–219.

Prain, D.
1903. *Bengal Plants* 1: 535–541.

Raizada, M. B. and R. Shahi
1951. The genus *Aegopodium* Linn. from India. *Jour. Indian Bot. Soc.* 30: 75–79.

Rao, R. S. and K. Hamadri
1976. *Pimpinella katraiensis* Rao & Hamadri—a new species of Apiaceae from the Sahydri Ranges, Maharashtra State. *Indian Forester* 102 (4): 232–236.

Rawat, G. S.
1987. On the occurrence of *Lalldhwojia cooperi* Farille (Apiaceae) in India. *Jour. Bombay Nat. Hist. Soc.* 34(3): 725–727.

Rechinger, K. H. and H. Riedl
1963. Umbelliferae. In: *Symbolae Afghanicae* V. *Biol. Skr.* 13(4): 27–135 (M. Koie and K. H. Rechinger, eds.).

Rheede tot Draakestein, H. A. van
1678–1703. *Hortus indicus malabaricus.* 12 vols.

Richard, A.
1820. Monographie du genre *Hydrocotyle. Ann. Gen. Sci. Phys.* 4: 145–225.

Ridley, H. N.
1942. *Distribution Areas of the Indian Flora.* 150th Annivers. Vol. Roy. Bot. Gard. Calcutta, pp. 49–52.

Roxburgh, W.
1814. *Hortus Bengalensis, or a Catalogue of the Plants Growing in the Honourable East India Company Botanic Garden at Calcutta,* 105 pp.
1832. *Flora Indica or Description of Indian Plants.* W. Grey (ed.)., 2: 89–97.

Royle, J. F.
1839. *Umbelliferae. Illustrations of the Botany and other Branches of the Natural History of the Himalayan Mountains and of the Flora of Kashmir,* p. 227.

Santapau, H.
1953. The flora of Khandala on the Western Ghats of India. *Rec. Bot. Surv. India* 16: 1–396.

Shan, R. H. and L. Constance
1951. The genus *Sanicula* (Umbelliferae) in the Old World and the New. *Univ. Calif. Publ. Bot.* 25: 1–78.

Shan, R. H. and M. L. Sheh (eds.)
1979, 1985. Umbelliferae. In: *Flora Reipublicae Popularis Sinicae* 55(1): 318 pp., 55(2): 282 pp.
1980. Xizang umbellifers: their geographical patterns and floristic characteristics. *Bull. Nanjing Bot. Gard.* 1980: 12–20.

Shan, R. H., M. L. Sheh, *et al.*
1980. New taxa of the Umbelliferae from Xizang (Tibet). *Acta Phytotax. Sinica* 3: 374–379.

Shishkin, B. K. (ed.)
1950–1951. [1973–1974]. Umbelliflorae. In: *Flora of the USSR* [Translated from Russian by R. Lavoot.] vols. 16 & 17, and pp. 478 & 285 respectively.

Singh, K.
1979. New plant records from Kheri District (U. P.) for the upper Gangetic Plain. *Indian Jour. Forestry* 2(2): 158–160.

Stewart, J. L.
1869. *Punjab Plants*, pp. 103–110.

Stewart, R. R.
1967. Check list of the plants of Swat State, North West Pakistan. *Pakistan Jour. Forestry* 17: 507–509.
1972. An annotated catalogue of the vascular plants of West Pakistan and Kashmir. In: *Fl. West Pakistan* i-xvii, 1028 pp. (E. Nasir and S. I. Ali, eds.).
1982. History and exploration of plants in Pakistan and adjoining areas. *Op. cit.*, 186 pp.

Strachey, R.
1906. *Catalogue of the Plants of Kumaon and of the Adjacent Portions of Garhwal and Tibet*, pp. 72–76.

Tardieu-Blot, M. L.
1967. *Umbelliferae: Flore du cambodge, du laos et du vietnam* 5: 1–77.

Tseng, C. C.
1967. Anatomical studies of flower and fruit in the Hydrocotyloideae (Umbelliferae). *Univ. Calif. Publ. Bot.* 42: 1–58, pls. 1–10.

Tutin, T. G. (ed.)
1968. Umbelliferae. *Flora Europaea* 2: 315–375.

Wagenitz, G.
1956. Die Gattung *Psammogeton* Edgew. (Umbelliferae). *Ber. Deutsch. Bot. Ges.* 69: 227–238.

Wallich, N. A.
1828–1829. A Numerical List of Dried Specimens of Plants in the East India Company's Museum, Collected under the Superintendence of Dr. Wallich of the Company's Botanic Garden at Calcutta. [Wallich Catalogue]

Wight, R.
1838–1853. *Icones plantarum Indiae Orientalis.* 3 vols.
1846. *Spicilegium Neilgherrense or a Selection of Neilgherry Plants* 1: 65–68, pls. 78–82.

Wight, R. and G. A. Walker-Arnott
1834. *Prodromus Florae Peninsulae Indiae Orientalis*, pp. 365–375.

Wolff, H.

1910. Umbelliferae-Apioideae-*Bupleurum*, *Trinia* et reliquae Ammineae heteroclitae. *Engl. Pflanzenreich* 43(4228): 214 pp.

1913. Umbelliferae-Saniculoideae. *Engl. Pflanzenreich* 61(4228): 305 pp.

1922. Umbelliferae; in Bot. Reisen in den Hochgebirgen Chinas und Ost-Tibets. *Feddes Repert.*, Beiheft, 447–453.

1925a. Neue Umbelliferen-Gattungen aus Ostasien. *Notizbl. Bot. Gart. Berlin-Dahlem* 9(84): 275–280.

1925b. *Pleurosperma* nova Tibetiae orientalis. *Feddes Repert.* 21: 241–244.

1926. Umbelliferae. In: Plantae Sinenses a Dre. H. Smith annis 1921–22 lectae. *Acta Hort. Gothoburgensis* 2: 289–328.

1927. Umbelliferae-Apioideae-Ammineae-Carinae, Ammineae novemjugatae et genuinae. *Engl. Pflanzenreich* 90(4228): 398 pp.

1929–1930. Umbelliferae Asiaticae novae relictae. *Feddes Repert.* 27: 112–128, 179–192, 301–335; 28: 109–112.

COLLECTORS CITED
(INDIA, BANGLADESH, BHUTAN AND NEPAL)

Abraham, V.	Bhutan
Agharkar, S. P.	West Bengal
Aitchison, J. E. T.	Jammu & Kashmir, Punjab
Almeida S. M.	Maharashtra
Anderson, T.	Sikkim, Uttar Pradesh, West Bengal
Ansari, M. Y.	Maharashtra ("Malabar")
Arora, C. M.	Madhya Pradesh, Uttar Pradesh
Babu, C. R.	Uttar Pradesh
Bacon	Uttar Pradesh
Balakrishnan, N. P.	Andhra Pradesh, Madhya Pradesh
Banerji, M. L.	Meghalaya, Nepal
Barber, C. A.	Karnataka, Tamil Nadu
Bartholomew, B.	Bhutan
Beddome, R. H.	Madhya Pradesh, Tamil Nadu
Beer, L.	Nepal
Bell, T. R. D.	Maharashtra, Punjab
Bhattacharyya, U. C.	Himachal Pradesh, Uttar Pradesh
Bhide, R. K.	Maharashtra
Blasco, F.	Kerala
Blinkworth, R.	Nepal, Uttar Pradesh
Booth, T. J.	Bhutan
Bor, N. L.	Himachal Pradesh, Manipur, Nagaland
Bourne, A. G.	Tamil Nadu
Bourrit, A.	Nepal
Bowes-Lyon, S. A.	Bhutan
Brandis, D.	Himachal Pradesh, Uttar Pradesh
Breganza, A. R.	Maharashtra
Bressers, J.	Bihar
Brothers, L. R.	Uttar Pradesh
Buchanan-Hamilton, F.	Nepal, Uttar Pradesh
Burkill, I. H.	Arunachal Pradesh, Himachal Pradesh
Burrit, A.	Nepal
Byers, A. C.	Nepal

Byers, E.	Nepal
Cameron, W.	Karnataka
Campbell, A.	Bihar
Cave, G. H.	Sikkim, West Bengal
Chakrabarty, A. K.	West Bengal
Champion, H. G.	Uttar Pradesh
Chand, T. R.	Meghalaya, Nagaland, Uttar Pradesh
Chatterjee, A. C.	Assam
Clarke, C. B.	Assam, Bangladesh, Bihar, Himachal Pradesh, Jammu & Kashmir, Meghalaya, Mizoram, Nagaland, Orissa, Punjab, Sikkim, Tamil Nadu, West Bengal
Cleghorn, H. C. F.	Tamil Nadu
Cooper, R. E.	Bhutan, Sikkim
Cousins, C. W.	Sikkim
Dalhousie, Countess or Lady	Uttar Pradesh
Dalzell, N. A.	Maharashtra
Das, R. E.	Uttar Pradesh
Deb, D. B.	Arunachal Pradesh, Kerala, Tripura
Debbarman, P. M.	Tripura
Dhwoj, Lall	Nepal
Drummond, J. R.	Himachal Pradesh
Dungboo	Sikkim
Duthie, J. F.	Himachal Pradesh, Jammu & Kashmir, Madhya Pradesh, Rajasthan, Uttar Pradesh
Edgeworth, M. P.	Himachal Pradesh, Jammu & Kashmir, Uttar Pradesh
Ellis, J. L.	Andhra Pradesh, Kerala
Falconer, H.	Jammu & Kashmir, Uttar Pradesh
Farille, M. A.	Nepal
Fischer, C. E. C.	Karnataka, Kerala, Tamil Nadu
Gage, A. T.	Mizoram
Gallar, W.	Jammu & Kashmir
Gamble, J. S.	Bihar, Himachal Pradesh, Madhya Pradesh, Punjab, Sikkim, Tamil Nadu, Uttar Pradesh, West Bengal
Gammie, G. A.	Arunachal Pradesh, Jammu & Kashmir, Sikkim
Ganguly, A. K.	West Bengal
Gardner, E.	Nepal
Gibson, A.	Maharashtra
Giles, G. M. J.	Uttar Pradesh
Gillett, G. W.	Uttar Pradesh
Godfrey	Mizoram
Goule, B. J.	Bhutan
Govan, G.	Uttar Pradesh

Griffith, W.	Assam, Bangladesh, Bhutan, Meghalaya, West Bengal
Gupta, K.	Uttar Pradesh
Haines, H. H.	Bihar, Madhya Pradesh
Hajra, P. K.	Arunachal Pradesh, Bhutan
Hamilton, F.	Nepal (cf. Buchanan-Hamilton)
Hamilton, H.	Jammu & Kashmir
Harsukh	Bihar
Hemadri, K.	Maharashtra
Henry, A. N.	Madhya Pradesh
Heyne, B.	Bangladesh, Tamil Nadu, "Peninsular India"
Hicks, J. H.	Bhutan
Hoffmeister, W.	"Himalaya"
Hohenacker, R. F.	Karnataka, Tamil Nadu (cf. Metz, Miquel)
Hooker, J. D.	Bhutan, Meghalaya, Sikkim, Uttar Pradesh, West Bengal
Hunter, A. G.	Punjab
Inayat Khan	Jammu & Kashmir, Uttar Pradesh
Jacob	Tamil Nadu
Jacquemont, V.	Jammu & Kashmir
Jaeschke, H. A.	Himachal Pradesh, Jammu & Kashmir
Jameson, W.	Himachal Pradesh
Jenkins, F.	Assam
Jones, W.	Nepal
Joseph, J.	Madhya Pradesh, Tamil Nadu
Kanjilal, P. C.	Meghalaya
Kanjilal, U. N.	Assam
Keshavanand, R. B.	Jammu & Kashmir
Khan, K.	Punjab
King, G.	Kerala, Sikkim, Tamil Nadu, West Bengal
King's collector	Meghalaya, Sikkim
Kingdon-Ward, F.	Arunachal Pradesh, Assam, Meghalaya, Nagaland
Koelz, W. N.	Himachal Pradesh, Jammu & Kashmir, Meghalaya, Nagaland, Punjab, Uttar Pradesh
Kuntze, C. E. O.	Sikkim
Kurz, S.	Sikkim, West Bengal
Lace, J. H.	Himachal Pradesh, Uttar Pradesh
Lall Dhwoj	Nepal (cf. Dhwoj, Lall)
Lancaster, C. R.	Nepal
La Touche, T.	Uttar Pradesh
Law, J. S.	Maharashtra
Lockwood	Bihar
Long, D. G.	Bhutan
Ludlow, F.	Bhutan
Mackinnon, P. W.	Uttar Pradesh

Madden, E.	Himachal Pradesh
Malhotra, S. K.	Himachal Pradesh, Uttar Pradesh
Malla, S. B.	Nepal
Mann, G.	Meghalaya
Mazumdar, R. B.	West Bengal
McCosh, D.	Nepal
Meebold, A. (K.)	Jammu & Kashmir, Kerala, Maharashtra, Tamil Nadu, Uttar Pradesh
Metz, F.	Karnataka
Miehe, G.	Nepal
Miquel, F. A. W.	Karnataka, Tamil Nadu
Mishra, O. P.	Uttar Pradesh
Mooney, H. F.	Bihar, Madhya Pradesh, Orissa
Morris, D.	Nepal
Mukerjee, S. K.	Bhutan, West Bengal
Mukherjee, P. K.	Arunachal Pradesh, Himachal Pradesh, Jammu & Kashmir, Karnataka, Kerala, Maharashtra, Meghalaya, Nagaland, Nepal, Sikkim, Tamil Nadu, Uttar Pradesh, West Bengal
Munro, W.	"Nilgherries" [Nilgiri Hills]
Nair, N. C.	Arunachal Pradesh, Himachal Pradesh, Jammu & Kashmir, Punjab, Uttar Pradesh
Nakao, S.	Nepal
Narayanaswami, V.	Andhra Pradesh
Nasir, E.	Jammu & Kashmir
Naithani, B. D.	Uttar Pradesh
Osmanton, B. B.	Uttar Pradesh
Pande, B. D.	Nepal
Pangtey, Y. S.	Uttar Pradesh
Panigrahi, G.	Arunachal Pradesh, Assam, Meghalaya
Pant, P. C.	Uttar Pradesh
Parmanand	Himachal Pradesh
Polunin, O.	Nepal
Prain, D.	Sikkim, West Bengal
Prain's collector	Sikkim
Quarashi	Madhya Pradesh
Raizada, M. B.	Himachal Pradesh, Uttar Pradesh
Raju, D. C. S.	Orissa
Ram, B.	Punjab
Rao, A. S.	Bhutan
Rao, R. S.	Arunachal Pradesh
Rao, T. A.	Jammu & Kashmir, Uttar Pradesh
Rau, M. A.	Himachal Pradesh, Uttar Pradesh
Rayan, G. M.	West Bengal
Reid, F. A.	Uttar Pradesh

Ribu	Sikkim, West Bengal
Rohmoo (sometimes cited as "Rhomoo" or "Rohmoo Lepcha")	Sikkim, West Bengal, "East Himalaya"
Roxburgh, W.	West Bengal, "India orientali"
Royle, J. F.	Himachal Pradesh, Jammu & Kashmir, Punjab, Uttar Pradesh, "N. W. India"
Russell, R. S.	Jammu & Kashmir
Sahni, K. C.	Arunachal Pradesh
Sakya, A. R.	Nepal
Santapau, H.	Maharashtra, Uttar Pradesh
Sarkar, R. K.	West Bengal
Saroja, T. L.	Tamil Nadu
Saulière, A.	Tamil Nadu
Saxena, H. O.	Uttar Pradesh
Saxton, W. T.	Himachal Pradesh
Schlich, W.	Himachal Pradesh
Schmidt, B.	Tamil Nadu
Sebastine, K. M.	Andhra Pradesh, Kerala, Madhya Pradesh, Tamil Nadu
Sedgewick, L. J.	Maharashtra
Sen Gupta, G.	Bhutan
Shakya, P. R.	Nepal
Sherriff, G.	Bhutan
Shetty, B. V.	Kerala
Shrestha, T. B.	Nepal
Simons, C. J.	Assam, Meghalaya
Sinclair, I. W. J.	Bhutan
Singh, N. P.	Uttar Pradesh
Singh, S.	Himachal Pradesh
Smith, W. W.	Sikkim
Stainton, J. D. A.	Jammu & Kashmir, Nepal
Stewart, J. L.	Himachal Pradesh
Stewart, R. R.	Jammu & Kashmir
Stocks, J. E.	Maharashtra
Stoliczka, F.	Himachal Pradesh, Jammu & Kashmir
Strachey, R.	Uttar Pradesh
Subbarao, G. V.	Andhra Pradesh, Himachal Pradesh
Subramanyam, K.	Andhra Pradesh, Tamil Nadu
Sunderaj, D.	Andhra Pradesh
Sykes, W. H.	Nepal
Talbot, W. A.	Karnataka, Maharashtra
Tanner, H. C. B.	Jammu & Kashmir
Taylor, G.	Bhutan
Thapi	Nepal

Thomson, T.	Bhutan, Himachal Pradesh, Jammu & Kashmir, Meghalaya, Sikkim, Uttar Pradesh, West Bengal
Thothathri, K.	Bhutan
Vaid, K. M. M.	Himachal Pradesh
Vicary, N. A.	Uttar Pradesh
Vivekananthan, K.	Tamil Nadu
Wadhawa, B. M.	Maharashtra ("Malabar")
Wald, E.	Nepal
Walker-Arnott, G. A.	Tamil Nadu
Wallich, N.	Bangladesh, Nepal, Tamil Nadu, Uttar Pradesh
Watt, G.	Manipur, Nagaland
Webster, G. L.	Jammu & Kashmir
White, J. C.	Bhutan
Wight, R.	Tamil Nadu, "Peninsular India Orient"
Wigram, C.	Nepal
Williams, L. H. J.	Nepal
Winterbottom, J. E.	Uttar Pradesh
Wollaston	Nepal
Wood, J. J.	Bihar, Orissa
Wyss-Dunant, E.	Sikkim
Younghusband, F.	Sikkim

INDEX OF SCIENTIFIC NAMES*

* Synonyms given in italics.